Methods and Goals
in Human
Behavior Genetics

Contributors

RAYMOND B. CATTELL
DANIEL FREEDMAN
JOHN L. FULLER
RILEY W. GARDNER
RUTH GUTTMAN
IRVING I. GOTTESMAN
DEWEY L. HARRIS
GABRIEL W. LASKER
GARDNER LINDZEY

JOHN C. LOEHLIN
SARNOFF A. MEDNICK
ROBERT C. NICHOLS
FINI SCHULSINGER
RICHARD T. SMITH
H. ELDON SUTTON
RICHARD E. STAFFORD
STEVEN G. VANDENBERG
RALPH L. WITHERSPOON

Methods and Goals in Human Behavior Genetics

Edited by
STEVEN G. VANDENBERG
CHILD DEVELOPMENT UNIT
DEPARTMENT OF PEDIATRICS
UNIVERSITY OF LOUISVILLE SCHOOL OF MEDICINE
LOUISVILLE, KENTUCKY

1965

ACADEMIC PRESS   New York and London

QH431
.V3
1965

Copyright © 1965, by Academic Press Inc.
ALL RIGHTS RESERVED.
NO PART OF THIS BOOK MAY BE REPRODUCED IN ANY FORM,
BY PHOTOSTAT, MICROFILM, OR ANY OTHER MEANS, WITHOUT
WRITTEN PERMISSION FROM THE PUBLISHERS.

ACADEMIC PRESS INC.
111 Fifth Avenue, New York, New York 10003

United Kingdom Edition published by
ACADEMIC PRESS INC. (LONDON) LTD.
Berkeley Square House, London W.1

Library of Congress Catalog Card Number: 65-24988

PRINTED IN THE UNITED STATES OF AMERICA.

List of Contributors

Numbers in parentheses indicate the pages on which the authors' contributions begin.

Raymond B. Cattell (95), *Department of Psychology, University of Illinois, Urbana, Illinois*

Daniel Freedman (141), *The Committee on Human Development, University of Chicago, Chicago, Illinois*

John L. Fuller (245), *The Jackson Laboratory, Bar Harbour, Maine*

Riley W. Gardner (223), *The Menninger Foundation, Topeka, Kansas*

Ruth Guttman (199), *Department of Psychology, The Hebrew University of Jerusalem, and The Israel Institute of Applied Social Research, Jerusalem, Israel*

Irving I. Gottesman (63), *Department of Psychiatry, University of North Carolina, Chapel Hill, North Carolina*

Dewey L. Harris (81), *DeKalb Agricultural Association, Incorporated, Sycamore, Illinois*

Gabriel W. Lasker (17), *Department of Anatomy, Wayne State University College of Medicine, Detroit, Michigan*

Gardner Lindzey (297), *Department of Psychology, University of Texas, Austin, Texas*

John C. Loehlin (163), *Department of Psychology, University of Texas, Austin, Texas*

Sarnoff A. Mednick (255), *Department of Psychology, University of Michigan, Ann Arbor, Michigan, and Psykologisk Institut, Kommunehospitalet, Copenhagen, Denmark*

LIST OF CONTRIBUTORS

Robert C. Nichols (231), *National Merit Scholarship Corporation, Evanston, Illinois*

Fini Schulsinger (255), *Psykologisk Institut, Kommunehospitalet, Copenhagen, Denmark*

Richard T. Smith (45), *Department of Chronic Diseases, School of Hygiene and Public Health, Johns Hopkins University, Baltimore, Maryland*

Richard E. Stafford (171), *Child Development Unit, Department of Pediatrics, University of Louisville School of Medicine, Louisville, Kentucky*

H. Eldon Sutton (1), *Department of Zoology, University of Texas, Austin, Texas*

Steven G. Vandenberg (29), *Child Development Unit, Department of Pediatrics, University of Louisville School of Medicine, Louisville, Kentucky*

Ralph L. Witherspoon (187), *Institute of Human Development, The Florida State University, Tallahassee, Florida*

Preface

In recent years there has been a rapid increase in interest in the role of heredity in human behavior. A number of psychologists, psychiatrists, and ethologists are engaged in studies designed to add to the presently meager information. A common characteristic of these investigators is their desire to move beyond the stage of the fruitless and outdated controversy concerning the importance of "nature" as *opposed* to "nurture." Instead they wish to study how heredity and environment jointly operate to produce the observable phenotypes. Several psychologists, collaborating with appropriate consultants, decided that the time was right for examining the trends in research in behavior genetics in man and for a critical review of methodology.

This volume is the first result of such a study. In planning the volume it was kept in mind that some of the methods and findings of biochemical genetics, anthropological genetics, and animal behavior genetics are highly relevant to human behavior genetics. Goals similar to those of this volume are met at a less advanced level by Fuller and Thompson's *Behavior Genetics*.* Much relevant research on humans has been done since that book was prepared, especially with twins, so that a consideration of research methods focused on *human* behavior genetics seemed worthwhile.

This volume was planned to be of interest to two types of readers. First of all it provides information for psychologists who are interested in the genetics of personality and ability. It is hoped that the volume will, in addition, be of some value to geneticists who are desirous of knowing about recent attempts by psychologists to study hereditary factors in human behavior. The emphasis throughout the book is on methods rather than on results. It is hoped that

---

* John L. Fuller & W. Robert Thompson, *Behavior Genetics*. New York: Wiley, 1960.

a later volume may be devoted to a presentation of findings from several research programs with an aim to integrate and interpret these findings.

The contributions to this volume are in some cases similar to papers presented during a meeting held in Louisville where this volume was planned, while the comments following these papers are based on tape recordings of the ensuing discussions. In editing these discussions, a compromise was sought which would preserve the lively flavor of the original exchange of ideas but avoid the lack of precision and occasional repetitiveness of such discussions. At times the order of specific comments was changed to enhance the logical progression of the ideas presented.

The Louisville meeting was supported by a grant from the National Institutes of Mental Health (MH 07708). I wish to acknowledge the help and thoughtful suggestions of Dr. Francis J. Pilgrim, both during the planning of the meeting and the preparation of this volume. I am indebted to the audiovisual service of the University of Louisville for the tape recordings. Mrs. "Mickey" Gliessner did a painstaking job in transcribing the tapes and patiently typed several versions of the papers and discussions. The many comments by the participants on their own and each other's papers were a great help in the editorial task. Drs. Gardner Lindzey and John Fuller served as discussion leaders and kept the participants oriented toward the goals of the meeting. Mrs. Elsie Long and Mr. Maurice LeCroy assisted with many arrangements, while Dr. Ann Schwartz assisted ably with the editing and correspondence with the authors, especially during my absence for several months due to illness. Thanks to her, progress was maintained in spite of this setback.

June, 1965   STEVEN G. VANDENBERG

Contents

LIST OF CONTRIBUTORS v

PREFACE vii

Biochemical Genetics and Gene Action

 H. Eldon Sutton 1

Studies in Racial Isolates

 Gabriel W. Lasker 17

Multivariate Analysis of Twin Differences

 Steven G. Vandenberg 29

A Comparison of Socioenvironmental Factors in Monozygotic and Dizygotic Twins, Testing an Assumption

 Richard T. Smith 45

Personality and Natural Selection

 Irving I. Gottesman 63

Biometrical Genetics in Man

Dewey L. Harris 81

Methodological and Conceptual Advances in Evaluating Hereditary and Environmental Influences and their Interaction

Raymond B. Cattell 95

An Ethological Approach to the Genetical Study of Human Behavior

Daniel Freedman 141

A Heredity-Environment Analysis of Personality Inventory Data

John C. Loehlin 163

New Techniques in Analyzing Parent-Child Test Scores for Evidence of Hereditary Components

Richard E. Stafford 171

Selected Areas of Development of Twins in Relation to Zygosity

Ralph L. Witherspoon 187

A Design for the Study of the Inheritance of Normal Mental Traits

Ruth Guttman 199

Genetics and Personality Theory
Riley W. Gardner 223

The National Merit Twin Study
Robert C. Nichols 231

Suggestions from Animal Studies for Human Behavior Genetics
John L. Fuller 245

A Longitudinal Study of Children with a High Risk for Schizophrenia: A Preliminary Report
Sarnoff A. Mednick and Fini Schulsinger 255

General Discussion
Gardner Lindzey (Chairman) 297

AUTHOR INDEX 339

SUBJECT INDEX 343

Methods and Goals
in Human
Behavior Genetics

Biochemical Genetics and Gene Action

H. Eldon Sutton

Biochemical genetics has become a very broad subject, and the concepts of gene action are changing rapidly. Consequently, certain aspects of gene action which could play a role in psychological variables will be outlined briefly although the connections may not be too obvious at the present time.

The action of genes may be roughly divided into two attributes—the quantitative and the qualitative. Knowledge of the qualitative effects of gene action date approximately from the demonstration by Ingram (1956) that mutant hemoglobins differ in their protein structure by single amino acid differences. This was a major milestone in understanding gene action. The specific amino acid sequence in one of the two identical half molecules of hemoglobin is 287 units long, and a single amino acid substitution accounts for the differences that have been found. Since the original discovery, the amino acid sequence of a number of hemoglobins has been analyzed in detail, and, with rare exceptions which actually are *not* exceptions to the hypothesis, all are found to differ by a single amino acid substitution as shown in Table I.

The method by which the primary structure of proteins is specified has also yielded to investigation during the past 6 or 7 years. There are still some mysteries surrounding protein synthesis, but the major outline of the scheme seems fairly clear. It is well established that DNA (deoxyribonucleic acid) is the chemical substance which transmits genetic information from one cell to another. Proteins, such as hemoglobin and enzymes, are the agents which ultimately express gene action by catalyzing biochemical reactions within cells, hemoglobin, for example, being concerned with oxygen transport. How does the information in DNA become expressed in terms of protein structure? Ribonucleic acid (RNA) appears to

TABLE I
AMINO ACID SUBSTITUTIONS OF HEMOGLOBIN A

Alpha chain

Hb Variant	1	2	16	30	57	58	68	116	141
Position:	Val.	Leu.	Lys.	Glu.	Gly.	His.	Asn.	Glu.	Arg.
I			Asp						
G$_{Honolulu}$				Gln					
Norfolk					Asp				
M$_{Boston}$						Tyr			
G$_{Philadelphia}$							Lys		
O$_{Indonesia}$								Lys	

Beta chain

Hb Variant	1	2	3	6	7	26	43	63	67	121	146
Position:	Val.	His.	Leu.	Glu.	Glu.	Glu.	Glu.	His.	Val.	Glu.	His.
S				Val							
C				Lys							
G$_{San\ Jose}$					Gly						
E						Lys					
G$_{Galveston}$							Ala				
M$_{Saskatoon}$								Tyr			
Zurich								Arg			
M$_{Milwaukee-1}$									Glu		
D$_{Punjab}$										Gln	
O$_{Arabia}$										Lys	

be the intermediary. There are several types of RNA, each of which has its unique function. The DNA of the nucleus directs the synthesis of an RNA fraction called "messenger" RNA (mRNA), which contains the genetic information specifying the amino acid sequences. The messenger RNA appears to be transferred to the cytoplasm where it forms complexes with ribosomes (Fig. 1). Ribosomes are particles composed of a second type of RNA. A third

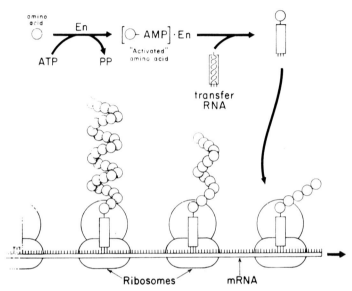

FIG. 1. Diagram of protein synthesis showing association of messenger RNA with ribosomes and alignment of the transfer-RNA-amino-acid complexes on the messenger RNA (from Sutton, 1965).

fraction of RNA is called "soluble" RNA (sRNA), or more commonly "transfer" RNA. The three fractions of RNA function approximately as follows: The messenger RNA carries the genetic information in the DNA to the cytoplasm where it serves as a template for the specific alignment of amino acids. The ribosomes appear to be production sites for protein synthesis. They do not seem to control the structure of the protein synthesized and may only provide the necessary configuration for the other elements and enzymes essential to the process. The soluble RNA is a small piece of RNA. There is a whole series of these soluble RNAs, each of them complexing with a specific amino acid. This activated complex is required for alignment of the amino acids on the messenger template. Once they are aligned, the amino acids are polymerized to form a long polypeptide chain, which folds to become the native protein. Ribosomes appear to have a long existence in the cytoplasm. (I might add that they also appear in the nucleus.) Their origin is not known, although they are present in every cell and their structure appears to be specified by the nucleus. Messenger RNA is ordinarily a very short-lived substance. It is thought to function perhaps for several polypeptide chains, but it is destroyed rapidly. Soluble RNA apparently recycles. Its origin is also something of a mystery, particularly since it contains some rather unusual components for RNA. (This picture, although very brief, includes much of what we know about the transfer of information from the genetic material to the final protein product.)

More relevant to psychologists, perhaps, are the quantitative effects observed in gene action. Several recent discoveries suggest models of how genes may act quantitatively. We know for example that most genes do not act most of the time. If this were not so we would not have the differentiation of cells necessary for the development of various organs and tissues. Hemoglobin, for example, is not synthesized in most cells; it is synthesized only in very special cells. Most other proteins are synthesized in cells of special tissues. Since all nuclei and their chromosomes are derived from a single zygote, it seems clear that there must be mechanisms for regulating the activity of genes. The clearest model for the regulation of individual genes is based upon studies by Jacob and Monod (1961) of two structural genes. Figure 2 summarizes the model which has emerged from these very elegant studies on the enzyme β-galactosidase from the bacterium *Escherichi coli*. This enzyme, which can be detected only if substrate or substrate analog is present in the growth medium, has its structure specified by what we might call structural gene A. This structural gene is analogous to the structural genes of hemoglobin, only in this case the information specifies the amino acid sequence of β-galactosidase. In the β-galactosidase system there is a second gene located adjacent to gene A. It specifies the structure of an enzyme which is responsible for the uptake of galactosides into the cell. I have designated this second gene structural gene B. These structural genes direct the synthesis of messenger RNA, which serves as a template in the cytoplasm for specific synthesis of these two proteins. Jacob and Monod also detected mutations influencing the ability to produce these enzymes which were *not* mutations in the structural loci. One of these sites they have called the regulator gene. The regulator gene is not closely linked to the two structural genes on the *E coli* chromosome. In other organisms, regulator genes or what appear to be regulator genes may be on different chromosomes than the genes they are regulating. The regulator gene apparently synthesizes or directs the synthesis of a repressor substance which can inhibit the

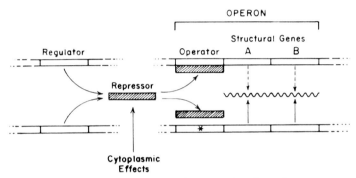

FIG. 2. Diagram of the relationships of structural and regulator genes according to the hypothesis of Jacob and Monod. A diploid cell is indicated. Mutation of an operator on one chromosome (indicated by asterisk) may lead to lack of repression of the structural genes which are part of the same operon. This results in uncontrolled synthesis of messenger RNA (from Sutton, 1965).

activity of the structural genes, A and B. If the repressor does have this very specific action there must be a site for this action. Mutations were located adjacent to the structural genes which showed the characteristics predicted for the site of action for repressor. This location has been termed the operator. Whether the operator is a separate gene or not is yet to be resolved, but at least it is part of the genetic region of the structural genes. The unit of operator plus structural genes has been given the designation "operon." The nature of the repressor substance is uncertain. Combination of repressor with operator turns the structural genes off; if the repressor is not combined with the operator the structural genes are turned on. This off-on switch is influenced from the environment (cytoplasm) by two types of substances called co-repressors and effectors. A co-repressor is a substance which enhances the activity of the repressor as part of a negative feed-back system. In some systems the co-repressor may be the end product of the metabolic sequence; this end product acts with the repressor to shut off the synthesis of inducible enzymes. Actually it is not known whether it is the product itself which does this, we only know that the level of end product enhances the activity of the repressor. The effector is a substance which inhibits the complex of the repressor and operator and therefore turns on structural genes. Such is the case with induced enzymes when a substrate or substrate analog is present. The effector combines with the repressor; the repressor is thereby inactivated; and the structural genes function to produce enzyme. This may seem a complex system, but it explains very well *how* genes may be turned off and on depending upon the environmental circumstances. This system has been studied in detail only in a few systems. However, there is evidence that it is valid for other microorganisms and it explains much of what we know about such systems in higher animals. It is, of course, easier to manipulate the environment of a microorganism than cells in a higher organism, so our knowledge of these systems is limited at the present time. However, using tissue culture, we may soon know how extensive such mechanisms are in the biochemistry of man and other higher organisms.

A second mechanism for influencing the quantitative action of genes has recently been discovered in higher organisms, including man. This mechanism involves inactivation of an entire, or almost entire, chromosome. Our knowledge is based solely on observations of the X chromosome. It has long been a matter of speculation why females with two X chromosomes and males with one X chromosome are relatively similar to each other in sex-linked traits which have nothing whatsoever to do with either primary or secondary sex characteristics. After all, dosage effects are a well-known genetic phenomenon. Those of us who are biochemists were really not too uncomfortable about this because the feedback mechanisms could account for it, but fortunately this did not deter some other experimenters from worrying. As a result we now have rather good evidence that there can be inactivation of a large portion of an X chromosome. Several lines of evidence support this hypothesis (Lyon, 1962). In cytological preparations from females, but not normal males, one can demonstrate a substance in the nucleus called the sex-chromatin body (Barr & Bertram, 1949). This is a DNA-containing structure next to the nuclear membrane. It has been followed rather carefully cytologically, and under some circumstances it appears to be a condensed X chromosome. A second line of evidence comes from studies of the action of genes controlling coat color in mice (Russell, 1961). Genes which behave quite regularly in dominance relationships while in autosomal positions show a variegation effect when translocated to the X chromosome. For example, in heterozygotes, while some tissues or areas show expression of the dominant gene as expected, others show expression of the recessive allele. An analogous effect has been found in man in the case of the enzyme glucose-6-phosphate-dehydrogenase in heterozygotes. There are persons who carry mutant forms of this enzyme. The mutant form is unstable while the usual enzyme is stable. Males who bear a single copy of the mutant gene show 100% unstable enzyme. Heterozygous females may show 100% unstable enzyme, almost 100% normal enzyme, or a "semi-stable" mixture. Examination of the kinetics of this enzyme in red cell suspensions has provided quite strong evidence that the mixture of enzymes is segregated into individual red cells (Beutler et al., 1962). That is, there are two different red cell types, one with normal enzyme and one with mutant enzyme, rather than a homogeneous mixture of red cells which contain both types of enzymes. In other words, we have some red cells which have 100% abnormal enzyme, the rest of them have 100% normal, and the percentage of abnormal enzymes really reflects the relative numbers of red cell types, *not* the amount of enzyme in each red cell. Since the trait is located on the X chromosome we can infer that in some red cells one of the X chromosomes is active and in the remainder the second X chromosome has had its gene expressed (see also the recent paper of Davidson et al., 1963). Further, there is the cytological demonstration by tritium labeling that one of the X chromosomes replicates late (Morishima et al., 1962). In a female, all of the autosomes plus one of the X chromosomes replicate in synchrony. The other X chromosome, however, replicates later. In males all chromosomes replicate at the same time. These various diverse lines of evidence combine rather nicely to support the notion of a mechanism for inactivating an entire chromosome, in this case, the X chromosome. No suggestion had been made concerning the nature of this mechanism, but it seems that all X chromosomes beyond the first are inactive. For example, in persons with the chromosomal constitution of XXXY, there appear to be two inactive X chromosomes and a single active one. Which X chromosome becomes active in each cell is a matter of chance very early in embryonic development. Once this decision is made it is apparently maintained. The inactive X chromosome is transmitted as an inactive chromosome in progeny arising from that cell. So we see two rather extreme mechanisms for quantitative gene action, one of them involving single structural loci (as in the β-galactosidase instance), the other involving entire X chromosomes. Possibly, intermediate types of gene inactivation exist. For example, histones are thought to influence the activity of genes considerably (Huang & Bonner, 1962). To what extent they are involved in either one of these mechanisms we do not know.

Figure 3 illustrates a system that we are currently investigating (Sutton & Karp, 1964), demonstrating the extent to which structural genes may be influenced in their activity. This particular system involves the blood proteins known as haptoglobins. So far as we can tell the structural elements which form the series of protein bands are identical. What does vary is the relative amount of the products which are synthesized by the structural genes in these heterozygous individuals. The structural genes themselves are transmitted as straight-forward Mendelian alleles showing codominance. The elements which determine how much each of the two alleles functions also are genetic. To what extent environmental forces may be involved here we do not know. However, if I were looking at a Caucasian population rather than a Negro population, as we have here, nearly all of the heterozygotes would be of type Hp2-1a. In Negroes, the other patterns occur commonly. At least some of the variation in heterozygous types is transmitted with the structural genes.

Some rather exciting information is the recent implication of RNA in nerve function. All of us have heard humorous stories of grinding up professors when they become senile and feeding them to their students so that the information could be passively transferred. This proposal had its origin in several lines of evidence which demonstrate changes in RNA as a result of certain types of learning processes and in planarian regeneration studies in the presence and absence of enzymes which destroy RNA. In the latter case, planaria which are first conditioned and then chopped in half will regenerate from either end. If they regenerate in normal pond water both retain the conditioned response. If they are allowed to regenerate in the presence of RNase (an enzyme which

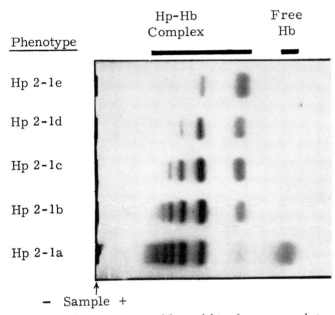

FIG. 3. Electrophoretic patterns of haptoglobins from persons heterozygous for the two structural alleles Hp^1 and Hp^2 but with different relative activities of the two alleles (from Sutton and Karp, 1964).

destroys RNA), only the head end retains the conditioned response, while the tail end does not "remember" anything. This is a rather striking experiment, but there are other lines of evidence which also point to a possible role of RNA in memory.

The RNA memory work is reminiscent of another possibly less familiar system, the synthesis of antibodies. For many years there has been much speculation as to *how* antibodies acquire their specific information. I am still unable to give you any concrete answer, but the evidence points rather strongly now to the Burnet hypothesis (Burnet, 1959) in which the amino acid sequences of antibodies, or gamma globulin molecules, are entirely specified by the genetic make-up of the host organism. Somehow either the coding sequence in individual cells becomes altered through mutation so that the mature organism has a variety of sequences to draw upon, or there are a number of genes transmitted which allow a variety of potential gamma globulin types. Against the latter hypothesis, it has recently been observed that gamma globulins are composed of two sub-units. Each of these two sub-units was shown to be the locus for certain inherited differences transmitted as simple Mendelian traits. This suggests that there is a single genetic locus being transmitted from parent to offspring for a particular type of sub-unit. In the presence of a foreign antigen the particular cell type bearing the appropriate antibody (and there may be a variety of appropriate antibodies) reproduces, giving the organism a clone of competent antibody producers. This particular clone persists, which explains the organism's memory for past antigenic experience. In this particular model, all of the information for antibody structure comes from the genome, while the antigen selects from among the potential antibodies those which are appropriate. This may be the type of effect which is observed in the case of RNA and memory. There may be a whole range of potential RNAs to be selected from. We would then have the RNA transmitted as a genetic trait (or traits) but expressed purely as a function of the environment. This would permit the genome of the organism to influence the cell in defining the total potential of the organism.

REFERENCES

Barr, M. L., & Bertram, E. G. A morphological distinction between neurones of the male and female, and the behavior of the nucleolar satellite during accelerated nucleoprotein synthesis. *Nature*, 1949, **163**, 676.

Beutler, E., Yeh, M., & Fairbanks, V. F. The normal human female as a mosaic of X-chromosome activity: Studies using the gene for G-6-PD-deficiency as a marker. *Proc. Natl. Acad. Sci.*, 1962, **48**, 9–16.

Burnet, F. M. *The clonal selection theory of acquired immunity.* Nashville, Tenn.: Vanderbilt University Press, 1959.

Davidson, R. G., Nitowsky, H. M., & Childs, B. Demonstration of two populations of cells in the human female heterozygous for glucose-6-phosphate dehydrogenase variants. *Proc. Natl. Acad. Sci.*, 1963, **50**, 481–485.

Huang, R. C., & Bonner, J. Histone, a suppressor of chromosomal RNA synthesis. *Proc. Natl. Acad. Sci.*, 1962, **48**, 1216–1222.

Ingram, V. M. A specific chemical difference between the globins of normal human and sickle cell anaemia haemoglobin. *Nature*, 1956, **178**, 792–794.

Jacob, F., & Monod, J. Genetic regulatory mechanisms in the synthesis of proteins. *J. Mol. Biol.*, 1961, **3**, 318–356.

Lyon, M. F. Sex chromatin and gene action in the mammalian X-chromosome. *Amer. J. Human Genet.*, 1962, **14**, 135–148.

Morishima, A., Grumbach, M. M., & Taylor, J. H. Asynchronous duplication of human chromosomes and the origin of sex chromatin. *Proc. Natl. Acad. Sci.*, 1962, **48**, 756–763.

Russell, L. B. Genetics of mammalian sex chromosomes. *Science*, 1961, **133**, 1795–1803.

Sutton, H. E. *An introduction to human genetics.* New York: Holt, R'nehart and Winston, 1965.

Sutton, H. E., & Karp, G. W. Variations in heterozygous expression at the haptoglobin locus. *Amer. J. Human Genet.*, 1964, **16**, 419–434.

DISCUSSION

Sutton: Perhaps some of you have contrary ideas about what was said about RNA and memory. We might inquire which fractions of RNA are involved in these memory experiments. There is no good evidence on this point; the magnitude of the RNA changes are hardly compatible with what we know about messenger RNA levels, and yet what we think we know about ribosomal RNA certainly does not suggest it as a very likely candidate for a place in the response to learning. Consequently, most of the theories do suggest messenger RNA as the responding agent, even though it is normally a very small fraction.

Freedman: I'd like to comment on this question of memory and RNA. I have given a certain amount of thought to it since it is such a popular notion right now, but I think it is another blind alley as far as an explanation of memory is concerned. Probably the best book on memory is Charles Bartlett's book on "Remembering" in 1932 which takes the Gestalt point of view and talks about schemata. Schemata are, of course, the essence of human memory and I bring this up, not to get into the discussion of memory, but to point up the fact that it has been shown anthropologically and physiologically that what we remember depends upon our general knowledge and perspective. To make a connection between the things that Proust talks about in "Remembrance of Things Past" in which he recalls things by smell associations or by the tilt of the street to the RNA level is quite a jump.

Sutton: Well, of course it's a jump and yet there must be a physical molecular basis for memory.

Freedman: Well, one would have to postulate that with the tilt of the street there's sort of a tilt of the RNA balance.

Sutton: I don't think that *what* one remembers has any relevance, at the present time anyway, to any proposed chemical mechanism, but we *must* postulate such a mechanism. Perhaps I should say you may not have to *postulate*, but you have to *assume* that a mechanism exists. For example, perhaps RNA is influenced or perhaps there is an intercellular reorganization. At the present time these ideas cannot in any way be associated with *what* is remembered or *how* an association is made at the conscious level.

Freedman: All right, we'll both agree that there is a jump.

Gardner: I'm ready to sympathize with Dr. Freedman's feeling that there is a jump and that it is a little frustrating at this point. Some of us who have studied memory know that the behavioral structures which, if you will, lie in the enduring patterns of behavior that are involved in memory processes are extremely complicated and undoubtedly hierarchically organized all the way down to the molecular level. I would agree that it is frustrating, but I do not consider that this concept is inadequate or inappropriate. It is very exciting to learn about some of the possible links in this chain largely of hierarchical unknowns. We know a few things about the superordinate controls and, now, a little about the molecular level of control and regulation. It is our job to fill in the steps in between. I find this very exciting.

Fuller: One of the problems at the molecular level is the unstable nature of messenger RNA. I also question whether the Burnet theory of a clone of cells is appropriate. In the first place the Burnet theory is controversial, but, if we do accept it, it implies to me that there may be a clone which can develop a nervous system for the cell division completed, perhaps not a clone of cells but a clone of ribosomes or something of this sort. There seem to be difficulties which don't concern the psychological level about which I have reservations as well as Dr. Freedman, but difficulties which are actually at the molecular level itself.

Sutton: Well, I think that it is very relevant to question the "clone of cells" concept. We don't know how many cells might be required to cause these changes, and, as you say, it may be either a clone or a group of cells which is involved. On the other hand, it may be that a given cell, once turned off or on, may induce the same system around it. In the experimental situations I am familiar with, it seems likely that more than one cell is required. Here, the presence of "on" substance may cause the adjacent cells to turn on and so forth. In contrast to the Burnet hypothesis, each cell may be capable of more than one function of the sort under discussion.

Gardner: Let me give one further reaction to this. Frankly, I am very naive about the molecular operations of genetics. I am a psychologist. I was really struck with this work on cellular regulatory mechanisms and the place of feedback in these systems. In considering controls at a much higher level, it seems to me that "behavioral controls" may account for individual consistencies and differences in certain aspects of remembering, thinking, and concept formation, etc. Some of the usable concepts at that level also are inhibitory or regulatory operations of the same kind, the same general patterns of feedback mechanisms, the importance of the negative feedback, etc. In spite of the difference between levels here, I am struck with the analogous ideas which seem useful as theoretical explanations. I think that there is a lot of empty ground here. Psychologically we are very ignorant of the memory process and we only have hope that in the next 50 years we can peel off some of the layers.

Sutton: It is gratifying that geneticists have finally been able to attack some of these more complex problems rather than dealing with the environmentally immutable attributes which are simpler to analyze, but which, perhaps, are not as important.

Gardner: I am sure I don't understand the implications, but now you are talking my language.

Gottesman: Before we leave RNA, I would like to ask a question about the inactive X chromosome. Are there any other differences at the molecular level between males and females that might account for sex differences and behavior? What differences in behavior are not explained by learning theory or sex typing?

Sutton: I'm not certain I really understand your question. Of course, in mammals, the presence or absence of the Y chromosome appears to be the primary determinant of sex regardless of the number of X's in a cell. No one really knows what the Y chromosome is doing. For a while it was believed to be devoid of genes, but with the discovery that it is the primary sex determinant we must postulate the presence of genes on the Y.

Fuller: There is possibly a histocompatibility gene on the Y chromosome of the mouse. Some people are worried about the dogma that there are no genes on the Y chromosome. I think that the evidence for this gene being on the Y is pretty good.

Guttman: Actually, what the Y chromosome does, it does for a very short time in embryogenesis, then it stops being active. Really, as far as we know, the Y chromosome just works at the time of the development of the testes and that is that, it has finished its job.

Lasker: That is true of most anatomical structural determination. For example, bone determination happens very rapidly. One moment there are no bones and suddenly the embryo has all the bones already species-specific. So I do not think a limited period of activity is limited to the Y chromosome; in anatomical determination, the period of organogenesis is very short, extraordinarily specific, and rapid.

Fuller: One thought occurred to me concerning the inactivation of the X (sometimes called the Lyon hypothesis) as this was mentioned in the discussion of red cell mosaicism in females with glucose-6-phosphate dehydrogenase deficiency. If we assume that there is a similar mosaicism in cells of the nervous system, then these cells differ in a very fundamental fashion from red blood cells or pigment cells since they are connected functionally throughout life. This would mean that any sex-linked factors affecting behavior would exist in a mosaic nervous system whose parts will still interact, a situation quite unlike the enzyme deficient red cells. I do not know if this has any meaning, but we might have to look for the mosaicism at the cellular level rather than at the behavioral one. We have some mosaic *Drosophila* which are partly male and partly female and it would be interesting to see what happens when part of the brain is male and part female.

Freedman: Dr. Vandenberg, I wonder if you could say a few words about twinning and the Lyon hypothesis. You had published some material regarding this.

Vandenberg: Yes, of course. The idea was as follows: if the Lyon hypothesis were true and held in a general fashion, you would expect to find greater variability within female twin pairs than within the male pairs. It looked quite exciting, and I published the data, but, on the suggestion of Dr. Sutton, I redid the calculations and found an error. Now there is not really any significant difference for all variables as a whole. It may still be of interest to check this in some other data, particularly with the variables showing greater female within-pair differences. The situation is not clear-cut, even though there may be some excess variability within female identical twin pairs.

Freedman: It was my observation also that you had as many pairs of male identical twins that had decisive differences as you had female.

Lasker: On this matter of variability in the sexes, the older anthropological literature is full of suggestions that females are less variable than males. Perhaps one needs a relative index of variability rather than an absolute one. I do not know whether there is any substantial basis for this statement which repeatedly occurs in the earlier literature.

Freedman: Witkin has observed that females are more oriented to the environment than males. If the environment is stable, you might consider this orientation to be the source of the invariability.

Lasker: I wonder where Witkin's psychological idea originated. In the physical-anthropological sense there was an old notion that females stay closer to the norm in some way, but they may have been considering only those characteristics that happen to be easily modified in males.

Sutton: Dr. Vandenberg, could you discuss the rationale behind the idea that females should be more variable? I think it may depend on your point of view.

Vandenberg: If the female is a mosaic of the two X chromosomes there is an opportunity for more variability. The assumption is that the decision, which of the two X chromosomes will be inactivated, is made after the twinning, and late enough in embryogenesis to have permitted a number of cell divisions. Then, each twin would possess different proportions of the two Xs. This mosaicism should result in greater variability within the female pairs than in the male pairs where there is only one X. In the extreme case, one twin would have cells with only one X chromosome while the other twin had cells with only the other X chromosome.

Sutton: This excess variability would occur only within twin pairs, not in the population of females.

Vandenberg: Yes, that is correct. It could be detected statistically only in identical twin pairs.

Lasker: But if in populations of females the variability is down for other reasons, then the direct male-female comparison in this manner would not be valid anyhow.

Sutton: That is correct.

Loehlin: Shouldn't this hypothesis suggest that females should be less variable in the general population because they are mixed of these extremes?

Sutton: It does suggest that to me. But this is the population of females versus the population of twin pairs.

Vandenberg: With X chromosomes, you would have to have marked differences between the twins in the proportions of the two X chromosomes to detect mosaicism. You would also have to have a marked discrepancy in the effect of the two X chromosomes on some variable. For example, one of the alleles producing greater size than the other, with one twin possessing 70% of the first allele and the other only 20%. Then you *could* get a marked discrepancy. This needs much more thought before it is a useful hypothesis for twin studies. Perhaps, as suggested by Dr. Sutton, the opposite approach would be better. It has occurred to me to look for identical twin females who show a marked discordance in measures that cannot be explained in any simple way by environmental factors.

Guttman: The inactivation of the X may not be a matter of chance; it is only a matter of chance because we don't know anything about it. Some day, and twins may help us in this, we may find that this is a nonrandom process.

Vandenberg: Inactivation might be influenced by other genes or by the region in which a cell is located.

Guttman: It could be influenced by the genome as a whole.

Fuller: The number of cells in the embryo at the time that inactivation occurs is also critical. If this is a fairly large number the differences in the mosaics are going to be minor.

Sutton: In several cases there is a great departure from a 50-50 type of distribution. I would like to mention in conclusion a twin pair showing a mosaicism of a small acrocentric chromosome. Each of the persons in this twin pair was mosaic but apparently mosaic for different amounts of tissue. They were both in an institution for mentally defective children. One had an I.Q. of about 30 and the other about 70. They varied in a number of other traits as well. With respect to the blood antigens there was no evidence of non-identity. The X chromosome was not involved, but it is an example of mosaicism inducing a great deal of variability.

Studies in Racial Isolates

Gabriel W. Lasker

This chapter will attempt to introduce some enlightenment from the field of anthropology on genetics. When I have worked in human genetics, I have worked in the United States, entirely in pedigree studies and the like. In anthropological field work there are so many factors which are difficult to bring under control, that genetic analysis has not been possible in most of my studies and the most I have been able to do is carry a genetic point of view into the interpretation of my results.

Recently, basic genetic theory has been making spectacular headway, and, if an attack on the important problems concerning the inheritance of behavioral characteristics of man is to benefit from these advances, we may well look at the conditions within human behavioral genetics first. This leads us to the question of the relationship of human genetics and the genetics of other organisms. The genetic material, DNA (deoxyribonucleic acid), is present in both, chromosomes are similar, crossing-over and linkage work in the same way; everything we know about chemical and molecular genetics is similar in man and other organisms. Now we come to the cultural level which we know is very different.

Before dealing with this I should mention in passing one point to keep in mind. Work in human genetics is subject to nonscientific appraisal. There are underlying assumptions by granting agents, editors, and sometimes investigators themselves that get in the way. These assumptions are not felt directly in scientific work or editing of individual papers, but more subtly through public reward for poor work whose conclusions seem to point in what are considered to be socially desirable directions. Sometimes, there is also a lack of public reward for good work if it could easily be misinterpreted to point the other way. Distortion of values is a risk in this field which is not felt in some other fields. Here, perhaps, anthropology has something to offer. In areas of knowledge in which value structure has plagued studies of our own society, work in very out-of-the-way places where quite different values prevail, sometimes permits escape and a certain degree of detachment. This is one thing anthropology can do for studies in behavioral genetics, that is, it can provide opportunities for a more detached look at some of these questions.

Culture, as a separate factor in human genetics, has a kind of inheritance. It is transmitted and modifies behavior, of course, and is modified in the transmission. The mechanism of chromosomal inheritance is, however, lacking for culture. Rather, there is a sort of blending inheritance which permits very rapid change and modification. Culture has a certain natural selection of its own since selection of culture-bearers for survival tends to select for culture. At both sides, culture is related to the biological base; that is, as a determinant of biological processes and by determinants of culture in biology.

Not much is usually said about how human culture modifies the field of action of the biological genetic mechanisms. Biochemists are now interested, for instance, in certain aspects of human activity which do modify natural selection, such as drug action. Natural selection for polymorphisms in man does occur even in respect to synthetic drugs which, as far as we know, do not exist in the natural world and have therefore been a factor only for the last generation. These reactions must have become established in response to other factors than the drug itself. In the case of G6PD (glucose-6-phosphate dehydrogenase) deficiency, it is well known that many drugs including certain sulfonamides produce anemia in the enzyme-deficient individuals. Other cases of this kind have recently come to light. This is not, I think, an area in which we are going to be able to predict progress.

There is, however, one area which seems very important to me in respect to the cultural base, i.e., the selection of mates. In this strongly culturally determined process man differs considerably from other organisms. If we are going to examine this distinction it must be done at the population level, where psychologists could make a unique contribution.

At this moment, I should like to digress and raise another point.

Dobzhansky and other geneticists note that man's genetic system is marked by world-wide range and he is a polytypic species, i.e., a varying species with local subgroups differing considerably. Dobzhansky (1962) has reasoned that, as a result, instead of a classical case of natural selection, in which a favorable homozygous type is selected, man presents a situation in which selection in many cases is for the heterozygote. The hemoglobin of sickle-cell anemia in man has been cited repeatedly as one good example of this. Several other instances of balanced polymorphism have been inferred in which some disadvantage to the individual of being homozygous for either of the alleles is known. In the case of sickling, the anemia of the homozygote sickler and the susceptibility to malaria of certain types of homozygous non-sickler leads to a favorable natural selection of heterozygotes. Under the balanced model, mankind represents an array of more-or-less normal genotypes for various functional capacities, differently modified and variably adaptive in different environments.

But to return to the population, I think the influence of the breeding structure would be germane under either theory—a balanced model or the classical model of selection for a type. Probably features of both models will be important if we take a world-wide view of the species. This has rarely been done, and it is what I wish to propose here. Studies of the human breeding structure at the population level have been conducted only in small local areas. Such studies are now fairly numerous, but there has been very little synthesis of the results. I know of only one determination of gene frequencies in men which deliberately considers a single species. McArthur and Penrose (1951) did this for the ABO blood groups. They sampled, where possible, each type of population of each country, multiplied by a factor representing the probable size of this population, and arrived at the probable gene frequencies in man. It seems surprising that there has been so little thought along this line, since it permits one to study individual populations and see their deviation from a central figure representing mankind. Brues (1954) showed that there are at least as many classes of ABO blood group distributions which are *not* found in any human population as there are which are found. There are no populations with a very high frequency of B; those in which B is relatively high, all have some A, and those in which the frequency of A is at its highest, there is no B. Brues suggested a natural selection against those distributions which are not found, and vectors of selection toward the average human figure. The deviations from this average she thought might be due to random factors. Some of the deviation from the central value could be explained as differential selection under different circumstances in different parts of the world. A third explanation, the "founding" effect, has not been proposed, but probably should be considered. The number of generations subsequent to the founding of the species may have been too short to permit evolution of all combinations of frequencies from those which occurred in original man. It is well to remember, however, that differences in blood group frequencies are found between the gorilla and the chimpanzee, the anthropoid apes which now definitely seem more closely related to man than the Asiatic apes. The study of population from the genetic point of view probably best begins with the models of the theoretical population geneticists such as Wright (1943), who have attempted to lay down some of the conditions. The human population may be considered to be composed of clusters of individuals. Mating is then essentially a random event occurring at a given rate within a cluster or "island," and at a reduced rate between "islands." In this model, the island can be considered a closed unit. I have in the past attempted to identify it with a community, at least in peasant and folk societies. The model also works fairly well for some tribal societies, in which the isolating mechanisms between the islands are often reasonably effective. For certain organisms, and to some extent man, the single individual is the island and the probability of the mating of two individuals depends on their places of birth. This assumes that the surface distribution of the mating population is more or less homogeneous, whether or not the space is considered as two dimensional. However, social factors may introduce a third dimension. In some societies, notably those of India, even with physical propinquity the social distance may be considerable. If the populations are large, the net effect of social factors may be small, but this dimension should be considered.

Darwin himself was concerned with sexual selection in man. The basis of selection which he considered was largely individual. We now know, for instance, that individuals who are ugly by our standards are about as likely to mate and reproduce as good-looking ones. Simple selection leaves the "unselected" to mate among themselves yielding as many or more offspring. These questions have been properly investigated in a variety of situations, in different countries and societies, and I think that simple exclusion from the breeding population under the classical model of selection is very unlikely. However, further investigation may reveal factors leading toward homozygosity when a breeding population is split by such practices.

How can one study the human breeding population on a worldwide basis? Obviously all one can do is extrapolate to larger entities from sub-samples. One would expect certain cultural characteristics to correlate highly with breeding practices. I already have some impressions from field work of my own and from other studies that, in Latin America for instance, a certain distance between birthplaces of mates may be more or less characteristic of peasant societies. To the extent that the type of society defines the size of the breeding population, one could take the number of individuals in the world in such societies and multiply by the characteristic distance. Other types of societies could be treated in the same manner.

There are some surprises. The most isolated populations, as you would expect, turn out to be on small islands; yet, if you look at mankind in general the populations of islands are not usually extreme. There are exceptions. For instance, off the north coast of Australia in the Gulf of Carpentaria, in that little square gulf that juts down between Arnhem Land and Cape York, there are some islands which are cut off from the mainland; these people have rafts and nothing else. According to Simmons *et al.* (1962), 34 people on about 15 rafts left one of these islands (in 1940 and 1947) because of lack of water and tried to reach another island. The island people can see other lands, they know they exist, they see smoke from the fires. But, of the 34 people on the rafts, only 17 ever reached their destination, the rest were lost at sea. The island is so isolated that with the mode of transportation available they could rarely get from one island to another. On one of these islands blood group B has the highest frequency found in Australian aborigines anywhere on the subcontinent. On another of the islands there is some blood group A and no B. Presumably, random factors led to elimination of one gene on the first island, the other on the second.

Probably, because of their circumscribed size and lack of opportunity for subsequent rapid expansion of a once limited population, islands do not show the extreme racial types found in continental groups. Therefore, I think, a satisfactory explanation of major differences in the human population will come through extrapolating from island situations to situations that were formerly widespread on continental land masses. On the mainland a group can expand very rapidly after a period of isolation because of possible introduction of favorable cultural developments such as better weapons or improved food acquisition techniques.

Hope of reconstructing past breeding populations through documentary evidence is slim. In a few countries there are church records, parish records, and official registers. For instance, the Mormons have been very assiduous in collecting records of this kind. These records have sometimes been made available to scholars, but one must still find the birthplace and some details of all ancestors for many generations. In six generations one must locate 64 ancestors. With bastardy rates of the order of 1 or 2% one runs into problems of reliability. If the ancestor is an outsider, the probability of recording error is increased. Usually, the best solution is dealing with the immediate situation. Here one can ask a person where his mother was born, where his father was born, and in a very high percentage of cases get the correct answer. Individuals in all societies usually know this even though marriage may not dictate paternity.

The extreme isolation that is sometimes presumed rarely occurs. In Australia, an ethnologist, Tindale (1953), was able to interview individuals belonging to groups which had had no contact with white men or native police. Even in such tribes matings outside the tribe amounted to 15% of the total, a higher proportion than one might have predicted. This means a genetically effective immigration across tribal boundaries of about 7% per generation.

In Australia, the groups are kept small by the very sparse vegetation, shortage of water, and rough living conditions. Where there are very large groups, mating across tribal barriers is reduced. Thus in Africa, where there are huge kingdoms, although the frequency of extra-tribal marriage for the whole group may be small, at the periphery it may be great so the gene flow across tribal boundaries is not impeded greatly in this case.

At first I thought that peasant societies were actually quite isolated since the individuals are tied to the land. This could offset the large size of these communities compared to those of tribal societies, even though peasants depend on a central authority, and are thus members of a large population. But the newer data, especially from the Australian islands, suggest that some primitive societies are more isolated than any peasant societies *ever* were. Furthermore, some peasant societies, notably in China, are not at all isolated—the peasant village is not endogamous. Even though the distance men go to seek spouses is limited, each village does go to the next, uniting the whole continental area into a single mass with individuals isolated by distances but not into discrete units. This is reflected in biological differences in the area. The distribution of certain characteristic physical features show gentle clines. Thus, for instance, shovel-shaped incisors reach their maximum frequency in the north, whereas individuals who lack this type of teeth increase as one moves further south in the Eastern Asiatic mainland. Stature also exhibits this cline decreasing more or less gradually from the north to the south of China.

To return briefly to the behavior of isolated peoples, I think there are especially great difficulties in ascribing any behavioral differences to genetic differences. Some of those places that I have listed as known to be reproductively very isolated and small remain isolated. In Bass Straight Islands off Australia, the offspring of Tasmanian aborigines, Australian aborigines, and white men have remained in isolation for some time in a very constricted space. On Pitcairn Island six of the mutineers of the *Bounty* with their Tahitian mates left their offspring with very little admixture from the outside. Tristan da Cunha in the South Atlantic was recently evacuated, but I understand that most of the people are going back. At the time of Napoleon, the British put a small garrison on the island, but since then there has been a more or less self-contained population. It consists almost entirely of the descendants of seven or eight men and a similar number of mulatto women from St. Helena Island ("traded" for potatoes by a whaling captain) plus a number of subsequent additions. The population was evacuated after a recent volcanic eruption but, in general, those few who married in England are not among those returning to the island. There are some isolated islands in Japan. Hosojima is one with lots of inbreeding (Ishikuni et al. 1960). In other parts of the world, allegedly isolated communities when adequately studied sometimes turn out to be less isolated than previously believed. For example, among 110 inhabitants of the Upper Xingu River in Brazil there were 13 to 15 individuals who came from other tribes.

Now it is true that these isolated remnant groups of people often represent genetic extremes. The Bushmen, the Australian aborigines, the Ainu, the Pigmies of Africa and the Andaman Islands, and the inhabitants of inland parts of New Guinea represent racial extremes, but they are also some of the most extreme cultural types so it is not possible to ascribe some of the "racial differences" in behavior to the genetic base.

There are, however, possibilities for collection of more direct genetic data in these places and I think doing this is a very urgent matter. These populations give us our last chance to examine those few items of data we can get to anchor the reconstruction by analogy of earlier periods in human history. Thus demographic data are very urgently needed. We ought to know what tendencies these groups have to breed among themselves and how small the breeding population of these people really is, so that we can interpret the probable influence of random genetic drift in the wider human scene. Specimens of blood would permit studies of blood groups, haptoglobins, transferrins, and the like. Correlation of data concerning the demography and the gene frequencies can then provide two independent approaches to a study of the genetic structure of these populations. Aspects of the cultural environment, such as the kinship system and the means of subsistence, may very well be among the determinants of effective population size and hence of the genetic process so this information should also be recovered. Geographic and climatological conditions are intimately related to cultural factors in determining the way of life. We can try to obtain the time dimension in order to check the probable rates of change in the amount of time available. We must rely on archeology to get any real notion of time depth. Skeletal remains can help in the problem of time depth, but what one is able to do with the skeleton itself is so limited that I think it is critical to use such archeological material only with good data on modern populations. Therefore, I think we must make a last fast effort to recover demographic and genetic data on those few peoples still remaining outside the influence of international technological society. This will put us in a better position to unravel the fabric of the human species as a single species, as a breeding species, not only as of now, but for some time in the past, and, by implication to plot the kind of changes going on here and now and in the future.

REFERENCES

Birdsell, J. B. Some population problems involving Pleistocene man. *Cold Spring Harbor Symp. Quant. Biol.*, 1957, **22**, 47–69.
Bonne, B. The Samaritans: a demographic study. *Human Biol.*, 1963, **35**, 61–89.
Brues, A. Selection and polymorphism in the A-B-O blood groups. *Am. J. Phys. Anthropol.*, 1954, **12**, 559–597.
Crow, J. F., & Morton, N. B. Measurement of gene frequency drift in small populations. *Evolution*, 1955, **9**, 202–214.
Dobzhansky, T. Genetics and equality: equality of opportunity makes the genetic diversity among men meaningful. *Science*, 1962, **137**, 112–115.
Glass, B., Sacks, M. S., Jahn, E. F., & Hess, C. Genetic drift in a religious isolate. *Amer. Naturalist*, 1952, **86**, 145–159.
Goodman, M. Serologic analysis of the systematics of recent hominoids. *Human Biol.*, 1963, **35**, 377–436.
Ishikuni, N., Nemoto, H., Neel, J. V., Drew, A. L., Yanase, T., & Matsumoto, Y. S. Hosojima. *Amer. J. Human Genet.*, 1960, **12**: 67–75.
Kucznski, R. R. *Demographic survey of the british colonial empire.* Vol. 3, London: Oxford Univ. Press, 1953.
McArthur, N., & Penrose, L. S. World frequencies of the O, A and B blood group genes. *Ann. Eugenics*, 1951, **15**, 302–305.
Porteus, S. D. *The psychology of a primitive people, a study of the Australian aborigine.* New York: Longmans Green, 1931.
Shapiro, H. L. 1936. *The heritage of the Bounty. The story of Pitcairn Island through six generations.* New York: Simon & Schuster, 1936.
Simmons, R. T., Tindale, N. B., & Birdsell, J. B. A blood group genetical survey in Australian aborigines of Bentinck, Mornington & Forsyth Islands, Gulf of Carpentaria. *Amer. J. Phys. Anthropol.*, 1962 [N.S.] **20**, 303–320.
Spuhler, J. N., & Clark, P. J. Migration into the human breeding population of Ann Arbor, Michigan, 1900–1950. *Human Biol.*, 1961, **33**, 223–236.
Tindale, N. B. Tribal and intertribal marriage among Australian aborigines. *Human Biol.*, 1953, **25**, 169–190.
Wright, S. Isolation by distance. *Genetics*, 1943, **28**: 114–138.

DISCUSSION

Freedman: Isn't the human species the only example of a species that is also the only member of its family?

Lasker: This is a matter of convenience in classification. There are two kinds of biological classifications and this insistence that man is so isolated is based on what we think about ourselves and so using a grade classification, rather than a genetic classification. The serum proteins of man are more similar to those of the chimpanzee and the gorilla, than those of the chimpanzee and gorilla are to those of the orangutang, Goodman, (1963). The gorilla and the orangutang are put in the family of the Pongidae, by virtually everyone today, and man is put in a separate family. So, I think there is no basic biological validity to this separate family for man. It is possible of course that the human specializations are so significant in themselves that man should have a family to himself. Then, the logical thing would be to give another family to the gorilla and the chimp and still another to the orang. One could do this, but it is a matter of convenience to lump them together, and extreme splitting in taxonomy is not usually viewed with favor. I do not think men are so different in fundamental biological respects, although most primatologists believe that the assumption of the erect posture and the specializations of the cerebal cortex for speech are so important that we should give man a very distinct name, that of a family. Does that answer your question?

Freedman: Yes. In a way it answers the other question which I have in mind when I think of island populations and the progressive speciation in the Darwin finches, on the Gallapagos Islands, which has never happened in man. In this context, would you care to comment on Coon's recent hypothesis in regard to the origin of races?

Lasker: Well, I do not know what I could say about it. I have not yet read the book, but I have read some reviews. The question seems to be whether there have been parallel separate evolutions over any considerable period of time? I do not know of any evidence favoring the view. There is a degree of discreteness of populations, of course, and when you look at the details, apparently human populations are sometimes, in some places, more isolated, so that it might take a considerable time for genes in one population in one part of the world to reach through the species to distant places. This was probably especially true under past conditions. Virtually all known mutants apparently did, surprisingly, spread to virtually all known populations, though with different frequencies.

Lindzey: Coon's argument really would be the reverse of that, wouldn't it? There was independent evolution, let's see, five times?

Lasker: This has been said. I have not seen the original.

Sutton: I have read half of the book. I do not think he goes to quite that extreme. He does not deny some gene flow. He maintains that racial groups have existed prior to what we consider the present time.

Lasker: There is some surprising evidence that people actually did move around, and passed from hand to hand some things like stone tools—very remarkable similarities are found over wide areas including the Americas and Asia for instance. Africa and Europe have the same kinds of sequences. Well, did they evolve those tools independently too, or did they borrow them from each other? The same paleolithic tool types occur in Africa and in Europe, and in western Asia and parts of India. In eastern Asia there are slight differences, but nobody is claiming that people did not cross Asia. People moved very great distances in historic times in a single generation just by walking or on horseback. There are the well-known migrations and raids of Tamerlane, the Huns, the Mongols, and Alexander the Great. People with small knowledge of technology and some method of carrying food were able to move considerable distances, and when people move they mate. The Iroquois Indians were fighting far away from their base at times. They came from Ontario and New York State, but originally had come from further south. I think that we cannot only consider how far an individual goes to get his mate in any one generation, but we have to bear in mind that with the food shortages, man could move quickly over long distances, rarely, but sometimes.

Fuller: All you need is one individual to move the genes in.

Lasker: Of course you have to account for that particular gene taking hold.

Fuller: Well, you remember that these people did not come as individuals, they came as a population.

Guttman: About the problem of isolates, I strongly agree with the urgency of studying them quickly and on a large scale and I would like to cite Israel as an opportunity which has almost been missed. I don't think we have missed the Samaritans because they are not going to marry anybody else, although they may die out in a few generations. I do not know whether you would call a 2000-year-old community an isolate, but I think you could, even on the human evolutionary scale. We have some villages in Israel which came *in toto* from Kurdistan and from Cochin in India and from places in Yemen, which really, as far as we know, had not married outside the community for about 2000 years, but now they are doing it very quickly.

Lasker: The religious isolates I didn't mention. Some of the highest rates of isolation in the world are found in these religious isolates, but they are possibly a special case as part of a wider cultural context. Dunkards have been studied in Pennsylvania by Glass *et al.* (1952), the Hutterites of northern United States and Canada by Steinberg, and the Samaritans of Israel and Jordan are being studied right now by Batsheva Bonné (1963) and Boyd and another group. Mourant in London is also interested in them. There are also some groups in Japan, do you recall the island where Neel's group studied the Buddhists and the Christians?

Sutton: Hosojima is the only one I know about.

Lasker: No this is a different island, the Michigan group worked on one of the islands of Japan on which they found religious isolates of Christians.

Sutton: I have done blood proteins on all the groups mentioned.

Guttman: I know, but in general only blood groups are studied. We would collect behavioral information as well. We're trying to do it now and we are hoping that NIH will be in as much of a hurry as we are. But we may miss our chance. In the Kurdish villages the rate of consanquinity was 17%, a very high figure, now it is down to about 5% and soon it will be down to that of the normal interbreeding population.

Freedman: I have a big question. How would you tease out the genetic aspects from the cultural aspects of the population? The only thing that I could think of is using very early appearing traits or early stages of development.

Guttman: Not necessarily. We have a scheme which shall be presented.

Lasker: My view is to get at those things we know we can attack first. There is one bit of information that can be obtained even after the communities have gone, one can still find out where people were born and who they married.

Guttman: Batsheva Bonné has done beautiful work on villages in Italy. These are not complete isolates but the gene frequencies sort of peter out as you move away from a certain focus. She also studied the migration patterns of some isolated villages in northern Italy.

Lasker: Here you get into the complex problem of selective pressures, drift, and migration. You have a multifactorial situation to interpret when you find a build-up of gene frequency in a small location and a fall-off away from it.

Sutton: I should like to make one comment about analyzing some of these populations. Most of the really elegant theory development is based upon the assumption of 2 or 3 alleles. In the hemoglobins and transferrins, we know of 20 or 30 alleles, and these are based only on electrophoretic differences. There may be a very large number of alleles. I am not a population geneticist, but it must influence the theoretical interpretation if we work with only 2 or 3 of the 20 or 30 alleles coexisting in populations.

Lasker: Yes. Classical geneticists will aver that a mutation has little effect on a population unless it is repeated. Unless you get repeated mutation, ordinarily it cannot become established in a population, unless it confers selective advantage. With a 1% selective advantage, it will generally wind up in the dump heap. Now what does that say about the mutations we do find which have occurred repeatedly? If you trace them back, the great majority of people with the mutant gene came from a single ancestor or two. The fate of most mutations to this allele has been extinction and only a few remain. The rarer the condition, the more reasonable is the assumption that all mutants came from the same event. I do not know what length of time it would take for the spread of a neutral mutant, but it may not be too great. Since some conditions such as Tay Sach's disease are generally restricted to the Jewish population it makes sense to consider these as relatively recent events, since it is hard to believe that the mutation in prehistoric times would remain confined to these populations. I was very interested in pentosuria, a condition in which a pentose sugar is excreted. All the cases which were carefully studied had been in Jews or people who might have been Jews. Recently some good pedigrees have turned up from Lebanon. In any case, distribution in non-Jews is restricted to Semitic peoples from the near East. The restricted distribution suggests that pentosuria is quite possibly a relatively recent mutation, although I do not know how many thousand years ago it occurred.

Multivariate Analysis of Twin Differences[1]

Steven G. Vandenberg

It is useful in thinking about psychological contributions to human genetics to adopt a scheme of *stages* or degrees of knowledge about hereditary mechanisms in man as a way to place a lot of facts in relation to one another.

In the most advanced stage, we would have knowledge about a condition due to one gene located in a known region on a given chromosome; in general, human genetics has not reached this state yet, although we are approaching it for sex-linked traits. We have not reached this stage because there are very few firmly established linkages, so we have not been able to construct chromosome maps in man.

The next best stage of knowledge is one in which the gene can be assigned to a specific chromosome although not to a definite location on it. Once there are several, widely spaced, markers for each chromosome one could determine the chromosome on which other single gene traits are located. In human genetics, this stage also has been reached only for genes on the sex chromosomes. Tentative locations for such genes will probably soon be suggested by studies of chromosomal abnormalities.

A stage of yet less exact knowledge is where we can only distinguish between autosomal and sex-linked traits. Varying degrees of ignorance might further be distinguished within this stage; for instance, we may suspect that a single gene is responsible for a given trait, but, because the condition is a rare one, lack enough cases to work out a precise mechanism; or, one gene is thought to be responsible but other diseases may lead to conditions closely resembling this one, so called phenocopies, resulting in a diagnostic

[1] The work reported here was supported by a grant from the National Institutes of Health (MH 7033) and the National Science Foundation (GB 466).

problem. I should emphasize that I am thinking especially about problems of neuropsychiatric or psychological conditions or phenomena.

Then we might distinguish situations where more than one gene is involved. If it turns out that there are only two or three genes, pedigree studies might still be useful, and we can try to build various models for combinations of genes and test how well such models fit actual data. We could do this right now, I think, on such things as height and handedness; perhaps even intelligence, if we are willing to regard intelligence as a general factor.

Finally we may have conditions controlled by many genes. Here we have to revert to the methods of biometrical genetics: analysis of variance, regression analysis, path coefficients, and heritability estimates and this is where twin studies come in. Regardless of the number of genes involved, we often have situations in which the effect of the gene can be so varied as a result of interaction with the environment that pedigree studies are, at the moment, not profitable. In such situations you can still do biometrical genetic studies, but conclusions from similar studies done in different locations may differ if the environmental conditions vary in the samples used. Here anything we can learn about environmental effects would make later studies of hereditary factors better, if not easier. As a matter of fact we do not have very good ideas yet about how theories derived from studies of environmental effect can be fitted into biometrical genetic studies. A demonstration with appropriate theoretical models could help clarify this whole problem and set the pace for future work. It would be good to take a particular phenomenon, such as height, which is understood fairly well, and work it through in as much detail as we can, to provide a model for studies of other variables. Such an approach can lead to more adequate models than the one in which heredity and environment are proposed as two independent sources of variance which together add up to 100%. The multiple abstract variance analysis, proposed by Cattell (1953), in which between-family and within-family hereditary and environmental variances are considered, as well as various interaction terms, is a good example along these lines. Finally it is well to keep in mind that complete knowledge of hereditary mechanisms is not necessary for doing useful applied work. In animal breeding, production of milk, bacon, or eggs can be considerably improved without precise knowledge of specific genes. It is true, however, that rather drastic selection is necessary to achieve such improvement. The mere elimination of poor producers would, in subsequent generations, alter production little or not at all; for this reason such methods would not work as eugenic measures with man.

My own background and temperament has led me to choose problems near the bottom of this list rather than near the top and this is dictated in part by previous work I have done, some familiarity with statistical techniques, and other things in my personal history, but in no way reflects a value judgment about what is more important. Furthermore, for what are socially the most significant and most important psychological problems, we are more or less forced to work at the bottom of this list, if we are going to make a start right now. A few more general remarks may be in order before I go to the area of twin studies. For neuropsychiatric problems it is still useful to get more adequate surveys of the incidences of various diseases in various parts of the world, because unless there are marked geographical differences in gene distribution, the presence of similar incidence rates under widely differing cultural conditions would tend to point toward hereditary factors. These studies of incidence rates could be made more profitable if they were combined with the collection of biochemical information. We can also do family studies and studies of incidence rates in relatives of patients, and work of this kind should certainly be done more widely.

Finally we come to twin studies. Twin studies are, perhaps, a more economical approach to human behavior genetics than family studies, as long as we are completely ignorant of hereditary factors. First of all, the twin-study method eliminates the age factor which we cannot deal with in many of our variables. Even in studying height, age is a serious problem; if you want to do pedigree studies you can start with adults which limits the accessible range of relatives somewhat, or if you start with children you must wait until they reach adult status before you can compare their measures with those of their relatives. In twin studies you do not have to worry too much about the sampling problems at this stage of the game. For the study of normal variations, twins can furnish quite a bit of information. We will not be able to work out precise mechanisms, but we can do a lot of screening and look for promising variables or areas in which variables might be found that can subsequently be studied by other techniques. I think that we need a lot more work on normal psychological variation before we are ready to do pedigree studies with existing measures of cognitive or personality variables.

Now I would like to review what can be done with twins. First of all, of course, there is the conventional comparison of the concordance rates of identical and fraternal twins, either for qualitative traits when you are looking for all-or-none phenomena, or for quantitative traits when you compare fraternal and identical within-pair variances and evaluate the statistical significance of their ratio with the usual F test. There must be replications of these studies because there can be marked fluctuation in the results of twin studies by different investigators. There have been a fair number of twin studies, but we are only now reaching some uniform conditions, such as the determination of zygosity by standardized methods. It is difficult to compare results that were based on other methods of determining zygosity. Furthermore, the samples are usually rather small, a fact for which no investigator can be blamed, but one which will lead to fluctuation of results; therefore replication is almost essential before we can base firm conclusions on twin studies. I hope that we will find a good deal of agreement when we do the replications. The second point I want to make is that it will be very interesting to try to understand why there are discrepancies between different studies and *why* a seemingly small change in a psychological test can lead to quite different results in heritability estimates. This may give us a new kind of experimentation in which we try to change heritability estimates by maximizing hereditary factors in a particular psychological variable, or maximizing environmental aspects, sometimes perhaps by a rather small change in the measure used. Still, within the framework of the conventional twin studies, we should be alert to the fact that when we collect such data, which, although economical in comparison with family studies, still are rather expensive, we have the opportunity to perform several other analyses on the same data. First of all, since we do collect the information on the blood types, it might be worthwhile to check routinely for linkage of any of the variables with the blood types using the fraternal twins as sibships of two cases. One of these studies alone will not give any firm information because the samples will be too small, but if there are consistent results between studies (and everybody is going to have to collect the information on blood types, anyway), it is possible that we would find something. I am not hopeful that this is going to lead to many discoveries, but it is not going to be very difficult to do. Secondly, when we have identical twins we can ask why some identical pairs are more divergent than others. While this is not going to tell us much about heredity, it will tell us a lot about environment. It may be a valuable complement to the study of hereditary factors particularly if this is in work with very young twins, where we can still secure accurate childhood histories. Next we can do conventional correlational studies. Finally, I would like to summarize a few reports to show further methods of getting more information from traditional twin studies than we get by the usual passive collection of twin data. We can introduce at least *some* experimental dimension. We normally consider variables either as all-or-none phenomena, where we look for the presence or absence of concordance, or as continuous and then we do analysis of variance. Dr. Stafford points out that it may sometimes be fruitful to break this set. For instance, PTC (ability to taste phenylthiocarbamide) was formerly considered an all-or-none phenomenon (this is the ability to taste a substance that tastes bitter to most people), and yet it is actually a continuous variable. Conversely, we might take some psychological variable known to be continuous, but just set an arbitrary cutoff point and classify people above a certain score as having this attribute and those below as devoid of the trait. This may be more economical and it may lead to clearer results. It is not impossible, for instance, that such a crude classification might be useful in studying the inheritance of pitch perception. We can also go the other way—color vision is traditionally considered an all-or-none phenomenon, but Pickford's work (1951) seems to indicate that it might be more accurate to call it a continuous variable. Most people are at one end of the distribution, but there is still enough spread within the normal range to give the usual genetic models interesting complications. Another approach has been tried by Loehlin (p. 163 of this volume). He takes information obtained on identical and fraternal twins, looks for items which give a high concordance in identicals and relatively low concordance in fraternals, then constructs a new test from items which will give you *high* heritability to see what sort of an animal you have. One can analyze the collection of items one has obtained to see if one dimension underlies the collection or whether there are more. Gottesman in his chapter will have something to say about that too. In this way an experimental approach is added to the traditional twin-study method. This is particularly emphasized because both the analysis of my data and comparison with other data have failed to yield a short easy generalization of my results. When I thought about this, an idea which is certainly not new struck me with great force; namely, some psychological traits may have a simple genetic mechanism controlling them, but these traits with high heritability are not necessarily variables which we are currently looking for, or even aware of. Most of the variables which are of interest to genetics today were discovered accidently. Even the "obvious" trait of color blindness is a rather recent discovery, yet people must have had this defect for centuries before science became aware of this polymorphism. We may have missed many differences because they have no applied usefulness at this time. In general, most of our psychological tests have been constructed for their usefulness in predicting success in graduate school, or in elementary school, or whether or not one is going to have problems when he enters the armed forces, etc. We have to start with these tests, but we shall need new instruments for studies of the genetic bases of psychological traits.

A few more methods will be mentioned that should be used to get needed information. First, I cannot make a strong enough plea for new co-twin control studies with samples larger than two or three pairs of twins and with clear aims. For instance, one could test the prediction that systematic training of one twin in concept formation at age 3 would give differences between the twins at elementary school age which would be relatively stable, or would *not* be stable depending on the type of training. Or, one could take a small number of identical twins and give one of each pair early training in reading to see if there were persistent, or even permanent twin differences. This would be a very valuable complement to studies of the role of heredity. Again, sometimes we can capitalize on accidents. In Sweden, a psychiatrist, Dencker (1958), studied the effect of brain damage, relatively independent of hereditary and early childhood factors, by comparing brain-damaged twins with their unaffected identical twins. We must be more original. There are probably many studies of this kind that could be done with various diseases if one could find enough identical pairs in which only one twin has the disease. A Swedish psychologist, Naeslund (1956), a former student of Husen, I believe, used a very ingenious way of studying the effectiveness of competing methods of teaching. He took 18 pairs of twins, including 10 identical pairs, and separated them into two classrooms. In one room, reading was taught by the phonics method and in the other by the sight method. Two teachers rotated between the two classes to control for the influence of the teachers. The phonics method was superior for average children, but for gifted children there was no real difference; so there is no clear-cut answer for all children as you would expect. I think this is a wonderful way to do research on teaching methods and one which should be used more. I would also like to make a plea for the study of identical twins discordant for mental illness. A start has been made; I was told that the National Institutes of Health is organizing a registry for such pairs all over the country.

I would like to conclude by presenting an idea for multivariate analysis of twin data. A formal presentation of this usually causes difficulties in understanding on the part of the audience, so I am going to retrace my own thinking instead. While not as elegant, it may be a more psychological presentation since in following my attempts to deal with the problem you will, I hope, arrive at an intuitive understanding of my solution.

In the Michigan twin study, we found statistically significant results for so many variables that people keep asking me: "Do you *really* think that all these variables are different with independent hereditary control for each?" Well, I did think that this was true for many of the variables, but there was no way I could prove it. If the conventional correlation coefficient obtained in some study of unrelated people shows a very low correlation between the scores on two tests, it does not say anything about independence of genetic mechanisms or genetic control over these two traits. Even if each of the variables had a high heritability and was the result of, or was influenced by the *same* hereditary sources of variance they could still have a low correlation if different environmental factors acted on each trait and vice versa. First of all I asked whether the twin that was taller was also the brighter twin, and I did this for identical and fraternal twins separately. For clarity I will consider height and intelligence as two very simple traits. Let me give you just two correlations now. I will go to psychological variables to make the discussion more concrete. If we intercorrelate the differences on the Information and the Vocabulary subtests of the WISC for elementary school age identical twins we get a correlation of .06; in other words, whatever it is in the environment *within* the family which produces differences or is associated with differences on the Information subtest is not related to whatever produces differences in the Vocabulary subtest. However, the intercorrelation for the differences on these same tests for fraternal twins gives a somewhat different picture—the correlation is .56. To keep the argument fairly simple, we have here the same within-family differences operating to produce (or associated with) differences on these tests, but in addition we have hereditary twin differences which can contribute to (or lower) the correlation. Here I would be tempted to conclude that this hereditary component has led to the greater correlation between the twin differences. I think this is essentially correct reasoning, but it is very awkward and we have lost some very significant information. In calculating correlations, we really lose sight of the *size* of the twin differences; each of the two correlations is calculated around its own mean with its own standard deviation, so we have really thrown away our most significant information, which is that differences tend to be greater for fraternal than for identical pairs. So we should compare covariances and we should do this simultaneously for a number of variables, because it is very awkward to treat successive pairs of variables and later reconcile findings on variables A and B, and variables B and C, and on A and C, etc. What we want is a method for treating a set of variables simultaneously, to determine whether or not they contain a number of relatively independent hereditary components. Well, given two tables of covariances of twin differences, one for identical and one fraternal twins, could an investigator subtract the identical twin difference covariance matrix from the fraternal twin difference covariance matrix? The covariance matrix for the fraternals is due to hereditary and environmental factors while the one for identical twins would only be due to environmental factors. Thus, if subtracted, one would have hypothetical hereditary covariances left. This is, of course, not permissable for a number of reasons. First of all, mathematically, if you subtract two matrices you may get a matrix that is no longer Gramian, that is, it no longer has properties such that you can obtain roots and vectors. Intuitively one might see that the implied relations between the variables obtained by the subtraction may be incompatible. Furthermore, this subtraction is incorrect genetically, because heredity and environment are not merely additive components; there is a strong interaction component that may actually be larger or more significant than either of the two separately. Yet we would like to do something of this kind, and it turns out that there is something we can do. Let me remind you what we do with just one variable: We calculate the ratio between the within pair variance for fraternals and for identicals and then test the significance of this ratio by the F test. Now we must generalize the F test to a multivariate test. The mathematical equivalent of getting a ratio between two covariance matrices is to find a matrix such that multiplying one of the original ones by *it* yields the second. Mathematically this is quite straightforward, although not easily calculated without a computer. Then we can ask how many roots are significant in the resultant matrix, which is the multiplier or transformation matrix that produces the fraternal from the identical covariance matrix, and this will tell us how many independent sources of hereditary variance there were in our data. Here I have to insert warnings about the size of the sample, about the fact that you cannot interpret mathematical models in terms of real phenomena without reservations, and so on. We can also ask whether certain variables have independent hereditary components, and I suspect that we may even, to some extent, be able to identify certain hereditary components of variance with particular variables, somewhat as you would in factor analysis. What we have been talking about is solving an equation of the form $|D - \lambda M| = 0$, where D is the covariance matrix of fraternal or dizygous twin differences and M the covariance matrix of the identical or monozygous twin differences. We are obtaining a multiple discriminant function on the differences of the scores of the two types of twins. We get a composite score from our differences that will discriminate identical and fraternal twins as well as possible, then we remove the effect of this first composite and find another composite that will again give us the best possible discrimination and continue until we have exhausted the variance or until we cannot make any further discrimination. This is really a very conventional technique but one has to grow accustomed to thinking in this way, at least it took me a long time to see the hereditary implication.

This method has been applied to the six scores of the Primary Mental Ability Test (PMA) (Thurstone, 1941) that was administered to 45 pairs of identical and 37 pairs of fraternal twins as part of the Michigan Twin Study.

The over-all design of this study has been described by Sutton, Vandenberg, and Clark (1962). The zygosity diagnosis was based on a decision function derived from the results of tests of blood group systems (see Sutton, Clark, & Schull, 1955 for details about this procedure).

The equation $|D - \lambda M| = 0$ was solved for the six PMA difference scores and the six roots shown in Table I were obtained.

TABLE I
Roots Obtained from PMA Difference Scores

Roots	$\|D - \lambda M\| = 0$
1	3.99972
2	2.23172
3	1.58655
4	1.00141
5	.64598
6	.38206

To test the significance of these roots Bartlett's chi squared test of the homogeneity of the remaining roots after extraction of 1, 2 . . . k roots was applied (Bartlett, 1950). The results are shown in Table II and indicate that at least four roots are significant, or in other words, that in this sample there are at least four independently significant hereditary components.

It is of interest to compare these roots with the F-ratios for the DZ over the MZ within pair variances for the six original scores. Table III shows these F-ratios; four of them are significant beyond the 1% level.

We concluded that there are at least four independent components in the six PMA scores of the subjects in this sample. Since the analysis of the single scores led to the conclusion that number,

TABLE II

CHI SQUARED VALUES FOR THE HOMOGENEITY OF THE REMAINING ROOTS AFTER EXTRACTION OF k ROOTS

k	Chi square	d.f.	p
1	30.698	20	between .10 and .05
2	17.483	6	less than .01
3	7.250	2	less than .05
4	2.147	.5	n.s.
5	0	0	

verbal, space, and word fluency were under a statistically significant degree of genetic control, it may be warranted to conclude that the four independent components represented by the significant roots of $|D - \lambda M| = 0$ are rather similar to the number, verbal, space, and word fluency abilities. More details about this analysis are given in another place, Vandenberg (1965).

TABLE III

F-RATIOS OF FRATERNAL OVER IDENTICAL WITHIN-PAIR VARIANCES OF SIX PMA SCORES FOR 37 AND 45 DEGREES OF FREEDOM

Number	2.583	.01
Verbal	2.651	.01
Space	2.419	.01
Word fluency	2.479	.01
Reasoning	1.401	n.s.
Memory	1.213	n.s.

I shall close with two more remarks. I overlooked, for quite a while, the fact that before one compares results between identical and fraternal twins there is an opportunity to look at identical twin differences only and their correlations. This provides a method for studying the extent to which within-family differences between twins on different variables are correlated. This is a way of studying in great detail at least a limited aspect of the environment, and with information on early childhood experiences, one has a really powerful way of studying how these experiences affect cognitive variables, personality variables, etc.

REFERENCES

Bartlett, M. S. Tests of significance in factor analysis. *Brit. J. Statist. Psychol.*, 1950, 3(2), 77–85.

Cattell, R. B. Research designs in psychological genetics with special reference to the multiple variance method. *Amer. J. Human Genet.*, 1953, 5, 76–93.

Dencker, S. J. A follow-up study of 128 closed head injuries in twins using co-twins as controls. *Acta Psych. Neurol. Scand.*, 1958, Suppl. 123, 33: 125 pp.

Naeslund, J. *Metodiken vid den forsta lasundervisningen; en oversikt och experimentella bidrag.* Uppsala: Almqvist and Wiksell, 1956.

Pickford, R. W. *Individual differences in colour vision.* London: Routledge and Kegan Paul, 1951.

Sutton, H. E., Clark, P. J., & Schull, W. J. The use of multi-allele genetic characters in the diagnosis of twin zygosity. *Amer. J. Human Genet.*, 1955, 7, 180–188.

Sutton, H. E., Vandenberg, S. G., & Clark, P. J. Hereditary abilities study: selection of twins, diagnosis of zygosity and program of measurements. *Amer. J. Human Genet.*, 1962, 14, 52–63.

Thurstone, L. L. *The primary mental abilities tests.* Chicago: Science Research Associates, 1941.

Vandenberg, S. G., Innate abilities: one or many? *Acta Genet. Med. Gemell.*, 1965, 14, 41–47.

DISCUSSION

Gardner: I'd like to say, Bravo! for this method of treating twin difference data to get at underlying relations. In practice this should give us answers we have long wanted.

Vandenberg: I think it does. I would prefer to work with larger matrices because I am worried about sample size. This is one reason why I have been holding up publication till I have repeated this with a larger sample.

Gardner: When you get down to the point of counting the number of significant latent roots you really have to worry about sample size. When you are concerned with replications you are on the right track.

Vandenberg: I think so. Next we should take a variable that has repeatedly shown evidence of a hereditary factor and subject it to a procedure of "refining" for the purpose of increasing heritability. Then you could finally go to family studies.

Freedman: I should like to stress that each identical twin pair has a different inheritance, so to speak, from every other identical twin pair. When we lump these different identical twin pairs together to extract what is inherited in common from among them, we lose so-called genetic individuality. We come up with something which is necessarily less than what is actually inherited, or *may* be inherited.

Vandenberg: I couldn't agree more. I have just two comments on that. First, I think there is a great deal of uniformity in the way we acquire the primary or most important cognitive abilities, so there may be less difference here than with personality variables. Second, this is merely a link in a chain that will, maybe 10 years from now, lead to some fairly precise knowledge about hereditary mechanisms in human behavior. This method allows us to take two variables that we have considered independent and test for a relation in the genetic component. If we find such a relation we can look for a way of constructing a new test, a measure of this new construct which has some characteristics of both variables. We would then be one step closer to the underlying hereditary component. In the personality area, everybody may be somewhat unique and it may be difficult to find measures which give a high heritability index, let alone a neat genetic mechanism. We may always be limited somewhat by this uniqueness. But I'm not pessimistic about the cognitive area and prospects may not be poor in personality studies either, particularly with certain gross characteristics as the schizoid personality.

Freedman: Yes, there are certain maturational events that are certainly common to the whole human race, such as smiling and sitting up.

Stafford: Your report about the Swedish psychologist studying learning is very apropos. We often try to find *the* way of teaching reading. In fact, the best way is undoubtedly going to differ depending upon the child. It is probably different at different maturational levels. If there are two discrete types, phonics and sight readers, these differences may be hereditary. One other question, did you mention linkage?

Vandenberg: Yes, I mentioned linkage with blood types as a by-product of twin studies. Eventually we are going to have a fair amount of information on the blood types of fraternal twins, their height, their shoulder widths, their scores on some tests, etc. and it seems a shame to have this unused. Other people might be going to great expense to collect these data.

Stafford: I wonder if it would expedite our research and possibly make communication a little easier if we took the chromosomes and, applying the knowledge that we have to date of linkage, assigned letters to the known linkage groups, hoping they did not exceed 23. Later, with the help of some chromosomal aberrations we might be able to tie C to chromosome 17, or A to 9, etc. I do not know if this would be helpful, and it might never come to pass, but certainly it would make communication easier.

Lasker: I am not sure that an arbitrary numbering system would be better than what we have. We have letters now for the three known linkages on autosomal chromosomes, one is with ABO, one is with Rh and one is with

MN, if I am not mistaken. An arbitrary system might only confuse matters as it has with numbering the chromosomes by their length and shape, and perhaps as Patau has claimed, the Denver system has already gone too far in trying to number even the morphology of the human chromosomes. One cannot always tell the difference between different chromosomes. For instance, in a preparation one cannot always distinguish the X chromosome from an autosome.

Sutton: The possibility of exact identification of particular chromosomes varies with both the preparation and the investigator. Some investigators claim that they can identify every chromosome, and others can as convincingly point out where they cannot. I think the hope still exists that one can by use of various accessory techniques, but there is danger in falsely precise labeling.

Vandenberg: I think some people may not be familiar with this. I wonder if Dr. Sutton is willing to say a little bit about how the technique works and so on.

Sutton: The technique for studying human chromosomes involves taking about 10cc of blood and culturing it through several divisions. Colchicine is added which halts cell division at the metaphase stage. There are then various techniques for spreading the white cells on a slide and examining metaphase figures. It is a fairly simple matter to count 46 chromosomes if all of them are there and perfectly spread. It is also a fairly simple matter to assign individual chromosomes to one of 6 or 7 groups, and within these groups several of the chromosomes can be identified quite clearly because of their total length and the position of the centromere. But within other groups identification is far from clear. If you take a whole series of objects of slightly different lengths, it is possible to array them so that the two longest match up, and the next two, and so forth giving a very convincing array going from long to short, which is the primary basis for arranging chromosomes. But, because there are slight differences in contraction, such arraying is subject to tremendous errors. A little bit of progress has been made in assigning linkage groups to specific chromosomes, and some suggestions are now in the literature, but we are very far from knowing any good associations.

Lindzey: In the chromosomal aberrations associated with mongolism, where multiple chromosomes are mentioned, would the fact that more than one is mentioned ordinarily be attributed to a measurement error, or is there some suspicion that there are multiple chromosomes involved?

Sutton: Well the similarity among persons with mongolism suggests that it is always the same chromosome. Now, if you get a variation in the chromosomal pattern you next try to interpret this variation in terms of a translocation which maintains the original extra chromosome 21. So far this has not always been possible, but I am not sure that the restrictions are very severe. I think this is a reasonable procedure, although there is some circular reasoning. The fact that so few viable trisomic conditions have been found, and the fairly uniform manifestations of mongolism rather convincingly suggest that the same supernumerary element is involved in all cases.

Fuller: The other trisomics have been identified with a different chromosome, and, though very rare, they also show a uniformity. This strengthens the argument, although only two other trisomics are known.

Sutton: Well, a whole variety of aberrations are known. There is the so-called isochromosome, where the division is apparently at right angles to the normal division, so that the daugher chromosomes consist of two non-identical chromosomes, one of which has both the long arms, the other which has both the short arms. Each daughter cell is thus a partial trisomic and partial monosomic. This has been observed on several occasions and the fact that a coherent picture can be made of all this suggests that there is some reasonable basis for it.

Gottesman: I am wondering if I might comment on the concept of heritability in twin studies, since people in animal genetics use the same term. They are using it in a different way than we are. Heritability in a twin study refers to within family variance and it underestimates by an unknown amount. If there were random mating we could multiply by two and get a better estimate, but we do not know how much assortative mating there is for each of the various behavioral characteristics. I would like Dr. Fuller to clarify the relationship between "heritability" in twin studies and the "heritability" that the animal geneticists talk about.

Fuller: Well, I think at this stage it is perfectly true that in a family, on an over-all family population basis, with the assumptions of random mating, half of your genetic variance will appear between, and half within the family, and if you make your comparison of the identical twins solely against their co-twins it is one-half the total. However, remember that only with the genetic variance do you get half the estimate. You are really measuring phenotypic variances, and here, of course, we do not really know the ratio of the within-family phenotypic components to the between-family phenotypic components. All we can say is that you always tend to underestimate heritability from the twin study.

Loehlin: It is also true that you underestimate the environmental variation in a twin study. So it is a question of which are you underestimating more, the environmental or the hereditary variance, when you confine yourself to within family analyses.

Fuller: That is something of course that we really do not know. We would like to say that there is more variation between families, but I think it is quite conceivable that some characteristics, for instance, being the older sib in a family, may really have more effect upon a specific area of psychological development, than does growing up in different families. This is why we speak very loosely when we try to say that growing up in different families is an important factor in the psychological development.

A Comparison of Socioenvironmental Factors in Monozygotic and Dizygotic Twins,[1] Testing an Assumption

Richard T. Smith[2]

This is an exploratory study of the relative intra-pair similarity of monozygotic (MZ) and like-sexed dizygotic (DZ) twins on selected socio-environmental factors. In studies of twins, the higher intra-pair similarity for MZ as contrasted with DZ twins is frequently attributed to hereditary influence. This conclusion is based, in part, on the assumption that the environment is common to both types of twins.

A number of writers have commented on the validity of this assumption. In Lilienfeld's review of problems in genetic-epidemiological field studies (Lilienfeld, 1961), he has questioned the logic of this assumption in twin studies. He cites as an example the greater similarity of MZ than of DZ twins with respect to serum cholesterol levels and makes the suggestion that "before deriving an inference that this is a reflection of genetic control of serum cholesterol levels, it would be important to determine whether MZ twins are more similar than DZ twins with respect to dietary or other environmental factors that may influence serum cholesterol levels." This observation would likewise apply in other twin studies in which the variables under investigation are as likely to be influenced by environmental factors as by genetic factors. Similar examples can be made of (1) risk of exposure and tuberculosis infec-

---

[1] This study was supported, in part, by U.S. Public Health Service Grant M-6637 A and conducted under a U.S. P.H.S. Fellowship, MF-8883.
[2] Department of Chronic Diseases, School of Hygiene and Public Health, Johns Hopkins University, Baltimore, Maryland. Paper read at the Genetics and Epidemiology Section of the American Public Health Association annual meeting held in New York City, October 5–9, 1964.

tion; and (2) the pattern of interpersonal relations and mental illness (Rosenthal, 1961; Simonds, 1957). Several other studies have indicated that environmental factors may contribute to the higher concordance for MZ twins (Jones, 1955; Newman, Freeman, & Holzinger, 1937).

The present study is an attempt to assess the degree of similarity and dissimilarity for MZ and DZ twins on selected factors in the environment. Various studies have shown that MZ twins are more similar than DZ twins on selected personal and behavioral characteristics. It has been observed that MZ twins in general form closer bonds of attachment than DZ twins (Shields, 1954); MZ twins tend to be treated more alike by their parents (Jones, 1955); MZ twins are more likely to exhibit a division of social roles involving a mutual interdependence of role relationships, e.g., leader-follower role (Jones, 1955). This study is a more detailed investigation of selected personal and social characteristics of MZ and DZ twins.

Study Group and Methods

Initially, 262 white, like-sexed, adolescent-aged twin pairs were selected for inclusion in the study. The twins were selected from the public schools in a large metropolitan area. Their ages ranged from 13 through 16 years at the time of the study.

Prior to actual contact with the families, 43 pairs (16.4%) were excluded for reasons ranging from family out-migration to institutional care for one or both members of a pair. Of the remaining pairs, 26 (11.9%) of the families contacted refused to participate in the study. Twenty-nine (13.2%) families participated but refused permission to have blood specimens taken from the twins. Blood samples were collected from 164 pairs which accounted for three-fourths (74.9%) of the primary study group ($N = 219$) under investigation. This latter group of 164 pairs is used in the analysis of the data presented here.

A questionnaire was developed for the purpose of obtaining information on the personal and social characteristics of each twin. Data were collected on: work, school, sports, and leisure activities; sleep, dress, and study habits; food preferences; and other information related to twinness. In general, twins were interviewed in the school. In addition, an interview was held with parents for the purpose of obtaining necessary background information on each twin pair. The schedule included such items as household composition, education, occupation, and income of the head of the family, maternal and twin birth information, and other items related to the twins' family environment. The parent interviews were conducted in the home.

Zygosity testing was carried out by an independent laboratory.[3] Blood typing for dizygosity was performed in a sequential manner. Tests for the minor factors were carried out only when differences failed to appear on the major factors.

The data from this study will be presented in the following order: first, background factors, and second, personal and social factors related to the degree of intra-pair similarity observed for MZ and DZ twins.

Results

In assessing the degree of similarity of difference between MZ and DZ twins, it is necessary to look at the background characteristics of the study group first. Of the 164 twin pairs who participated in this study, 90 pairs were classified as MZ and 74 pairs were DZ. Table I shows the distribution of these cases by age and sex for each type of twin. Females outnumber males in each of the two classes of twins. This finding is in accordance with most twin studies. In general, the age distributions for male and female twins, for both MZ and DZ groups, are sufficiently similar as to rule out any selective age bias.[4]

Another factor related to the twins' background and family environment is that of parental recognition of the twins' zygosity. In this study, the parent was asked whether or not the twins were identical. This information was compared to the zygosity findings determined by blood typing. The extent of misclassification of twins is shown in Table II. The order of magnitude of misclassifi-

[3] Five blood group systems were used: ABO, Rh, MN, Kell, and Duffy. The probability of establishing a diagnosis of dizygosity is approximately 90% using the five blood groups (Juel-Nielsen et al., 1958). The error of dizygotic misclassification is considered to be approximately 10%.

[4] Tests for differences in age distributions between MZ and DZ twins, for independent samples, were made using the Kolmogorov-Smirnov test. No significant differences were observed.

TABLE I
Age Distribution of Twin Pairs by Sex and Zygosity

Age	Males				Females				Both			
	MZ		DZ		MZ		DZ		MZ		DZ	
	No.	%	No.	%	No.	%	No.	%	No.	%	No.	%
13	10	22.5	8	23.5	11	22.0	8	20.0	21	23.3	16	21.6
14	9	30.0	7	20.6	16	32.0	9	22.5	25	27.9	16	21.6
15	12	22.5	12	35.3	10	20.0	15	37.5	22	24.4	27	36.5
16	9	25.0	7	20.6	13	26.0	8	20.0	22	24.4	15	20.3
	40	100.0	34	100.0	50	100.0	40	100.0	90	100.0	74	100.0

TABLE II

Distribution of Twin Pairs by Serologic and Stated[a] Zygosity, and Tests for Differences in Proportions Misclassified, by Sex

Serologic zygosity and sex		ID		Stated Zygosity FRAT		D.K.		Total		Difference in Proportions misclassified (excl. D.K.)	z	P<
		No.	%	No.	%	No.	%	No.	%			
Males	MZ	29	72.5	6	15.0	5	12.5	40	100.0	−0.05	−0.50	ns
	DZ	7	20.6	25	73.5	2	5.9	34	100.0			
		36		31		7		74				
Females	MZ	41	82.0	6	12.0	3	6.0	50	100.0	−0.24	−2.58	.05
	DZ	14	35.0	24	60.0	2	5.0	40	100.0			
		55		30		5		90				
Both	MZ	70	78.0	12	13.0	8	9.0	90	100.0	−0.15	−2.44	.05
	DZ	21	28.0	49	67.0	4	5.0	74	100.0			
		91		61		12		164				

[a] Stated by parent.

cation observed in this series varied from 12 to 15% for MZ twins, females and males respectively, to 20.6 and 35.0% for DZ twins, males and females respectively. The major error is observed among females where a high of 14 pairs are presumed by the parent to be identical and are defined by blood typing as DZ. This could affect the degree of environmental variation in terms of intra-pair similarity patterns between MZ and DZ twin pairs, espe-

TABLE III

PERCENT DISTRIBUTION[a] BY BIRTH ORDER FOR MONOZYGOTIC AND DIZYGOTIC TWINS, BY SEX

Birth order	Percent distribution					
	Males		Females		Both	
	MZ	DZ	MZ	DZ	MZ	DZ
01	29.9	35.9	32.6	31.5	31.5	34.3
02	34.2	23.8	26.7	23.2	29.4	23.9
03	20.4	22.4	16.4	16.2	18.9	15.1
04	8.6	11.4	9.5	13.1	9.3	11.6
05	1.4	—	—	9.9	0.7	9.1
06	4.1	2.4	14.8	1.1	9.6	1.5
07	—	4.1	—	2.8	—	2.0
08	—	—	—	1.1	—	0.6
10	—	—	—	1.1	—	1.9
11	1.4	—	—	—	0.6	—
Total % :	100.0	100.0	100.0	100.0	100.0	100.0
Total N :	40	34	50	38	90	72

[a] Adjusted for total number of pregnancies of mother (including stillbirths and miscarriages).

cially for the female group. That is, there may be a likelihood of greater similarity among segments of female DZ twin populations because of parental misunderstanding with regard to the twins' zygosity status. This would bias the analysis of female twin concordance rates on environmental factors in the direction of minimum differences thus erroneously leading to the possible conclusion that, for all practical purposes, the environment for MZ and DZ twins is the same.

The next two tables deal with background factors related to birth order and maternal age. Previous observations have shown that birth order, i.e., rank order of birth among all offspring, and maternal age of the parent at the birth of twins vary directly with the frequency of DZ twinning. These relationships do not seem to obtain for MZ twinning (Yerushalmy & Sheerar, 1946). In this study, DZ twin births do not occur more frequently in the higher birth order ranks than MZ twin births as shown in Table III. The average birth order rank for MZ pairs is 2.3 and for DZ pairs it is 2.8. In Table IV, the frequency of DZ twins tends to increase with

TABLE IV

PERCENT DISTRIBUTION[a] BY MATERNAL AGE FOR MONOZYGOTIC AND DIZYGOTIC TWINS, BY SEX

Maternal age at birth of twins	Percent distribution					
	Males		Females		Both	
	MZ	DZ	MZ	DZ	MZ	DZ
< 20	4.6	5.1	3.9	—	4.3	1.4
20–24	33.2	26.8	30.7	21.0	32.4	20.6
25–29	38.0	25.4	27.4	33.3	31.7	32.5
30–34	16.1	32.3	23.9	26.0	21.0	27.5
≥ 35	8.1	10.4	14.1	19.7	10.6	18.0
Total % :	100.0	100.0	100.0	100.0	100.0	100.0
Total N :	40	34	50	38	90	72

[a] Adjusted for total number of pregnancies of mother (including stillbirths and miscarriages).

maternal age in comparison with the MZ distribution. The mean maternal age for the MZ group is 26.8 years and for the DZ group 28.8 years.

Another factor likely to be of importance in assessing the degree of intra-pair similarity between these two groups is the socioeconomic status of the family. Families of differing social status usually reflect differences in styles of life including child-rearing practices, living standards, and leisure-time activities. The twins' environment, especially during their formative years, may be strongly influenced by the relative social status of the family.

A measure of family socioeconomic status (SES) was derived from the occupation, income, and education of the head of the household.[5] This is defined as a family's SES score. The distribution of SES scores for the families of MZ and DZ twins is shown in Table V. Families of DZ twins are more likely to be characterized by a low SES than families of MZ twins. This difference is statistically significant.[6] This finding is consistent with the results obtained by Lilienfeld and Pasamanick (1955) in their analysis of variations in the frequency of twin births by socioeconomic status. In that

TABLE V
DISTRIBUTION OF MZ AND DZ FAMILIES BY SOCIOECONOMIC STATUS[a]

SES score	Twin families					
	MZ		DZ		Both	
	No.[b]	%	No.[c]	%	No.	%
< 7	3	3.5	2	2.8	5	3.2
7–9	9	10.5	13	18.3	22	14.0
10–12	16	18.6	27	38.1	43	27.4
13–15	23	26.6	10	14.1	33	21.0
16–18	20	23.3	11	15.5	31	19.7
19–21	9	10.5	6	8.4	15	9.6
22–24	6	7.0	2	2.8	8	5.1
	86	100.0	71	100.0	157	100.0
Unknown:	4	—	3	—	7	—
Total	90		74		164	

[a] Low score = low SES; high score = high SES.
[b] Mean = 14.5; median = 14.4.
[c] Mean = 12.9; median = 11.4.

study, it was observed that the dizygotic frequency of twin births was highest in the lowest socioeconomic stratum. One possible explanation for this variation in MZ and DZ twins by SES levels may relate to family size. The larger the size of the family, the greater the likelihood of DZ twin births since DZ twinning increases with

[5] A hierarchy of scale values was assigned to each range of categories for each variable. The sum of the three scale values, one for each variable, resulted in a total SES for each family. A low score reflects a low socioeconomic status.
[6] Kolmogorov-Smirnov test: $D = -.27$; $P < .01$.

maternal age and to some extent birth order. Since this relationship does not obtain for MZ twins, it is conceivable that the excess of DZ twins in families in the lower SES levels is due to the number of offspring produced.

The second part of this paper concerns the findings obtained with respect to intra-pair similarities for MZ and DZ twins on a number of personal and social characteristics. A number of items in the questionnaire dealt with everyday habits of the individual. These included food, sleep, dress, and study habits. The twin was

TABLE VI

Proportions[a] Concordant with Selected Habits for MZ and DZ Twin Pairs, by Sex

| | Males | | Females | | Both | |
Habit	MZ	DZ	MZ	DZ	MZ	DZ
Eating between meals	72.5	66.7	77.1[b]	57.5[b]	75.0[b]	61.6[b]
Snack before bedtime	50.0	66.7	60.4	62.5	55.7	64.4
Time usually go to bed	59.0	72.7	72.9	57.5	66.7	64.4
Time usually get up	60.0	53.1	72.9	65.0	67.0	59.7
Dressing alike	59.0	57.1	68.7[b]	33.3[b]	64.4[b]	40.3[b]
Study together	23.1	9.1	54.2[b]	20.5[b]	40.2[b]	15.3[b]

[a] Proportions are based on varied number of pairs depending on response item.
[b] $P \leq .05$

asked a specific question on each of these items such as "Do you eat between meals?" The response to each of these questions was compared to the co-twin's response. In the matrix resulting from the twin and co-twin responses to an item, concordance ratios were computed using the combined diagonal cell frequencies of mutual agreements. In this manner, intra-pair similarity ratios were obtained for each subgroup, i.e., by sex and zygosity, for each of the items.

With respect to these habits, concordance rates are highest for female MZ twins as shown in Table VI.[7] In addition, the differences in concordance rates between MZ and DZ twins in general

[7] In this report, the normal test for equality of proportions in independent samples was employed; a difference at the .05 level is called "significant."

are more marked for females than for males. The over-all results indicate that there is a somewhat consistent pattern of greater intra-pair similarity among MZ as contrasted to DZ twins with respect to everyday habits.

The members of a twin pair were asked also to indicate the extent to which each engaged in school, extracurricular, and leisure activities. Information was obtained on such activities as participation in school and outside groups, involvement in athletics, extent

TABLE VII

PROPORTIONS[a] CONCORDANT WITH SELECTED ACTIVITIES FOR MZ AND DZ TWIN PAIRS, BY SEX

Activity	Males		Females		Both	
	MZ	DZ	MZ	DZ	MZ	DZ
Active in school groups	28.2	32.3	58.7	57.9	44.7	46.4
Active in other clubs	37.5	41.4	43.5	30.0	40.7	34.8
Number of sports played	23.1	25.0	20.8	25.0	21.8	25.0
Active in other outdoor sports	60.5	59.4	57.4	45.0	58.8	51.4
Attend sports games	79.5[b]	59.4[b]	66.7	60.0	72.4[b]	59.7[b]
Attend movies together	55.0	39.4	63.8	55.0	59.8	47.9
Play musical instrument	15.0	14.7	34.0[b]	17.5[b]	25.6	16.2
Same close friends	55.0[b]	30.3[b]	59.6[b]	35.0[b]	57.5[b]	32.9[b]

[a] Proportions are based on varied number of pairs depending on response item.
[b] $P \leq .05$

of other leisure-time pursuits, and type of friendship patterns. For each twin pair, those who indicated the same degree of involvement in a specific activity were defined as concordant.

The results are shown in Table VII. There are noticeable variations in paired agreements as one examines the different activities presented in this table. For example, twin pairs tend to show high concordance rates for attendance at sports games whereas for the number of sports played, the rates drop considerably. It is interesting to note that a markedly higher proportion of MZ twins as compared to DZ twins have the same close friends. This holds true for both sexes. In viewing the intra-pair similarities of MZ and DZ twins with respect to each of these activities, it is apparent that, in general, paired agreements are higher for MZ than for DZ twins.

Another activity investigated in this study was household chores. Six household chore items were included in the questionnaire. These items as well as the response options, and scores used for each item, are listed in the Appendix. In order to obtain a combined measure for comparing MZ to DZ paired responses, absolute mean intra-pair differences were computed for each group. This was accomplished by taking the paired absolute difference in total

TABLE VIII

TEST OF MEAN DIFFERENCE ON FREQUENCY OF HOUSEHOLD CHORE ACTIVITIES[a] BETWEEN MONOZYGOTIC AND DIZYGOTIC TWIN PAIRS, BY SEX

Sex and zygosity		N	$\bar{d}$	$Var._p$	$S.E._p$	t
Males	MZ	34	2.50	4.41	0.54	−0.33
	DZ	28	2.68			
Females	MZ	46	2.07	3.76	0.42	−0.90
	DZ	38	2.45			
Both	MZ	80	2.25	4.01	0.33	−0.89
	DZ	66	2.55			

[a] See Appendix for the items included and the response options and weights used for each item.

scores as derived from the sum of the values of all six items for each member of that pair. These differences were then totaled for all pairs and divided by the number of pairs in that group to arrive at an absolute mean difference. The results of these calculations as well as the tests of mean differences are presented in Table VIII.

As shown in this table, the relative degree of intra-pair differences is not much greater among DZ than among MZ twins. This observation is the same for both sexes. In general, both types of twins seem to be more or less equally influenced by factors within the family environment.

Twin data were also gathered on food preferences and patterns of daily beverage consumption. From a listing of staple foods, fruits, and vegetables on the questionnaire, the twin was instructed to indicate his degree of preference for each food item by checking one of a number of response categories. In a similar manner, information was obtained concerning each twin's daily consumption of beverages. The items included under each of these headings are listed in the Appendix.

TABLE IX

Tests of Mean Differences on Food Preference Items[a] between Monozygotic and Dizygotic Twin Pairs, by Sex

Preference factor, sex, and zygosity		N	d	Var.$_p$	S.E.$_p$	t
Staple food preferences						
Males	MZ	25	1.88	3.32	0.52	−1.43
	DZ	24	2.63			
Females	MZ	38	2.13	2.61	0.39	0.16
	DZ	30	2.07			
Both	MZ	63	2.03	2.90	0.32	−0.90
	DZ	54	2.31			
Fruit preferences						
Males	MZ	25	1.92	3.95	0.57	−0.29
	DZ	24	2.08			
Females	MZ	37	1.38	2.34	0.38	−2.64[b]
	DZ	29	2.38			
Both	MZ	62	1.60	3.02	0.32	−2.00[b]
	DZ	53	2.25			
Vegetable preferences						
Males	MZ	25	1.76	3.96	0.58	−0.57
	DZ	23	2.09			
Females	MZ	36	1.06	1.58	0.32	−2.42[b]
	DZ	28	1.82			
Both	MZ	61	1.34	2.62	0.31	−1.94[b]
	DZ	51	1.94			

[a] See Appendix for items included in each food preference category and the response options and weights used for each item.
[b] $P \leq .05$

To simplify the arrays of items included under these two general categories, as well as to assess the extent of paired similarities between MZ and DZ twins, absolute mean intra-pair differences were calculated for the various categories for each group. The findings on the food preference items are shown in Table IX while those on the beverage factor are shown in Table X.

Among males, the mean intra-pair differences observed for DZ twins are consistently higher than those observed for MZ twins for all food preference categories. That is, male DZ pairs tend to show greater variations in their food preferences. For females, mean differences are markedly greater for DZ twins than for MZ twins with respect to fruit and vegetable preferences. However, a slight reversal in direction occurs among female twins concerning staple foods.

TABLE X

TEST OF MEAN DIFFERENCE ON DAILY BEVERAGE CONSUMPTION ITEMS[a] BETWEEN MONOZYGOTIC AND DIZYGOTIC TWIN PAIRS, BY SEX

Sex and zygosity		N	$\bar{d}$	$Var._p$	$S.E._p$	t
Males	MZ	30	2.77	4.69	0.58	1.45
	DZ	26	1.92			
Females	MZ	41	1.98	6.73	0.61	−1.04
	DZ	33	2.61			
Both	MZ	71	2.31	5.90	0.43	0.00
	DZ	59	2.31			

[a] See Appendix for the items included and the response options and weights used for each item.

In Table X, a comparison of mean differences among males indicates that MZ twins are less similar with respect to beverage consumption than DZ twins. On the other hand, female DZ twins show greater variation on this factor than female MZ twins.

Other factors related to the degree of intra-pair similarity for MZ and DZ twins were investigated. One interesting facet pertains to the psychological factor of self-evaluation by means of comparison to the co-twin. Responses to this self-with-other appraisal was obtained by asking each twin the following type of question: "Compared to your twin, who does more work around the house?" Three response options were listed: (1) "myself," (2) "my twin," or (3) "no difference between us." Questions were asked with respect to eating and sleeping habits, household chores, school and sports activities, and parental treatment. Concordance rates were tabulated using the paired agreements to the response of "no difference." Among both males and females, concordance rates are higher among MZ than among DZ twin pairs. In general, MZ twins show a greater tendency to agree mutually that there is no difference between them.

As a final note on the data examined in this study, brief mention can be made of the findings obtained concerning the similarity of dress habits of twins. This information was obtained from the parent. One question asked of the parent was, "Up to what age did you dress the twins the same way?" Parents of female MZ twins tended to dress their twins in the same way to an older age level than did parents of DZ twins. However, parents of male MZ twins tended to treat their twins in much the same manner as parents of DZ twins. In addition to information about past practices, parents were asked how often the twins dressed alike at their present age. Among females, MZ twins are regarded by their parents as "frequently" or "almost always" dressing alike while, in contrast, DZ twins are considered to dress alike "rarely" or "never." This observed difference in the two distributions is statistically significant. For males, the direction is the same but the finding is not significant.

These factors as well as others not discussed in detail in this paper tend to give some support to the notion that MZ and DZ twins are treated, and thought of, differently by their parents—a situation which may, in turn, have a further impact upon the twins' perceptions of themselves. These findings may be important in twin studies investigating various aspects of mental illness.

Discussion and Summary

The implications with regard to twin studies may be considered in view of the observations presented in this paper. It would appear that there is a relatively greater degree of intra-pair similarity among MZ as compared to DZ twins with respect to habits, activities, personal preferences, parental treatment, and self-images. In addition, DZ twin families are more frequently observed to be of lower socioeconomic status than are MZ twin families. This finding supports the notion that there is a difference in the overall environment of the two types of twins which will, in turn, influence intra-pair differences.

If a given condition is likely to be subject to environmental influence, irrespective of the role of genetic influence, to the extent that these environmental factors are not taken into account, one is less certain that the greater concordance observed among MZ twins is not a manifestation, in part, of the greater environmental similarity among monozygotics.

These observations may be more important in those instances in which one is concerned with a particular condition where one is using a female twin study group. The direction and extent of difference is more pronounced among female twins than among male twins.

Based on the over-all findings presented in this study, it seems evident that the assumption of a common environment for MZ and DZ twins is of doubtful validity and, therefore, the role of environment needs to be more fully evaluated in twin studies.

REFERENCES

Jones, H. E. Perceived differences among twins. *Eugenics Quart.*, 1955, **2**, 98–102.
Juel-Nielsen, N., Nielsen, A., & Hauge, M. On the diagnosis of zygosity in twins and the value of blood groups. *Acta Genet.*, 1958, **8**, 256–273.
Lilienfeld, A. Problems and areas in genetic–epidemiologic field studies. *Ann. N.Y. Acad. Sci.*, 1961, **91**, 797–805.
Lilienfeld, A., & Pasamanick, B. A study of variations in the frequency of twin births by race and socio-economic status. *Amer. J. Human Genet.*, 1955, 7:204–217.
Newman, H. H., Freeman, F. N., & Holzinger, K. J. *Twins: a study of heredity and environment.* Chicago: Univer. of Chicago Press, 1937.
Rosenthal, D. Problems of sampling and diagnosis in the major twin studies of schizophrenia. *Psychiat. Res.*, 1961, **1**, 116–134.
Shields, J. Personality differences and neurotic traits in normal twin schoolchildren. *Eugenics Rev.*, 1954, **45**, 213–246.
Simonds, B. Twin research in tuberculosis. *Eugenics Rev.*, 1957, 49:25–32.
Yerushalmy, J., & Sheerar, E. Studies in twins. I. The relation of the order of birth and age of parents to the frequency of like-sexed and unlike-sexed twin deliveries. *Human Biol.*, 1946, **12**:95–113.

Appendix

For each household chore item (listed below) the following response options and weights were used.

Response Option	Weight
Every Day	4
Two to 6 times a week	3
Once a week	2
< Once a week	1
Never	0

Household chore items:

a. Help with the dishes
b. Make your own bed
c. Dust the furniture
d. Run errands to nearby store
e. Help with grocery shopping
f. Take out the trash or garbage

For each beverage item (listed below) the following response options and weights were used.

Response Option	Weight
> 3 glasses	5
3 glasses	4
2 glasses	3
1 glass	2
< 1 glass	1
None	0

Beverage items:

a. Milk
b. Chocolate Milk
c. Buttermilk
d. Orange Juice
e. Coffee
f. Tea

For each food item (listed below) the following response options and weights were used.

Response Option	Weight
Like very much	3
Not very much but eat it	2
Dislike, never eat it	1
Don't know, never tasted it	0

Staple food items:

    a. Chicken
    b. Ham
    c. Pork Chops
    d. Steak (beef)
    e. Liver
    f. Fried Eggs
    g. Boiled Eggs
    h. Mashed Potatoes
    i. Raw Potatoes
    j. Seafood (fish)

Fruit items:

    a. Apples
    b. Oranges
    c. Peaches
    d. Pears
    e. Grapes
    f. Plums
    g. Bananas
    h. Apricots
    i. Watermelon

Vegetable items:

    a. Raw Carrots
    b. Ripe Tomatoes
    c. Raw Celery
    d. Raw Radishes
    e. Raw Cucumbers

Personality and Natural Selection

Irving I. Gottesman

New ideas in human behavior genetics may be set in motion by construing psychological groups as Mendelian populations (Tryon, 1957; Thompson, 1957). The continuity, complexity, and fluidity of behavior (Thompson, 1964) together with the hazards of interpretation of phenotypic correlations impose barriers on those who do research in this area which may impair their vision of solutions to the particular problems of human behavior genetics. A major purpose of this paper is to entertain the notion that it may be fruitful to look at personality and natural selection simultaneously. Bridging the gap between population genetics on the one hand and social anthropology on the other may be one of the roles for the human behavior geneticist.

Demonstration of a heritable component in the variation observed in several personality traits and types (Gottesman, 1962, 1963b; Vandenberg, 1962) has immediate implications for the evolution of our species stemming from the theory of natural selection. The one nonrandom genetic process that accounts for the adaptive orientation of evolution is reproduction (Simpson, 1958, pp. 18-19): "If reproduction is differential, if there is a correlation between distinctive genetic factors in the parents and their relatively greater success in reproduction, then there will be an increase in the frequencies of those genetic factors (and combinations of them) within the population from one generation to another. . . . That, in brief, and shorn of numerous complications, is the modern concept of natural selection." The evolution of personality can be explained by this process only to the extent that personality is heritable; differential reproduction based upon superior adaption which has been transmitted culturally, complements that adaptation which was associated with genetic variation. Simpson said that the modern theory of evolution reinstates behavior, not merely as something to which evolution has happened, but as something which is itself one of the essential determinants of evolution (Simpson, 1958, p. 9).

Tables I, II, and III present data on the heritability of personality traits and on assortative mating for some of these traits. Only those traits measured by the Minnesota Multiphasic Personality Inventory (MMPI) have been selected for illustration (Dahlstrom & Welsh, 1960). For nonpsychologists who are unfamiliar with the MMPI I would like to digress for a moment and say that it is

TABLE I

MINNESOTA STUDY MMPI SCALE HERITABILITY INDICES FOR TOTAL GROUP, FEMALES, AND MALES

Scale	Total Group		Females		Males	
	H	F^a	H	F^a	H	F^a
1 Hypochondriasis	.16	1.19	.25	1.33	.01	1.01
2 Depression	.45	1.81	.22	1.28	.65	2.83
3 Hysteria	.00	.86	.00	.56	.43	1.74
4 Psychopathic deviate	.50	2.01	.37	1.60	.77	4.35
5 Masculinity-femininity	.15	1.18	.00	.99	.45	1.83
6 Paranoia	.05	1.05	.00	.70	.52	2.09
7 Psychasthenia	.37	1.58	.47	1.89	.24	1.31
8 Schizophrenia	.42	1.71	.36	1.56	.50	2.00
9 Hypomania	.24	1.32	.33	1.50	.00	.81
0 Social introversion	.71	3.42	.60	2.49	.84	6.14

^{a} The three values of F required for significance at the .05 level are 1.78, 2.04, and 2.72 for the total group, the females, and males.

not *just* another questionnaire test of personality. Each item on the personality scales of the MMPI was derived by an item analysis which contrasted the frequency of endorsement of a response, *true* or *false,* by some criterion group which had been defined after careful study by psychiatrists and psychologists, with frequencies from a control group. Take one scale, for example the one labeled "Depression"; the originators of the test, Starke Hathaway and J. C. McKinley, took an item, computed the frequency with which the normal group said *true,* and then computed the frequency with which the depressed patients under psychiatric study said *true.* If there was a significant difference between these percentages of about three times the standard error, the item was selected for that scale. I do not really want to go into the details of the construction further, but I thought that it ought to be known that each scale was derived empirically by this method of contrasting "normals" with diagnosed criterion groups. This is one reason why the MMPI is different from personality tests derived by factor analysis.

TABLE II

HARVARD TWIN STUDY MMPI SCALE HERITABILITY INDICES
FOR TOTAL GROUP, FEMALES, AND MALES

Scale	Total group		Females		Males	
	H	F^a	H	F^a	H	F^a
1 Hypochondriasis	.01	1.01	.00	.89	.09	1.10
2 Depression	.45	1.82	.48	1.91	.37	1.58
3 Hysteria	.30	1.43	.44	1.79	.11	1.12
4 Psychopathic deviate	.39	1.63	.46	1.47	.46	1.84
5 Masculinity-femininity	.29	1.41	.30	1.43	.29	1.42
6 Paranoia	.38	1.61	.40	1.67	.40	1.67
7 Psychasthenia	.31	1.46	.60	2.53	.00	.99
8 Schizophrenia	.33	1.49	.37	1.58	.29	1.41
9 Hypomania	.13	1.15	.19	1.23	.04	1.04
0 Social introversion	.33	1.49	.35	1.53	.29	1.40

[a] The three values of F required for significance at the .05 level are 1.47, 1.66, and 1.78 (preliminary analysis).

The heritabilities in Table I are derived from the study of 34 pairs of identical (MZ) adolescent twins and 34 pairs of fraternal (DZ) same-sex adolescent twins from the public high schools of Minneapolis and St. Paul, Minnesota (Gottesman, 1962, 1963b). It is worth noting at this point that zygosity was always diagnosed by extensive blood typing. The Smith and Penrose (1955) procedure was used which I would like to recommend to you. If just concordances for the blood group genes are taken, this is not as precise as the Smith and Penrose method which allows one to assign the probability of dizygosity to each pair. The blood group gene frequencies which were used are given in Race and Sanger (1962) and are based on the Caucasian English population blood-typed during World War II.

About 90% of the Minnesota sample reported one or two parents of Scandinavian extraction and something like three-fourths of the twins were Lutheran. When the data of Table I are compared with the data of Table II some differences will be noticed which might provide a basis for speculating about a different gene pool

TABLE III
Harvard Twin Study
Father x Mother Correlations
on MMPI Scales
(N = 66 pairs)

Scale	r
Hs	.26[a]
D	.08
Hy	.00
Pd	.26[a]
Mf	−.05
Pa	−.11
Pt	.06
Sc	−.11
Ma	−.25[a]
Si	.01

[a] Significant at the .05 level (preliminary analysis).

around Minneapolis and St. Paul from the gene pool around Boston, but that kind of speculation can be saved for the Discussion. Tables II and III present data from a large twin-family project in progress. The sample consists of 82 pairs of MZ twins, 68 pairs of DZ same-sex twins, 58 siblings of the twins the same sex as the twins, and 66 pairs of mothers and fathers of these twins, all from the greater Boston area. Some 29 twin pairs were removed from the data analyses because their tests were invalid. When self-reports are used from twins or any other population, one has to check whether the subjects understand the test items. The MMPI is one of the few tests that has a built-in indicator of lack of cooperation or poor reading comprehension. The results in Tables II and III are submitted as a promissory note on a more thorough analysis (e.g. Gottesman, 1963b). For the time being, it is hoped that the data will serve as a ticket of admission to the realm of speculation.

Darwinian fitness or adaptive value refers to the reproductive efficiency associated with a particular genotype and not necessarily to a strong back or a deadly aim. Any parental personality factors which would lead to celibacy, early divorce, or aversion to diapers —you can think of other things—would contribute to a lack of fitness. Selection against genotypes carrying a substantial number of genes associated with such personality syndromes (operationally defined by configurations of those traits with substantial heritabilities) should lead to a decrease in the numbers of those genotypes. In other words, discrimination against schizoid, highly aggressive, or paranoid types, and other varieties of neurotic and psychotic individuals as suitable marriage partners should decrease the frequencies of the associated genes in the next generation. For a variety of reasons, about two million males and two million females over age 45 are unmarried in this country and therefore contribute no genes to the gene pool; that may be a gratuitous assumption. Other genes are selected against by biological and psychological sterility. The personality correlates of family size in cultures with access to contraception are unknown, as far as I know. Kallmann's (1946) nuclear group of schizophrenics, i.e., hebephrenics and catatonics, had a marriage rate 55% that of the general population; their birth rate per marriage was 42% (1.4 versus 3.3) that of the general population. If this line of reasoning has any validity, why is mental illness still a characteristic of our species? Why has schizophrenia, for example, not been eliminated from the gene pool? It is precisely these kinds of questions which the theory of polygenic inheritance explains, or at least can handle. Fuller's work on audiogenic seizures in mice and Wright's work on polydactyly in guinea pigs would seem to indicate that only those genotypes in which a sufficient number of the genes are manifested in the phenotype as pathological will be selected against. Individuals on the continuum of pathological behavior who are below the "cutting score" established by their culture will then be "carriers" whose *offspring* may manifest the full-blown illness should there be a supra-threshold accumulation of the associated genes. When I speak about a cutting score established by the culture, I mean that different segments of our culture tolerate different amounts of psychopathology, so that an individual in one stratum of society might be hospitalized earlier and therefore prevented from marrying and perpetuating his genes. Is there, then, some adaptive value in moderate proportions of pathological genes in a polygenic system? If the schizophrenia scale on the MMPI, for example, can be construed to be a dimension of flexible behavior, as opposed to rigidity, as well as of irrational behavior, the adaptive value of the trait becomes apparent. Numerous studies have elicited the adjectives descriptive of the behavior of nonhospitalized individuals, e.g., college students and Air Force officers, who obtain the major configural patterns on the MMPI (Dahlstrom & Welsh, 1960). High Sc, or schizophrenia scores, for example, have been described as versatile, ingenious, inventive, and imaginative. In moderation the depression scale may reflect caution and control; the psychopathic deviate scale may reflect courage and vigor; paranoia may reflect sensitivity and awareness; some psychasthenia may correspond to conscientiousness; and social introversion may reflect inhibition and conformity. These speculations are only meant to convey the notion of an adaptive value of moderate amounts of what is ordinarily called pathology.

In one sense, the unit of evolution is the male-female dyad which cooperates to propagate its genes. Interpersonal behavior *vis a vis* the opposite sex then takes on added evolutionary significance in a civilization which can control its birth rate and in which reproduction is preceded by courtship and marriage. While we may lament the present state of the world, it might have been worse had there been no selection in favor of a modicum of cooperation or what Henry Murray (1961) chose to call *dyadism*.

Even the smallest selective advantage or disadvantage of a heritable characteristic is important for the gristmill of evolution. If a genotype produces on the average $1-s$ surviving offspring for every survivor produced by other genotypes, the former is said to be discriminated against by a selection coefficient s. Given enough time, the favored genotype will become established and the unfavored genotype will be eliminated from the population. The speed of the selection process will depend upon the size of s, the frequencies of the genes selected for, and the size of H, heritability. In a simple case, selection against a recessive gene with an initial frequency of 50% and weak selection, $s = .1$; the incidence of the gene diminishes at the rate of about 1% per generation (Dobzhansky, 1962). The predictions for polygenic systems are very speculative. The speed with which selection accomplishes specific results decreases with increasing number of loci (Stern, 1960). With the intense artificial selection of the laboratory, Tryon (1942) obtained virtually complete separation of his maze-bright and maze-dull rats after eight generations of inbreeding for this polygenically determined behavior.

Class Differences and Societal Structure

Class differences are differences between populations, not between individuals. Whenever there is a sizable degree of reproductive isolation between populations, the relative frequencies with which the different forms of genes occur in their gene pools will differ. In view of the demonstrated heritability for some aspects of personality, and the possibility of assortative mating (even though Table III does not show it to be important), it is possible that some social class differences observed in the incidence of mental illness and other aspects of personality may be partially on a genetic basis rather than the wholly environmental basis so popular at this time. Because this raises the spectres of Social Darwinism or of Aryan mythology, scientists must be careful not to throw out the baby of empirical behavior genetics with the bath water of race and class prejudice. Differences in gene frequencies for the blood groups are well established in such groups as the Jews in Rome contrasted with their Italian Catholic neighbors, the Basques, endogamous Indian castes, and Celtic-speaking Welshmen (e.g. Dobzhansky, 1962; Stern, 1960).

Behavioral differences among social classes which may have a partially genetic origin have only a speculative basis. The speculation will not have been in vain if it leads to more effective treatment procedures for physical and mental illness than those extant. Granted that the differences observed among social strata in the United States with respect to mental illness are quantitative and not qualitative, since when can a prevalence rate of the psychoses in the lowest stratum 800% greater than that in the upper two strata be dismissed as merely quantitative? That fact emerges from Hollingshead and Redlich's (1958) epidemiological study in New Haven. In this same study, the prevalence rate for the neuroses was some 360% greater in the *upper* two strata than in the lowest, suggesting that polygenes associated with the neuroses (Gottesman, 1962) may be accumulating toward the top of the social scale and those associated with the psychoses accumulating toward the bottom. At least it suggested that to me. In a more recent midtown Manhattan study of Srole, Langner, Michael, Opler, and Rennie (1962), social psychiatrists reported sick-well ratios for the mental health status of a representative sample of untreated adults in midtown New York based upon an intensive home survey. For every 100 well adults in the highest of six social strata they found 46 sick; for every 100 well adults in the lowest stratum they found 470 sick. While these authors do not rule out the possibility of constitutional factors in the etiology of the reported differences among social status groups, their preferred explanation focuses upon "certain specific forms of sociocultural processes operating within the framework of the social class system" (Srole *et al.*, 1962, p. 236). The gradient observed for the frequency of the psychoses across social classes suggests an analogy to the gradients of allele frequencies for type B blood and for some morphological characters observed across geographical areas and accounted for by a model of gene flow (Stern, 1960, pp. 727–731). The present analogy equates geographical distance with social distance, both leading to a decreased probability of mating.

A positive correlation exists between social status and IQ and between social status and the prevalence of psychoneuroses. A negative correlation exists between social status and the prevalence of high-grade mental retardation and between social status and the prevalence of psychoses. The possible adaptive value of subthreshold amounts of psychopathology has been outlined. All of these threads of ideas are assembled for the purpose of another speculation. To the extent that variation in intelligence and personality has a polygenic component, and to the extent that phenotypically deviant individuals represent relatively greater collections of homozygosity at the various loci in the polygenic systems for each trait, the threads are tentatively submitted in support of the balance theory of population structure (Dobzhansky, 1955,

1962). According to the balance theory, a balanced genetic load consists of genes and gene combinations which are advantageous in heterozygotes as compared to their effects in homozygotes. Sickle-cell anemia has been presented as a simple and clear illustration. Higher frequencies of heterozygotes for this trait are found in areas infested with malaria. It appears that the reproduction of the parasite *Plasmodium falciparum* is greatly impaired in the (smaller) sickle cell, so that these heterozygous individuals are more resistant to malaria than individuals with normal erythrocytes. If the speculation advanced here is generally correct it could explain the perpetuation of polymorphism in man's behavior and the constant but infrequent appearance of maladapted homozygotes. It may be that Nature not only abhors a vacuum, she also abhors a homozygote. Cattell (1960) has suggested that genotypes are negatively correlated with their environments. This led him to formulate the hypothesis that there is a cultural coercion to a bio-social norm. It would be of interest to pursue the possible overlap between the balance theory and the bio-social norm hypothesis. Both ideas bear on the perplexing mystery of the cause of Gaussian or normal distributions for human characteristics.

Behavior Genetics Model of Society

A behavior genetics model of society has been proposed by Tryon (1957; Hirsch, 1958). In a society that provides for social mobility, the varieties of genotypes migrate to different strata or social ecological niches by social selection. In this schema, the strata are ordered by the single major variable, money-reward. "Individuals receiving the same money-reward but for different kinds of ability tend to gravitate to the same social area. The hierarchy of social strata is determined by the hierarchy of money-rewards characteristic of all occupations. The abilities requisite for performance in the different occupations depend upon different sensory-motor and personality components, which are in turn determined by different independent polygenic combinations. Most matings occur within strata so that a correlation among abilities is developed not because there is one general factor underlying achievement in all fields, but because of the selective influence of the common denominator, money-reward, which collects comparable levels of various abilities within the same social strata. [Tryon has found a correlation of about .6 between the social area ratings of spouses even when reared in different cities.] The picture being drawn is a statistical one. It does not assert that *all* of the genetically controlled constitutional factors responsible for high achievement are confined to the highest social stratum or that *all* of the factors responsible for low achievement are to be found in the lowest stratum" (Hirsch, 1958, pp. 2–3).

It must be emphasized that the closer the approach to equality of opportunity, the more the variation observed in a society will be associated with genetic differences, and the more society will be able to utilize the resources of all its members.

Concluding Remarks

Human behavior genetics is an area of research which cannot remain in the academic ivory tower. Almost any datum generated by our efforts can be used as ammunition by proponents of various malignant propagandas. I would submit that we all retain ethical responsibility for our findings which extends beyond scientific reporting. This requires alerting the public whenever our hypotheses or data are misinterpretated or over-generalized (cf. George, 1962; Gottesman, 1963a).

There is a natural reluctance to accept the supposed determinism that is associated with views that human behavior is genetically influenced. The former is especially true when one thinks of oneself. The ego defenses aroused are in part due to the values placed on free will and equality which are part and parcel of our democratic way of life. The word "supposed" above was used intentionally. Allport (1937, p. 105) pointed out that the doctrine of genetic determination does not state that personality is inherited, but rather that no feature of personality is devoid of hereditary influences. This means that if the genes are altered the personal characteristics are altered and *not* that they are determined only by the genes. More recently Kallmann and Baroff (1955) noted that the belief that genetically determined disorders are unalterable finished entities was related to the early impression made by congenital anomalies. Considerable progress has been made in demonstrating that some gene-specific disorders are neither congenital nor unchangeable. Another source of reluctance to become involved in behavior genetics research is the lurid and disquieting history of the eugenics idea in some countries. The reader is referred to Dunn's (1962) presidential address to the American Society for Human Genetics on the differences between genetics and eugenics. It is a pleasure to call attention to the fact that many of the contributors to behavior genetics, including those in the present symposium, are just as interested in exploring areas such as early experience, prenatal effects, socialization, and psychodynamics as they are in genetics. Progress in understanding the complexities of human behavior necessitates such appreciation of complementary vantage points.

REFERENCES

Allport, G. W. *Personality: a psychological interpretation.* New York: Holt, 1937.
Cattell, R. B. The multiple abstract variance analysis equations and solutions for nature-nurture research on continuous variables. *Psychol. Rev.*, 1960, **67**, 353–372.
Dahlstrom, W. G., & Welsh, G. S. (Eds.). 1960. *An MMPI handbook.* Minneapolis: Univer. Minn. Press, 1960.
Dobzhansky, T. *Evolution, genetics and man.* New York. Wiley, 1955.
Dobzhansky, T. *Mankind evolving.* New Haven: Yale Univer. Press, 1962.
Dunn, L. C. Cross currents in the history of human genetics. *Amer. J. Human Genet.*, 1962, **14**:1–13.
Edwards, A. L. The prediction of mean scores on MMPI scales. *J. consult. Psychol.*, 1964, **28**, 183–185.
Fuller, J. L., & Thompson, W. R. *Behavior genetics.* New York: Wiley, 1960.
George, W. C. *The biology of the race problem.* New York: National Putnam Letters Committee, 1962.
Gottesman, I. I. Differential inheritance of the psychoneuroses. *Eugenics Quart.*, 1962, **9**, 223–227.
Gottesman, I. I. Science or Propaganda? *Contemp. Psychol.*, 1963, 381–382(a).
Gottesman, I. I. Heritability of personality: A demonstration. *Psychol. Monogr.*, 1963, **77**, No. 9 (Whole No. 572).(b)
Hirsch, J. Recent developments in behavior genetics and differential psychology. *Dis. nerv. System*, 1958, **19**, No. 7 (Monogr. Supp.).
Hollingshead, A. B., & Redlich, F. C. *Social class and mental illness.* New York: Wiley, 1958.
Kallmann, F. J. The genetic theory of schizophrenia. *Amer. J. Psychiat.*, 1946, **103**, 309–322.
Kallmann, F. J., & Baroff, G. S. Abnormalities of behavior (in the light of psychogenetic studies). In *Ann. Rev. Psychol.* Palo Alto, Calif.: Ann. Reviews, Inc., 1955. Pp. 297–326.

Marlowe, D., & Gottesman, I. I. The Edwards SD scale: a short form of the MMPI? *J. consult. Psychol.*, 1964, **28**, 181–183.(a)

Marlowe, D., & Gottesman, I. I. The prediction of mean scores on MMPI scales: a reply. *J. consult. Psychol.*, 1964, **28**, 185–186.(b)

Murray, H. A. Unprecedented evolutions. *Daedalus*, 1961, **90**, 547–569.

Race, R. R., & Sanger, R. *Blood groups in man.* (4th ed.) Oxford: Blackwell, 1962.

Simpson, G. G. The study of evolution: methods and present status of theory, In A. Roe and G. G. Simpson (Eds.), *Behavior and evolution.* New Haven: Yale Univer. Press, 1958. Pp. 7–26.

Smith, S. M., & Penrose, L. S. Monozygotic and dizygotic twin diagnosis: *Ann. Human Genet.*, 1955, **19**, 273–289.

Srole, L., Langner, T. S., Michael, S. T., Opler, M. K., & Rennie, T. A. *Mental health in the metropolis.* Vol. 1. New York: McGraw-Hill, 1962.

Stern, C. *Principles of human genetics.* (2nd ed.) San Francisco: Freeman, 1960.

Thompson, W. R. The significance of personality and intelligence tests in evaluation of population characteristics. In *The nature and transmission of the genetic and cultural characteristics of human populations.* New York: Milbank Foundation, 1957. Pp. 37–50.

Thompson, W. R. Multivariate analysis in behavior genetics. In R. B. Cattell (Ed.), *Handbook of multivariate analysis,* 1964. (Unpublished manuscript).

Tryon, R. C. Individual differences. In F. A. Moss (Ed.), *Comparative psychology.* New York: Prentice-Hall, 1942. Pp. 330–365.

Tryon, R. C. Behavior genetics in social psychology. *Amer. Psychol.*, 1957, **12**, 453. (Abstract)

Vandenberg, S. G. The heritability of selected psychological traits, a report on the Michigan Twin Study. *Amer. J. Human Genet.*, 1962, **14**, 220–237.

Discussion

Gardner: I have a psychologist's question about your handling of the MMPI. I hope you will forgive me for bringing up a whole thicket of issues that you very pleasantly did not want to get into here, with people who are not so involved in this. I can put my question as a suggestion with a little preliminary statement. For those of you who are not familiar in detail with the Minnesota Multiphasic Personality Inventory, it is a test which was originally constructed against criterion groups of patients and nonpatients as Dr. Gottesman described; it is scored in terms of scales by the original authors, and their successors, which Dr. Gottesman, in analyzing his data, apparently has taken as such. There has been a great deal of controversy about how many response factors are actually involved in this test. Recent evidence by Edwards, Messick, Jackson, and a number of other people indicates that some responses set attributes of the subjects such as his tendency to agree with items that are socially desirable, or his tendency to be compliant and to agree with any positive statement, introduced powerful correlations among the scales. Some of the correlations due to response set in the MMPI are of the order of .65 to .80.

The most sophisticated MMPIer's I know do not deal so much with the individual scale scores, which they rather distrust, but develop a body of clinical experience with the test and use this as a practical clinical tool, based primarily on relationships amongst the scores, that is, patterns of scores. They feel that a high score on a certain individual scale does not necessarily mean that the person has a great intensity of that attribute. In one context of other scores you might be interpreting it correctly, in another context not. To give an example, there is a homosexuality scale, and in certain constellations a high score would be interpreted as an exaggerated form of denial, and in another context as a high degree of intensity of this trait. It occurs to me that one can approach data empirically; if one factor-analyzes the data, one may obtain a great deal of information about the relationship between empirical factors and their heritability estimates and the traditional scales. One could get multiple forms of information from such an approach. I wondered if you have thought of doing this.

Gottesman: Yes, this is already in the computer.

Gardner: If we do not control for the response-sets we won't get anywhere. Unless we do, we cover all the scores as if with a paint brush.

Gottesman: This is really probably not the place for me to defend the MMPI against the response-set notions of Edwards and Messick. Marlowe and I (Marlowe & Gottesman, 1964a,b) have compared the use of Edward's (1964) ideas as applied to the MMPI and have found them to be of little practical value.

Gardner: Yes, I would suspect that is true, but that is because the clinical interpreter does not use the scale scores as if they were readings on a thermometer.

Gottesman: Right, this is a first approximation. I could have used index numbers which are what both Paul Meehl and Starke Hathaway suggested using, instead of these clinical labels I have given you, but that would have been less clear in this symposium. What we teach our students is that each of the MMPI scales or their configurations represent new constructs to be validated by the nomological network generated by research on both sick and healthy samples.

Lindzey: I think that Dr. Gardner's question is directed less toward the issue whether to use numbers rather than names, and more toward the question of profile analysis or some kind of multivariate analysis. It may be better to hold discussion until Dr. Cattell has presented his approach. Because it seems to me that you do have a rather nice comparison of two different approaches to measuring a similar domain of the phenotype—and you know they really are quite opposed in terms of basic assumptions—so it would be a shame to criticize the Multiphasic when we cannot yet criticize the factor pure test.

Lasker: For how long a period of time, do you think, has human society fulfilled, to any appreciable extent, the model of Tryon and Hirsch, that is presented here? And to what extent is it possible that this could have been in existence long enough and consistently enough to have any effect whatsoever in human genetic polarization?

Gottesman: I would guess that selection has been going on for a long time, but not necessarily with respect to money; I would guess that the caveman who was the strongest and the most cunning would have his first choice of a mate, so that in terms of adaptive behavior we get the same results.

Lasker: But then it would be a totally different ability with different genes responsible for it, that was being selected and each revolution, of which there are a series throughout human history, led to some extent to new polarizations, which requires a new orientation of values.

Gottesman: I was thinking in terms of adaptive behavior in general, rather than the traits that we can measure at the moment. Men have different channels for expressing adaptive behavior; one person may earn a lot of money by being very smart in the stock market, and another person may earn a lot of money by working very hard physically on a farm. Each would be the basis for the accumulation of wealth which then leads to the social stratification. There would have been selection for adaptive behavior rather than some specific occupation. Actually maybe that is a hedge.

Lindzey: Halsey, the English sociologist, considered this in 1958.[1]

Gottesman: I am familiar with a summary of it.

Lindzey: It seems to me, that your question really is whether we can make a specific model and estimate the parameters. Agreement about the upper and lower bounds of the parameter would lead to some agreement about whether or not this sort of process is likely to have been significant. Halsey is by political belief a socialist, and is really looking for this kind of process. With Savage, a mathematical statistician, and Caspari, a geneticist, he set up a model for the social distribution of intelligence. He made certain assumptions with regard to heritability, making the assumption of a single-gene-Mendelian mechanism, which is obviously wrong. He also made some estimates about the rate of gene flow from one class to another, and knowing from population figures what the sizes of the classes are, he arrived, as I recall the article, at very discouraging conclusions in regard to the likelihood that social class differences in intelligence could lead to any change in the gene frequencies in terms of a genetically determined range of intelligence. In other words, the phenotype as it is influenced by the genotype would not change over time given his ranges of heritability and given the rate of gene flow, so he really arrived at the same conclusion Dobzhansky reached in regard to a closed society, namely that it has no movement from one social class to another.

Lasker: If the communalities between genetic basis for success, in whatever way you want to define success, are very different in one kind of society from those in other types of society and if peoples have been going through changes in their types of social organization, then the effect of such a factor as this must be of a low order.

[1] Halsey, A. H. Genetics, Social Structure and Intelligence. *Brit. J. Sociol.,* 1958, 9, 15–28.

Lindzey: Dr. Fuller, you argued against cultural and race differences and for common selection factors, have you not?

Fuller: Yes, I thought that perhaps common factors in selection are more plausible than cultural and race differences, although I did not rule out the latter.

Guttman: I would like to add a question here. Has what is today a deviant personality always been considered one so that we can really speak of the evolution of a deviant personality in terms of polymorphism? I do not think we can.

Gardner: There is also the problem of the remarkable increase in the types of identifications or, if you wish, of types of sociocultural and occupational classifications.

Lindzey: In terms of Dr. Guttman's question, we do not have good evidence over really long periods of time about the stability of psychosis, but the evidence available is remarkably constant, across cultures and across time. About 100 years is the longest span of time for which we have records. It is at least conceivable that there is a form of psychotic process or mental disorder which in all times and all cultures would have been maladaptive and so had a low Darwinian fitness.

Gottesman: It is difficult to think of any culture in which prolonged loss of contact with reality is adaptive, in spite of great variation between cultures.

Lasker: There are some exceptions: the Eskimos go on spirit quests, the Chukchee do the same, the Indonesians get out of contact for specific reasons, it is done in Africa; there are widespread examples of behavior which we would consider out of contact.

Guttman: It seems to me that those must be their most prolific periods too, in terms of reproduction, but I do not think much is known about this aspect.

Lasker: Priesthood is often associated with sexual restraint, and celibacy; I imagine that it varies, and I expect that there would be examples in some places for almost any combination of these behaviors.

Gottesman: You are talking about processes which are under control but I am talking about psychosis, not a culturally accepted game. I do not completely accept the "myth of mental illness."

Lindzey: There is a general problem which is particularly related to remarks made about schizophrenia. What you said about schizophrenia is perfectly reasonable. Of course, a lot of other theories or models have been advanced in an attempt to account for the paradox of its persistence. Among those who are primarily concerned with mental disorder, a sizable number question whether schizophrenia is an entity. I, myself, would side with you on the entity issue intuitively, but others, e.g., Karl Menninger, believe in a unitary dimension of mental disorder, and that schizophrenia simply is a way of talking about a certain intensity or degree of severity. If you introduce this conception into the schizophrenic studies you have quite a different variable and the kinds of genetic theory that you need would be different. There is a related factor which you know very well, i.e., the possibility of contamination of data in genetic studies of schizophrenia. Possibilities for contamination are so extreme, that there is no need to blame experimenter bias. This is why persons like Gordon Allen, who has been much concerned with this kind of research, are pessimistic about conclusions in regards to genetics and schizophrenia.

Stafford: I would like to ask a question about this if I might. In your correlation of father-mother, if you ran those on a computer you would also have the correlations which might indicate cross-homogamy. Did you look at these?

Gottesman: I did not. I have so many data coming out of this study.

Stafford: I know. It is difficult to digest it all immediately. I feel that this is one place where, especially with personality traits, you have to look closely, because there may be occasions where the mate high in one trait meets someone high or low in another trait and there may be some real assortative mating. I am always taking a look at cross-correlations. I analyzed the cross-homogamy of 10 variables, which gave me, if I excluded the homogamy, 90 correlations. Four were significant at the .05 level and two at the .01 level, which is about what you would expect by chance. So I threw up my hands and figured that there was nothing there; but I still think it is worth looking every time you do one of these studies, and make sure that you intercorrelate all variables.

Gottesman: I might add that one of the nice things about doing a twin family study including some of the sibs is that you generate data which are of interest to so many people with varying backgrounds. Some of my students who are really against genetics in psychology have all kinds of raw data to use for their own particular ideas about parent-child relations, identification with the same sex parents, and resolution of whatever complex you want to talk about. I am very happy to entertain all those ideas.

Lindzey: This question really involves the usefulness of heritability estimates. Dr. Gottesman pointed out that between the Minnesota sample and the Harvard-Cambridge-Boston sample, there were appreciable differences in terms of the estimates for the same instrument using the same procedure, and suggested this might reflect gene pool differences which indeed it might. But this may just reflect the instability of the index, the extent to which it is sensitive to the genotype-environment interactions, with nothing in the design to control or sample this interaction. I would like to ask about Blewett's Primary Mental Abilities study. I recall he got rather different heritability coefficients.

Vandenberg: Not too different from mine, although, there are differences. I would not expect any of these figures to be final because they are based on small samples. They should never be taken without a considerable band of error around them.

Lindzey: Was his math as high as yours, the numerical, Dr. Vandenberg?

Vandenberg: No, his was .07, while I obtained .61 in Ann Arbor and .72 in Louisville. Thurstone's Chicago data give a value of .34. These should all be combined if possible. I could not agree more on the need for repeating these studies and I expect that going elsewhere is more important and better strategy than obtaining larger samples in the same geographical area.

Freedman: I have a question along this same line, Dr. Gottesman. I notice on this first sample, the Minnesota sample, you have some H coefficients as high as .71, .65, .77, and .84. This gives a stereotype picture of Swedes, because here they are social introverts, they are depressed, and they tend to be paranoid. But in the Harvard twin study, you have only one at .6 while the rest are all about the same, which does not give you much of a picture. Does it not bother you when you get all your heritabilities at about the same level?

Gottesman: It does bother me, and in the next step of the data analysis I am going to divide my sample by social class. There is a big difference in the proportion of males and females in the two samples. I had about twice as many females as males in the Minnesota sample but in the Harvard sample I had more nearly equal numbers, although still more girls than boys. Everyone here who is doing twin studies gets that imbalance, did you notice that? There is a good section in Fuller and Thompson (1960) that I read whenever I get depressed about the instability of heritabilities. They talk about the various things that can contribute to heritability changes and various ways of manipulating estimates of heritability. This is fine for everyone concerned, both environmentalists and hereditarians. How important the genetic factors are depends on how much trouble one wants to take to manipulate the environment.

Sutton: One question I would like to raise about comparing studies has to do with how the sample is collected, particularly when psychological variables are studied. Everyone begins with the restriction that both members of a twin pair have to be available. Next they usually restrict themselves to the normal population, which means both members have to be within the defined normal population. They are collected from cooperative families by necessity, and while you are cutting down your sample, you are also losing much variability. Yet, from one sample to another the cut-off point may be very different, so I see no reason why you should expect to get the same values. This kind of variation between studies is neither environmental nor genetic as far as these variables are concerned.

Gottesman: It was easier to do good sampling in Minnesota than it was in the Boston area, and my sample in Minnesota may turn out to be one of the better samples that people have been able to acquire in a twin study. I had access to the school records of every public school child in grades nine through twelve, and I contacted everyone of those twins and their families. I ended up with 75% of all possible female adolescent twins, and 43.4% of all possible male adolescent twins. I do not know what I have in the Boston area yet, because there are so many different school systems.

Sutton: How did you find the twins?

Gottesman: I went through the files, there were 31,307 children in grades nine through twelve and I took all those individuals with the same last name, same sex, and same birthdate from the roster.

Sutton: They still had to be bright enough to get to grade nine. This is grade, not age is it not?

Gottesman: Yes. Also because there are relatively few mental hospitals serving the twin cities, I could contact all the hospitals and there were no twins in the age range at the time that I did my study.

Lindzey: Is that not rather remarkable?

Gottesman: No, maybe not between the ages of 14 and 18. The population of Minnesota in general is not that big, perhaps 3 million for the whole state. The twin cities, Minneapolis and St. Paul have about three-quarters of a million which is almost one-third the population of the state. I have yet to evaluate the sampling in the Boston study.

Vandenberg: This is a problem that we have thought about too, of course. We are aware of just about all the twins in this area, but you really cannot use certain twins. We have several pairs in which either one or both are so deficient in reading that there is no point in including them in a study of this kind. Yet any conclusion which you draw would include consideration of these cases. That is one thing that I have not resolved. What we shall do, which is not really going to help on that issue, is study these cases individually and arrive at some kind of determination of ability. I do not see how these data will fit into the final analysis. My feeling is, if you have just a few of these cases and a clear-cut criterion below which they are excluded it should be possible to repeat these conditions somewhere else and minimize the difference that would result from such factors. If one person excludes a subject because he could not do some very difficult things and another excludes a subject only if he could not read at all, one would introduce some variation, but with agreement on some very low minimum performance, one has comparable studies and should obtain comparable results.

Sutton: When you have cut-off points that can be clearly stated, things are not quite so bad, but I think frequently that experimenters themselves are not very sure where the cut-off point is.

Lindzey: When you state a cut-off point, it is typically a character with both genetic and environmental determinants. This means, as pointed out earlier, one really does not know how much of either source of variation may be missing.

Vandenberg: Here, in Louisville, at the beginning of the group testing, we have been using a very short test that is very easy as a criterion for whether or not to excuse children. We have excused some who could not do this, but we can go back and give them individual examinations. Eventually I think we could collect the data, if we wanted to do a separate analysis of these pairs. It turns out that there have not been very many so far, perhaps ten pairs in which either one or both were so low that we did not want them in the study.

Biometrical Genetics in Man[1]

Dewey L. Harris[2]

Introduction

A logical starting point for this paper seemingly involves an explanation of the meaning of the expression "biometrical genetics." Although this term, which incidentally was coined by Mather (1949), is often used in the literature, there does not seem to be a clear definition. Thus, the following is my interpretation of its meaning. The roots are *bio* (from the Greek bios meaning life which is the root word of biology), *metric* (referring to measurement), and genetics (the science of inheritance). *Bio* and *metric* are often combined into biometry, the name given the science of statistics as applied to biological observations. So biometrical genetics refers to the application of statistics to the biological area known as genetics. As commonly used, biometrical genetics refers more specifically to the statistical aspects of quantitative genetic traits in relevant populations. Quantitative genetic traits are those traits of individuals which are characterized by continuous distributions when populations are considered. All informative work about the nature of the basic mechanism of inheritance has involved rather simple traits. Observations on traits involving presence or absence of certain pigments in the skin or in the eye, or the presence or absence of a certain amino acid often may be classified into distinct phenotypes. Most of these types of traits are controlled by the genetic constitution of the individual in a manner which allows, after a bit of study, an accurate and complete description of the

[1] Journal Paper No. J-4883 of the Iowa Agricultural and Home Economics Experiment Station, Ames, Iowa. Project No. 1448.
[2] Present address: DeKalb Agricultural Association, Inc., Sycamore, Illinois.

relationship between genotype and phenotype, i.e., identification of specific genotypes.

On the other hand, one is often interested in complex traits that are measured on some quantitative scale and seemingly are the result of the action and interaction of numerous genes and numerous environmental influences. This complexity and variation usually results in the inability to relate directly observed phenotypic values to specific genotypes. These are the kind of traits in which plant and animal breeders are largely interested because of economic considerations. Seemingly, these are also traits that are involved in most behavioral studies. Certainly, a few abnormal psychological conditions may involve enough discontinuity so that exact modes of gene action may be assessed. But most traits are a reflection of such complex functioning of body processes that they undoubtedly are influenced by so many hereditary factors that it seems futile (to me, at least) for an investigator to spend much time attempting to identify simple modes of inheritance.

Nevertheless, the underlying genetic mechanism of the more complex traits is basically the same, the difference being in the number of relevant loci and in the quantitative measurement of the trait rather than classification.

Biometrical Genetic Theory Relevant to Human Studies

Certain statistics, such as correlations between relatives and variances between and within families, may be calculated from a study of a quantitative trait in a population even when one has no knowledge of the nature of the genetic mechanism underlying the expression of this trait. This is precisely what was done by Galton and others, prior to the rediscovery of Mendel's work. However, the basic premise behind what is termed biometrical genetics seems to be that knowledge concerning the nature of the mechanism of inheritance, the so-called Mendelian mechanism, should be helpful in the understanding and interpretation of statistics that one might observe from quantitative studies. More exactly, this knowledge of the mechanism of inheritance should lead to a more basic parametrization for the population. We can consider covariances between full sibs, covariances between half sibs, and so on as parameters descriptive of the population; but, perhaps there are more basic parameters which are indicative of the genetic structure of the population and which illustrate the relationship among the various covariances between relatives. A mathematical description of the probabilistic genetic mechanism and the mating structure of the population should suggest this more basic parametrization.

The basic mechanism of inheritance is quite simple. However, the mathematical description of traits involving complex metabolic and biochemical reactions interposed between the basic action of the genes and the final phenotypic expression in a population with a complicated mating structure becomes quite difficult. Because of this complexity and in an effort to attain freedom from restrictive assumptions about the nature of gene action, most theoretical biometrical genetics has been limited to populations with rather simple mating structures. The applicability of the resulting parametrization to actual biological populations is thus highly dependent on how close the mating structure of the actual population is to the theoretical mating structure. There is considerable biometrical genetic theory relating to inbreeding, but this does not seem to be relevant for the considerations of this paper. The primary bit of biometrical genetic theory relevant to human studies is the development of the parametrization for genotypic covariances between relatives in random mating populations. Since variances are special cases of covariances and since correlations and regressions between relatives and the genotypic components of variance between families and within families are all functions of these basic covariances between relatives, the parametrization is involved in a basic representation of nearly all the descriptive quantities about the population.

The term random mating refers to the situation in which every individual in the conceptual population has an equal probability of mating with any individual in the population of the opposite sex. The conceptual population considered will be quite large and will be mating at random with no differential fertility of matings and no differential viability of gametes or zygotes; or, at least the differences in fertility and viability will not be associated in any systematic way with the genotypes involved. All of these assumptions are collectively termed panmixis.

The basic work underlying this parametrization is in the paper of Fisher (1918). Several other workers have been concerned with this since then; among these are Mather (1949), Cockerham (1954), and Kempthorne (1954, 1957). One should refer to these papers for the complete derivation, but I shall attempt to illustrate the development for a single locus affecting the quantitative trait with two alleles occuring at this locus.

Although it is generally impossible to identify genotypes for a quantitative trait, we may *conceptually* consider the three sub-populations of the over-all population corresponding to the three genotypes, AA, Aa, and aa for a single locus with two alleles. A possible distribution for these is shown in Fig. 1. The means of these sub-populations are represented by the symbolic values, d, h, and r, respectively. These values are termed the genotypic values for those genotypes. Under the assumptions of panmixis, the three genotypes AA, Aa, and aa will have frequencies p^2, $2pq$, and q^2, respectively, where p is the gene frequency for A and q is the frequency of a with $p + q = 1$. Thus the population mean is

$$\mu = p^2 d + 2pqh + q^2 r,$$

and the deviations of genotypic values from the mean are $i = d - \mu$, $j = h - \mu$, and $k = r - \mu$. These quantities are illustrated in the two-dimensional diagram in Fig. 2. Letting α represent

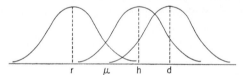

FIG. 1. Illustration of genotypic values d, h, and r.

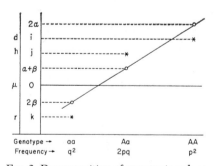

FIG. 2. Decomposition of genotypic values.

the "average effect" of gene A and β the "average effect" of gene B, we have the further decomposition of genotypic values as follows:

$$AA : d = \mu + i = \mu + 2\alpha + (i - 2\alpha) = \mu + 2\alpha + d_{AA}$$
$$Aa : h = \mu + j = \mu + \alpha + \beta + (j - \alpha - \beta) = \mu + \alpha + \beta + d_{Aa}$$
$$aa : r = \mu + k = \mu + 2\beta + (k - 2\beta) = \mu + 2\beta + d_{aa}$$

The average gene effects, α and β, are defined as the values which minimize

$$p^2(i - 2\alpha)^2 + 2pq(j - \alpha - \beta)^2 + q^2(k - 2\beta)^2$$

the average squared deviation of the coded genotype values, i, j, and k, from the sum of the average effects of the two genes involved in the genotype. The values which accomplish this are $\alpha = pi + qj$ and $\beta = pj + qk$. Note that the average effects of the genes depend on their frequency in the population. The deviation $i - 2\alpha = d_{AA}$, $j - \alpha - \beta = d_{Aa}$, and $k - 2\beta = d_{aa}$ are termed the dominance deviations since they are non-zero when there is dominance on the quantitative scale, i.e. $h \neq (d + r)/2$. The total contribution of this locus to the variance of the population is

$$\sigma_G^2 = p^2 i^2 + 2pq j^2 + q^2 k^2$$

This variance is composed of two parts

$$\sigma_G^2 = \sigma_A^2 + \sigma_D^2$$

where

$$\sigma_A^2 = p^2(2\alpha)^2 + 2pq(\alpha + \beta)^2 + q^2(2\beta)^2 = 2p\alpha^2 + 2q\beta^2$$

and

$$\sigma_D^2 = p^2(d_{AA})^2 + 2pq(d_{Aa})^2 + q^2(d_{aa})^2$$

σ_A^2 is termed the additive genetic variance for this locus and σ_D^2 is the dominance variance. Note that σ_D^2 is the quantity which was minimized by the choice of α and β.

If pairs of individuals are related in a certain manner in this population, the covariance between members of the pairs will involve certain fractions of σ_A^2 and σ_D^2 with the fractional coefficients involving the probabilities of genes possessed by the individuals being "alike by descent." Representing the individuals as X and Y

with X having the arbitrary genotype ab and Y having genotype cd, the covariance between the genotypic value of X, G_x, and the genotypic value of Y, G_y, may be found to be

$$\text{Cov}(G_x, G_y) = 2r_{xy}\sigma_A^2 + u_{xy}\sigma_D^2$$

where r_{xy} is Malécot's (1948) "coefficient de parenté"

$$r_{xy} = 1/4\,[P(\text{a} = \text{c}) + P(\text{a} = \text{d}) + P(\text{b} = \text{c}) + P(\text{b} = \text{d})]$$

and

$$u_{xy} = P[(\text{a} = \text{c}) \text{ and } (\text{b} = \text{d})] + P[(\text{a} = \text{d}) \text{ and } (\text{b} = \text{c})]$$

where $P(\text{a} = \text{c})$

is read "the probability that gene a is alike by descent to gene c" and so on. The term "alike by descent" means the two genes are copies in the reproductive process of the same gene in some common ancestor or one is a copy of the other. Only one of the two probabilities in u_{xy} may be non-zero under the assumption of panmixis.

Up to now we have only considered one locus with only two alleles. If the procedures are extended to include an arbitrary number of independent loci with an arbitrary number of alleles, the total genotypic variance may be partitioned as

$$\sigma_G^2 = \sigma_A^2 + \sigma_D^2 + \sigma_{AA}^2 + \sigma_{AD}^2 + \sigma_{DD}^2 + \sigma_{AAA}^2 + \dots$$
$$= \sum_{r=0}^{n}\sum_{s=0}^{n} \sigma_{A^r D^s}^2$$
$$1 \leq r + s \leq n$$

where n is the number of segregating loci influencing this trait; σ_A^2 now represents the sum over loci of the additive genetic variance for each locus; σ_D^2 is now the sum of dominance variances for individual loci; and the other terms represent variances due to interactions among additive effects and dominance deviations at two or more loci. For example, $\sigma_{AAD}^2 = \sigma_{A^2D^1}^2$ would represent the sum over sets of loci of the variance due to additive $\times$ additive $\times$ dominance interactions. For this general situation, the genotypic covariance between X and Y would be

$$\begin{aligned}
\operatorname{Cov}(G_x, G_y) &= 2r_{xy}\,\sigma_A{}^2 + u_{xy}\,\sigma_D{}^2 \\
&\quad + (2r_{xy})\,\sigma_{AA}{}^2 + (2r_{xy})(u_{xy})\,\sigma_{AD}{}^2 \\
&\quad + (u_{xy})^2\,\sigma_{DD}{}^2 + \ldots \\
&= \sum_{r=0}^{n}\sum_{s=0}^{n}(2r_{xy})^r(u_{xy})^s\,\sigma_{A^rD^s}^2 \\
&\quad 1 \le r + s \le n
\end{aligned}$$

The $\sigma^2_{A^rD^s}$ parameters form a more basic set of parameters for representing the covariance between relatives in the population. The values of $2r_{xy}$ and u_{xy} for certain simple relationships between X and Y are given in Table I (Kempthorne, 1957).

TABLE I
Values of $2r_{xy}$ and u_{xy} for Certain Relationships

Relationship of X and Y	$2r_{xy}$	u_{xy}
Identical twins	1	1
Full sibs	1/2	1/4
Parent-offspring	1/2	0
Half sibs	1/4	0
Parent-k^{th} degree offspring	$(\tfrac{1}{2})^k$	0
Uncle-nephew	1/4	0
First cousins	1/8	0
Double first cousins	1/4	1/16

The assumptions stated or implied in the above are:

1. panmictic population of infinite size
2. no mutation
3. no sex linkage
4. no linkage
5. no position effects
6. no sex influence upon genotypic values.

Also, the assumption that all non-genetic (environmental) influences upon the phenotype of an individual were independent of genotype and independent of family structure was implied in the definition of genotypic values as the means of sub-populations composed of certain genotypes. The failure of these assumptions to be true limits, to an unknown degree, the utility of this parametrization for a specific population.

Some modifications of this parametrization have been suggested by Bohidar (1960) for the presence of sex linkage, by Schnell (1963) and van Aarde (1963) for the presence of linkage, and by Griffing (1961) for position effects. Sex influence may be accounted for by defining the trait as actually being two traits; one trait in female individuals and another trait in male individuals. Of course, phenotypic values for female traits do not exist in males, and vice-versa, but the genotypic values may be considered to exist conceptually.

When an analysis is made of the variation among and within families such as full sib families, an analysis of variance is obtained by utilization of the following tabulation:

Source of variation	Degrees of freedom	Mean squares
Among families	$a - 1$	MS_A
Within families	$a(n - 1)$	MS_W

where a is the number of families and n is the number of members of each family. Since the variance of true family means is equal to the covariance between family members, the genotypic contributions to MS_W will be $\sigma_G^2 - \text{Cov}(F.M.)$ where $\text{Cov}(F.M.)$ is the genotypic covariance between two members of the same family. The genotypic contributions to MS_A will be $\sigma_G^2 + (n - 1) \text{Cov}(F.M.)$. If the environmental contributions of phenotypic values are independent of the genotypic values or the family membership of the individual, then both of these mean squares will have an additional component of variance, σ_E^2, due to environmental influences. Thus we have the following tabulation:

Mean square	Expectation of mean square
Among families	$\sigma_E^2 + \sigma_G^2 + (n - 1) \text{Cov}(F.M.)$
Within families	$\sigma_E^2 + \sigma_G^2 - \text{Cov}(F.M.)$

It seems worthwhile at this point to reemphasize that these considerations involve population studies. Thus, it is quite important that sample data upon which these considerations are used be representative of some population and, in particular, of the population to which one wants to make inferences. Of course, this is true of all types of population studies.

Various quantities are found in the literature for measuring the relative importance of genetic influences to environmental influences. These are the heritability estimates. Lush (1945) defines two forms of heritability values, heritability in the broad sense and heritability in the narrow sense. Actually the concept of heritability was used prior to Lush's work, but his definitions form the basis for most considerations today. Symbolically,

$$h^2(\text{broad}) = \frac{\sigma_G^2}{\sigma_G^2 + \sigma_E^2}$$

and

$$h^2(\text{narrow}) = \frac{\sigma_A^2}{\sigma_G^2 + \sigma_E^2}$$

Most estimates based upon correlations and regressions between related individuals estimate quantities intermediate between these. We thus have

$$h_{H.S.}^2 = \frac{4\,\text{Cov}\,(H.S.)}{\sigma_G^2 + \sigma_E^2} = \frac{\sigma_A^2 + \frac{1}{4}\sigma_{AA}^2 + \frac{1}{16}\sigma_{AAA}^2 + \cdots}{\sigma_G^2 + \sigma_E^2}$$

$$h_{F.S.}^2 = \frac{2\,\text{Cov}\,(F.S.)}{\sigma_G^2 + \sigma_E^2}$$

$$= \frac{\sigma_A^2 + \frac{1}{2}\sigma_D^2 + \frac{1}{2}\sigma_{AA}^2 + \frac{1}{4}\sigma_{AD}^2 + \frac{1}{8}\sigma_{DD}^2 + \cdots}{\sigma_G^2 + \sigma_E^2}$$

$$h_{P.O.}^2 = \frac{2\,\text{Cov}\,(P.O.)}{\sigma_G^2 + \sigma_E^2} = \frac{\sigma_A^2 + \frac{1}{2}\sigma_{AA}^2 + \frac{1}{4}\sigma_{AAA}^2 + \cdots}{\sigma_G^2 + \sigma_E^2}$$

and

$$h_{I.T.}^2 = \frac{\text{Cov}\,(I.T.)}{\sigma_G^2 + \sigma_E^2} = \frac{\sigma_G^2}{\sigma_G^2 + \sigma_E^2} = h^2\,(\text{broad})$$

where *H.S.* represents half-sibs, *F.S.* represents full sibs, *P.O.* represents parent-offspring, and *I.T.* represents identical twins. Similarly the quantity defined by Holzinger (1929) would be

$$\frac{\sigma_{WDZ}^2 - \sigma_{WMZ}^2}{\sigma_{WDZ}^2} =$$

$$\frac{\left[\sigma_E^2 + \sigma_G^2 - \text{Cov}(F.S.)\right] - \left[\sigma_E^2 + \sigma_G^2 - \text{Cov}(I.T.)\right]}{\sigma_G^2 - \text{Cov}(F.S.) + \sigma_E^2} =$$

$$\frac{\sigma_G^2 - \text{Cov}(F.S.)}{\sigma_G^2 - \text{Cov}(F.S.) + \sigma_E^2}$$

where *WDZ* indicates within dizygotic twin pairs, *WMZ* indicates within monozygotic twin pairs, and

$$\sigma_G^2 - \text{Cov}(F.S.) = \frac{1}{2}\sigma_A^2 + \frac{3}{4}\sigma_D^2 + \frac{3}{4}\sigma_{AA}^2 + \frac{7}{8}\sigma_{AAA}^2 + \cdots$$

The *F* ratio used by Vandenberg (1962) would be

$$\frac{\sigma_{WDZ}^2}{\sigma_{WMZ}^2} = \frac{\sigma_E^2 + \sigma_G^2 - \text{Cov}(F.S.)}{\sigma_E^2 + \sigma_G^2 - \text{Cov}(I.T.)}$$

$$= \frac{\sigma_E^2 + \frac{1}{2}\sigma_A^2 + \frac{3}{4}\sigma_D^2 + \frac{3}{4}\sigma_{AA}^2 + \cdots}{\sigma_E^2}$$

Another bit of theory which should be of interest to human behavioral geneticists is the work of Fisher (1918) and Wright (1921) relative to assortative mating. These workers consider a simple form of assortative mating that can be described by a simple phenotypic correlation between mates for the trait of interest. Fisher develops a parametrization of Cov(*P.O.*) in the absence of epistasis.

One of the few attempts to relate these biometrical genetic concepts to behavioral traits is the work of Burt and Howard (1956, 1957a,b). Their work, although a bit primitive, should be applauded for indicating a meaningful direction for genetic studies of human behavioral traits to follow.

Problems in Applying Biometrical Genetic Theory to Human Behavioral Studies

Most of the difficulties in applying this theory to human studies centers around the failure of some of the numerous assumptions involved in the development of the theory to be true. Among these inadequacies is the complexity of the mating system in human populations. There is a considerable sub-population structure in the human population because of race, geographical distribution, economic and social barriers, etc. However, on the other hand, few of these sub-populations are distinct enough to be considered as separate populations. Besides these considerations, there are probably numerous forces of an assortative nature influencing marriages. However, the complexity of these factors makes it difficult to describe the nature of assortative marriages. In fact, this would constitute a behavioral study in itself. Many of these forces would involve the traits of interest to the behavioral scientist. Another way in which human populations differ from the conceptual panmictic populations is through differential fertility of marriages. How much this is related to behavioral traits is not clear (to me, at least).

However, the greatest difficulty in human data seems to be the environmental correlations due to family structure. Although related individuals may be similar because they possess genes "alike by descent," they may also be alike because of similarities of environment. These sources of similarity also contribute to the observed covariances between relatives (and, thus, correlations and regressions). The genetics of this is further complicated because the environment imposed upon a set of sibs, for example, is influenced by the genotypes of the parents. In other words, phenotypic similarity of related individuals as measured by $Cov(x,y)$ is indicative of $Cov(G_x, G_y)$ only when other possible causes of $Cov(x,y)$ are absent.

The best discussion of the reasons for similarity of monozygotic twins is in the paper of Kempthorne and Osborne (1961). In this paper, a grouping of the common forces influencing monozygotic twins is attempted. Many of the considerations are relevant to other types of relatives as well. Even for twin data, it seems that monozygotic pairs might receive more similar treatment from their parents than dizygotic pairs.

There is one type of datum which eliminates part of these reasons for similarity of the related individuals; that is, datum on twins reared apart. This has been recognized by many workers and much effort has been made by some human geneticists to obtain such data. Of course, the number of twin pairs of this type is quite small.

In a natural population of other species, many of these limitations are also present but a properly designed experiment with an experimental population can eliminate these. Of course, procedures, such as diallelic crosses and control of environmental influences on siblings, are not possible with human populations.

Another problem in human behavioral studies is the measurement of behavioral traits. In some cases, there seems to be difficulty in the definition of the relevant traits. Certainly, measurement of learning ability or temperament is not as simple as measuring weaning weight of pigs or yield of corn. Measurement is usually made of what can be termed an "indicator" trait such as performance on an examination. Ideally, the measurement on the indicator trait should be equal to the true value of the primary trait for the individual plus a random error with this error being relatively small. However, this might not always be the case. When two traits such as a primary trait and an indicator trait are considered, there are two gross sources of the relationship between the two traits, genetic effects upon the two traits and environmental effects on the two traits. The inter-relationships involved here could be quite troublesome in the interpretation of behavioral genetic data.

A further problem is that we are concerned with variances and covariances, parameters for which the errors of estimation are usually quite large unless a considerable volume of data is used.

Overcoming these Difficulties

A very difficult question that now logically presents itself is: How should the human behavioral geneticist attempt to overcome these difficulties? My suggestions are necessarily superficial and do not immediately lead to definite courses of action. My suggestions are:

(1) to use data on relatives reared apart when possible,
(2) to attempt to improve the theory and understanding of assortative mating and of the sub-population structure of human populations,
(3) to keep clearly in mind the population being sampled and to which inferences are to be made, and
(4) to attempt to obtain sufficient data for reasonably accurate estimation.

Of course, points (1) and (4) are strongly antithetical. The situation I have discussed seems very discouraging. I sincerely hope that the contents of this paper will not discourage behavioral scientists attempting to study quantitative genetics in human populations but will alert them to the difficulties involved. To do effective research in this area, the behavioral scientist must realize both the utility and the inadequacies of biometrical genetic theory and thus use it intelligently and effectively in his research.

REFERENCES

Bohidar, N. R. Role of sexlinked genes in quantitative inheritance. Unpublished doctoral dissertation, Iowa State Univer. Sci. Tech. Library, Ames, 1960.
Burt, C., & Howard, M. The multifactorial theory of inheritance and its application to intelligence. *Brit. J. Statist. Psychol.*, 1956, **9**, 95–131.
Burt, C., & Howard, M. Heredity and intelligence: a reply to criticisms. *Brit. J. Statist. Psychol.*, 1957, **10**, 33–63.(a)
Burt, C., & Howard, M. The relative influence of heredity and environment on assessments of intelligence. *Brit. J. Statist. Psychol.*, 1957, **10**, 99–104.(b)
Cockerham, C. C. An extension of the concept of partitioning hereditary variance for analysis of covariance among relatives when epistasis is present. *Genetics*, 1954, **39**, 859–882.
Fisher, R. A. The correlation between relatives on the supposition of mendelian inheritance. *Trans. Royal Soc. (Edinburgh)*, 1918, **52**, 399–433.
Griffing, B. Accomodation of gene-chromosome configuration effects in quantitative inheritance and selection theory. *Australian J. Biol. Sci.*, 1961, **14**, 402–414.
Holzinger, K. The relative effect of nature and nurture influences on twin differences. *J. Educ. Psychol.*, 1929, **20**, 241–248.
Kempthorne, O. The correlations between relatives in a random mating population. *Proc. Royal Soc. (London)*, Ser. B. 1954, **143**, 103–113.
Kempthorne, O. *An introduction to genetic statistics.* New York: Wiley, 1957.

Kempthorne, O., & Osborne, R. H. The interpretation of twin data. *Amer. J. Human Genet.*, 1961, **13**, 320–339.

Lush, J. L. *Animal breeding plans.* Ames: Iowa State Univer. Press, 1945 (See also earlier editions).

Malécot, G. *Les Mathématiques de l'Hérédité.* Paris: Masson et Cie, 1948.

Mather, K. *Biometrical genetics.* London: Methuen, 1949.

Schnell, F. W. The covariance between relatives in the presence of linkage. In W. D. Hanson and H. F. Robinson (Eds), *Statistical genetics and plant breeding.* Washington, D.C.: Natl. Acad. Sci.–Natl. Res. Counc. Publ., 982, 1963.

van Aarde, I. M. R. Covariances of relatives in random mating populations with linkage. Unpublished doctoral dissertation, Iowa State Univer. Sci. Tech. Library, Ames, 1963.

Vandenberg, S. G. How "stable" are heritability estimates? A comparison of heritability estimates from six anthropometric studies. *Amer. J. Phys. Anthrop.*, 1962, **20**, 331–338.

Wright, S. Systems of mating I, II, III, IV, V. *Genetics*, 1921, **6**, 111–178.

Methodological and Conceptual Advances in Evaluating Hereditary and Environmental Influences and their Interaction

Raymond B. Cattell

As recent writings by Anastasi (1959), Burt and Howard (1956, 1957), Fuller and Thompson (1960), Isaacs (1962), Loehlin (1965), Vandenberg (1962) and others show, the relative importance of environmental and hereditary influences has become once more a controversial topic in psychology. But it has become so in a new sense. For, whereas the philosophically and politically tendentious debates of 30 years ago characteristically begot more heat than light, the serious technical attention of today is generating methodological and conceptual improvements. A growing awareness among psychologists generally, that advances in fields so diverse as learning theory and clinical personality theory are impeded by a bottleneck of knowledge due to half a century of neglect of psychological genetics, has also sufficed to call good minds to the problem. The present chapter expands what has been presented elsewhere (Cattell, 1963b) in article form and relates it to the views of a number of contributors in this symposium.

The central problem in evaluating the inheritance of mental traits has always been that psychological measurement in general deals with *continua*, frequently normally distributed, in which hereditary and environmental influences seem hopelessly intermingled. Furthermore, in many physical traits, as in Mendel's peas, the essential characters of the genotype can be seen in the phenotype. But direct analysis of behavioral inheritance, as seen through definite phenotypic, all-or-nothing characters, as required by Mendelian principles, is certain for only a handful of oddities, e.g., Huntington's chorea, color blindness, and perhaps, if Kallman (1946) and others are right, for some disorders like schizophrenia and manic depressive disorder.

Everywhere else, through the length and breadth of normal personality, psychologists have had to turn to the alternative approach of determining the contributions to observed variances of genetically and environmentally determined variances in the given population. From such obtained nature-nurture ratios, and differences, a second stage of research may try, as discussed below, to infer the nature of polygenic Mendelian mechanisms. This is where the population geneticist's models, as discussed in this symposium by Harris (1965) would be most helpful. But even if we have to be content with only the descriptive ratios, a very great contribution has been made to personality theory, to sociology, and to applied psychology in school and clinic. For here, except for the few studies by Burks and Roe (1949); Newman, Freeman, and Holzinger (1937); Cattell, Blewett, and Beloff (1955); Eysenck and Prell (1951); Gottesman (1963) and a few others, psychology is entirely in the dark.

A decade ago, I proposed what seemed to me two radical improvements in conceiving and experimentally determining these nature-nurture ratios (Cattell, 1953). The first consisted only in substituting the measurement of functionally unitary factors, in place of single, narrow and arbitrary variables—a step which has implications for reliability and genetic relevance, as discussed below. The second and probably more important proposition was the suggestion that we abandon dependence on the classical twin study method and utilize a new method called the Multiple Abstract Variance Analysis Method (Cattell, 1953, 1960). This method (which will henceforth be designated MAVA, for short, just as writers now abbreviate analysis of variance to Anovar) has resemblance to a method developed by geneticists in animal husbandry. However, it would be a mistake to consider them one, because this development in psychology was not only historically independent, but conceptually different. The analysis and formulae are different because in human beings we cannot control the independent genetic and enviromental influences and because humans produce social as well as biological families, with all the rich behavioral learning situations of the former, whereas cows do not!

Since the purpose of this paper is to discuss further advances in this area of method, and to supply additional formulae, some knowledge of the main MAVA design must be assumed, and only the briefest sketch of its essentials should be recapitulated here. Essentially it derives certain required abstract or hypothetical variances, such as "between-family hereditary variance" and "within-family environmental variance" contributions from a set of simultaneous equations. In each of the latter some actually observable variance, such as that between sibs (taken in pairs) reared apart, unrelated children reared together, etc., is equated to the appropriate abstract, hypothetical sources and their covariance. For example, the observed variance for sibs normally raised together is written:

$$\sigma^2_{ST} = \sigma^2_{wh} + \sigma^2_{we} + 2r_{wh.we}\sigma_{wh}\sigma_{we} \quad (1)$$

where σ^2_{ST} is the observed variance within sib pairs reared together. The abstracted or inferred variances are σ^2_{we} standing for "within-family environmental" variance, and σ^2_{wh} which is the within-family hereditary variance. The term $r_{wh.we}$ is the correlation of these sources of variance.

Several other equations like (1) can be set up, for sib pairs whose members are reared apart, unrelated foster children reared together, and other diverse combinations of hereditary and environmental contributions (Cattell, 1953, 1960). Incidentally, in principle, the method can be broadened to include other hereditary units than the family, such as kith groups and races and also other environmentally uniform groups, e.g., cultures, social status strata, etc. The choice depends on the ratio of nature-nurture influences in which one happens to be interested. However, even if, as usually happens, we are primarily interested in the family unit, there are more possible constellations that the eight main ones so far used, which have been:

1. Identical twins together
2. Identical twins apart
3. Sibs together
4. Sibs apart
5. Half sibs together
6. Half sibs apart
7. Unrelated children reared in the same family
8. Unrelated children reared apart

One should consider, for example, adopted and unadopted child comparisons in the same family and adoptions from near relatives, e.g., cousins, compared with unrelated adoptions. These interesting extensions on the fringe have not been used, as far as I know.

From the above simultaneous equations, each having on the left a value empirically determinable from a particular family constellation, and on the right the abstract,[1] hypothetical contribution variances, one can solve for the unknown desired "abstract" contributory variances. Compared with the MAVA method, the only other design used by psychologists—the twin method—is relatively powerless. For it yields (1) no information on the correlation of hereditary and environmental influences, and (2) it finally gives us the ratio only for the quite specific "within" family hereditary variation and the quite peculiar environmental difference between twins (which numerous psychologists and psychoanalysts have pointed out must be distinctly different from that obtaining between ordinary sibs in typical families). It is not so much that the MAVA method is more *accurate* than the twin method, but that it yields entirely new kinds of information which provide the necessary broad basis for making inferences about Mendelian mechanisms.

Among the critical comments which the MAVA design, like any new method, has rightly had to face, are three which deserve serious consideration. First, it is objected that the formulae sometimes hypothesize certain family relation values to be identical which in fact could be different. Indeed, such values might ideally justify distinct unknowns instead of a single "class" value. For this reason MAVA has been worked out (Cattell, 1960) (*a*) in a "full design" with eleven unknowns, and also (*b*) a "limited resources design" with only ten or fewer equations in which the benefit of the doubt is taken on certain alternatives. For example, (*b*) assumes the correlation between hereditary and environmental influences to be the same for fraternal twins as for ordinary brothers. Such initial economy of assumption is necessary because the number of unknowns for which we wish to solve—basically four unknown variances and six unknown correlations—treads

[1] "Abstract" is better than "hypothetical" here, because it is a postulate virtually of fact that these sources of variance *do* exist, and that the total variance must be their derivative.

very closely on the number of independent equations we can hope to set up for getting empirical observations and solutions.

Here, as in any armchair formulation of possible scientific concepts it needs no great skill in casuistry to assert ever finer possible differences in one's shade of meaning for this or that concept, e.g., a kind of environment. What matters is whether these finer definitions correspond to any demonstrably real classes and whether too much hair splitting finally defeats the scientific aim of an effective economy of explanatory concepts. Of course, the ideal procedure is to *enter* research initially with as many terms as one can conceive to be possibly necessary, even erring on the side of having a few extra, luxury assumptions, and to leave the empirical verdict to decide whether what one has treated as a variable can rightly be regarded as a simple constant. Actually, a whole series of degrees of economy have been reviewed (Cattell, 1960) for the MAVA model, with equation systems ranging from eight to twelve unknowns. The design with twelve unknowns involves getting equations and data from groups of half-sibs and identical twins reared apart—difficult to get in sufficient numbers. But even when reduced to the simplest form of finding eight unknowns the MAVA method is dealing in vastly more refined conceptions and comprehensiveness of solution than is the classical twin method.

A second shortcoming suggested is that children in foster families (or certain other classes) may not be biologically and culturally typical of the nation's children as a whole. Perhaps the onus of proof for this should be returned to the critic. But, if we assume he is right, one answer would be to match our control "normal" families on sociological parameters of origin, status, etc., with those containing the sibs reared apart and the unrelated reared together, i.e., to bring about simultaneously across our classes some match in biological and cultural range.

Parenthetically, let us note that this suggestion of selection of some kind acting on the adopted child variance is not the same as the assertion that *social placement agencies place adoptive children in chosen families in such a way as to bring about nature-nurture correlations*. The latter is not likely to amount to much except in the trait of intelligence. It is asserted that some placement agencies choose the educational level of the adoptive home to fit the level of intelligence of the child, as expected from his biological parentage. Though such non-random placement is unlikely except in intelligence, the existing MAVA method introduces and solves for a possibly non-zero placement correlation. Another way of avoiding the objection that foster children are a biologically abnormal sample would be to drop those equations and observations which deal with children raised outside their own biological families. This could be done, but only at the cost of laming the design by cutting it down to a relatively coarse set of equations and unknowns.

The third difficulty raised is that the standard errors of the estimates of within and between family hereditary and environmental variances, etc., obtained by the MAVA method are likely to be large, through the nature of the equations and derivations, so that the sample sizes required for a reasonably steady estimate may be expensively large. This is a complex question better answered later in the presentation.

Environmental-Genetic Relationship as Covariance and Interaction

A normal sequence of exposition might seem to require that we discuss more fully the meanings of the previously mentioned four named sources of variance (within and between family, etc.) before discussing their interaction. But the former, having the usual meaning and mode of calculation of within and between variances, present no mysteries for the average reader which cannot be safely postponed to later discussion. The correlations and interactions of heredity and environment, however, have some unexpected and intriguing features, whose clarification is essential to the primary overview of the model. In particular one must ask which correlations deserve to be considered.

The simplest MAVA model begins with the assumption that hereditary and environmental influences are additive and linearly related to their effects. Unlike the twin method, however, it does not assume they are uncorrelated. Both the linearity and the correlation are illustrated by equation (1) above for the sibs reared together. All penetrating discussions of the problem, and especially the considerations in the writings of Burt and Howard (1956,

1957), Fisher (1930), Penrose (1959), Wright (1960) and recent contributors make it extremely probable that significant correlations of hereditary and environmental influences will exist both within and between families.[2]

The *a priori* arguments that significant correlations of hereditary and environmental influences must exist are now supported and illustrated in a variety of personality aspects by the researches with MAVA by Cattell, Blewett, and Beloff (1955) and Cattell, Stice, and Kristy (1957). These pioneer studies have the approximate character of pioneer studies, but they indicate, for example, an r of $+.25$ between hereditary and environmental effects on intelligence in terms of between-family variances—a result, incidentally, which is remarkably close to the values of .22 to .30 repeatedly obtained for correlation (likewise when families are used) between intelligence and social status. Other intriguing correlations found in these two substantial studies are a correlation of $-.51$ between heredity and environment in their effects on the general level of inhibition, U.I. 17 and of $-.53$ of the same with regard to the factor U.I. 23, which Eysenck called "the neuroticism factor" and we have identified as *regression*. Possible mechanisms are, in the first case, that individuals of constitutionally low general inhibition, U.I. 17, tend to encounter more traumatically inhibiting pressures from the environment. In the second case, the negative correlation with ergic regression suggests a self-correcting "tempering of the wind to the shorn lamb" in which a protective atmosphere reducing regression-producing conflicts is developed in families constitutionally sensitive to such conflicts. Incidentally, the

[2] A volume of opinion to the contrary exists, however, among those less professionally immersed in the problem. For example, Eells *et al.* (1951) in America, and Isaacs (1962) in Britain assert there is no reason to believe that native intelligence is in any way correlated with social status. Indeed, the former went so far as to reject from an alleged intelligence test any items which yielded a social status differential! Incidentally, this did a disservice to public confidence in culture fair intelligence tests (Cattell, 1949). For the aim of the latter must be to meet the criterion first of being measures of g_f and secondly of being culture fair. The Eells-Davis tests checked only on the latter, and in a situation in which partialling out culture almost certainly partialled out much intelligence. The study of MacArthur and Elley (1963) suggest that the Culture Fair intelligence measure correlates with social status about $+.22$ to $+.24$ compared with $+.30$ to $+.34$ for traditional intelligence tests.

unforeseen but systematic finding that heredity-environment correlations are more frequently and largely negative has become one of the two main supports for the *law of coercion to the bio-social mean* (Cattell, 1957). This law suggests a variety of special studies to exploit its possibilities further.

Although the MAVA model thus respects and allows for the determination of heredity-environment correlations in a way which the twin method, for example, does not, it has not hitherto been extended to include also interaction effects. Interaction, in the usual statistical meaning, sometimes gets conceptually confused with covariance. Obviously it is independent of whether the two "effects" (influences) are correlated in the population or not, and its existence simply means that some dependent variable cannot be accounted for solely in terms of an additive, linear composite of the two influences. Assuming that the hereditary and environmental influences on a trait are usually correlated, we have next to consider that they may not produce their effects on a linear, additive fashion. Let us first note that the formulae so far used, as in (1) above, require a non-interactive relation as follows:

$$d_{ST} = k_{wh}h_w + k_{we}e_w \qquad (2)$$

where d is the observed deviation [here exemplified between sibs reared together (ST)] and e_w and h_w are the standard scores respectively of environmental and hereditary influences operating on the individual differences within the family. [Note that subscripts for an individual pair, i, are omitted, but that instead of including the k's as hidden in the score scaling, as is done in equation (1), we here consider the influences from environment and heredity to be in standard scores and the magnitudes of their relative influences to be shown by k_{wh} and k_{we}.]

If now we hypothesize that interaction as well as covariance of (1) exists, then any kinds of product, raised to any powers, of e_w and h_w, would need to be correspondingly modified. In the *embarras de richesse* presented by so many possible hypotheses and formulae, and where experiment bids us begin with the one most likely hypothesis it is reasonable to take the simplest. Thus we should perhaps begin simply by adding to (2) a product term, $e_w h_w$, weighted by a new coefficient k_{hew}, which is quite independent of k_{we} and k_{wh}. This product term could be interpreted, for example, as meaning in the illustrative case of intelligence, that a good environment operates more powerfully to improve intelligence when the native intelligence is already high.

The effect of this modification of the assumption in (2), while retaining correlated effects as in (1), is as illustrated in (3). (To disencumber the formula, the subscript ST is omitted and we have simply ow to designate *observed* variance *within* the family.)

$$\sigma^2_{ow} = k^2_{wh} + k^2_{we} + 2r_{wh.we}k_{wh}k_{we} + k^2_{hew}\sigma^2_{hew} \\ + 2r_{we.hew}k_{we}k_{hew}\sigma_{hew} + 2r_{wh.hew}k_{wh}k_{hew}\sigma_{hew} \quad (3)$$

(The σ_{ew} and σ_{hw} terms vanish because they are in standard scores.)

It might at first appear that we have complicated our solutions by the introduction of as many as four new unknowns, but actually σ^2_{hew}, $r_{wh.hew}$ and $r_{we.hew}$ can be derived from σ_{hw}, σ_{ew} and $r_{wh.we}$, so that essentially only one is introduced, namely, k_{hew}, the weight to be given to interaction. Consequently, in the ten or so equations covering all family constellations, only two unknowns would be added; k_{hew}, the weight for within family interaction, and k_{heb}, the weight for between family interaction effects.

The prospects for experimentally employing a MAVA solution for an interaction as well as a covariance model are therefore reasonably good. It would mean either finding two new, independent simultaneous equations from new family constellations, e.g., for half sibs reared apart, or dropping two of the correlations discussed in the next section, on grounds that they must be very small, or adopting the approximation or assumption that interactions within and between families are much the same, thus reducing k_{hew} and k_{heb} to k_{he}.

The Possible Varieties of Heredity-Environment Correlation

Having extended our model to include heredity-environment interaction as well as correlation it perhaps behooves us to be more specific about the varieties of psychological, social or physiological hypotheses which might be tested by calculations of such correlation and interaction. Such a psychological examination of correlation sources, for example, might also affect experimental design by suggesting that extra family constellations might not be needed in some cases since r is almost certainly zero.

In an earlier discussion of these correlations (Cattell, 1953; Cattell, Blewett, & Beloff, 1955) we discussed the usual "mathematical expectation that among four variances there could be six correlations to consider. These would consist of four involving relations of one "within" with one "between" family deviation (of individuals—see - - - in Fig. 1), and two—$r_{bh.be}$ and $r_{wh.we}$—confined to terms both of which are either between or within (see —— in Fig. 1). However, by our usual habits of thought on *statistical* rules, the first reaction, from Anovar practice, would be that the first four must vanish because of the assumption of non-correlation of any individual's within group deviation with his family's between group deviation. The fallacy in this simple reaction is that we are dealing with abstract variances, not concrete variances. True, it is a mathematical impossibility in groups consisting of random samples to get within and between variance correlated. But when this is true of the *total, concrete* variance *it is still possible for within group hereditary variance to be correlated with between group environment, or environment in the former with heredity in the latter, for these are special abstractions from the concrete whole.* That is to say, significant correlations in all six senses shown in Fig. 1 are possible.

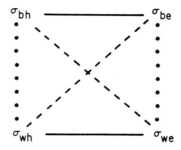

Fig. 1. Possible correlations by simple consideration of mathematical combinations.

Consequently, our question is no longer whether within-between correlations are *possible,* but whether certain of them are likely enough, from systematic causes, to justify entering a non-zero term for a numerical solution. Any psychologist can think of several theories which would require correlation *within* the family, i.e., of deviation on heredity with deviations on environmental treatment, as when parents toss more intellectually demanding problems to their more gifted child. A demonstrated instance is the $r_{wh.we}$ of $-.50$ which we found (Cattell et al., 1955) for the dominance factor, E, in the 16 P. F., suggesting that sibs or parents must be encouraging more dominance in the less dominant and fixing "trash for overtopping," in Hamlet's terms, upon the more dominant. Similarly, any sociologist can generate hypotheses requiring family differences in heredity to become associated systematically with differences in environment, to yield a significant $r_{bh.be}$. A rather special instance of this is the tendency of bright children to have been brought up in the environment provided by less bright parents, and vice versa, due to regression to the mean.

These "homogeneous" correlations (the solid lines in Fig. 1) are fairly obvious, but one must take a more sophisticated look in order to see the instances of within and between correlations. An instance of $r_{wh.be}$—an individual's likelihood of greater genetic deviation within his family being related to greater (or lesser) deviation of the environment of his family from the mean of families—is provided, for example, by the fact that genetically less assortative matings will tend to yield greater within family hereditary variance in the children. Since a large difference between the parents is something which contributes to the nature of the common family environment or atmosphere surrounding the children, a child more deviating from his family hereditary mean (as occurs when parents are genetically very different) is likely to be brought up in a family in which the environmental effects of having more "discordant" parents are also greater. Thus a relation (by eta if not by r) of a $wh.be$ kind, could become significant here. A systematic relation in the converse direction, to lead to a significant $r_{bh.we}$, would arise from the theory that families with a high manic-depression genetic endowment will tend to create an environment more uneven for different children (depending on the age at which they happen to experience parental instability), so that the genetic constitution of the family (as an average) becomes related to the environmental variability within any family.

The "within-between" relations of within-environment with between-environment, $r_{we.be}$, and of within-heredity with between-heredity (broken lines in Fig. 1) would not, mathematically, be able to yield any *linear* correlation, for the usual reason in Anovar. But they *could* yield relations of the kind shown in Fig. 2. In 2(a) for example, the families which deviate positively in environment could systematically contain individuals who deviate more in both directions than do families below normal. For example, a family high in interest and capacity in giving educational aid to its children might discriminate among them more, according to native capacities, in the amount of education given (the bright child to the most challenging college), whereas in a family with low educational interests a general mediocrity of educational pressures might

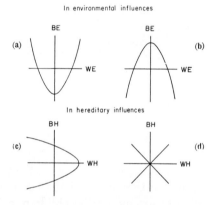

FIG. 2. Some major examples of association of within and between deviations (and, therefore, variances) without linear correlations appearing. (Substantive examples of these are discussed in the text.)

prevail. The opposite direction of association, in 2(b), could be illustrated by the tendency of parents with greater and more enlightened affection for their children to bestow it more evenly, whereas careless favoritism might prevail in parents less conscientiously interested in their children. An instance of a related pattern in heredity (σ_{wh} and σ_{bh}) could result (Fig. 2 (c)) if matings are less assortative at lower than at higher levels. [The data, (Table II) in Higgins, Reed and Reed (1962), suggest this may be true in the trait of intelligence.] Then greater within family hereditary variance would prevail at lower hereditary family levels.

Effects of the kind in (a) and (b) could yield significant curvilinear correlation, in etas, and in (d) could result in a correlation coefficient never before described, but which could be formulated as a "double-eta." However, none would yield a linear covariance term. Consequently, in the review of evidence to this point we would conclude that there is need to enter the MAVA solution with only four linear correlations in anticipation of possibly significant values, namely, with $r_{wh.we}$, $r_{bh.be}$, $r_{bh.we}$, and $r_{wh.be}$.

At this point, however, one has to take note of two, possibly three, quite peculiar *psychological* correlations. First, as we have already noted, when children are adopted there may be a tendency (see following section) for agencies to place children of certain prospective genetic maturations in families environmentally selected to fit those expectations (in the interests of harmony), e.g., the child of a dead genius to the adoptive family of higher education and the less intelligent to the less intelligent foster parents. There is little documentation of how far this goes on, consciously or unconsciously, but it is generally believed by sociologists to be a serious possibility. Accordingly we introduced into the MAVA design the principle of *placement correlations*, indicated by a double prime, $r''_{wh.be}$, representing the correlation tied to the *within* family heredity differences of adopted children, and $r''_{bh.be}$, representing the correlation produced *between* families by the above mechanism. It may well be questioned whether the former, which implies that placement agencies distinguish, in say, the first year of life, between a sib to go to a "better" and another to a "poorer" family environment, should be seriously entertained, but we have played for safety.

Another class of possible additional correlations is that created by any need to change a correlation that has already been designated when it is likely to become systematically different in size because of a new range of variance being created by special circumstances. For example, we have agreed in the standard MAVA formulae to a $\sigma^2_{we''}$, which accepts the presumed fact that differences of environment between twins are different from those between ordinary sibs—σ^2_{we}. Within this narrower range for fraternal twins the correlation of environmental and hereditary influences could well be different from that among ordinary sibs. (This raises the unknown variances to five, unless as in our earlier exposition, one also likes to distinguish the environmental variance of identical from fraternal twins, by $\sigma^2_{we'}$, giving six.) The question now arises whether the value $r_{wh.we'}$ is likely to be different from $r_{wh.we}$. Against this it might be urged that although $\sigma^2_{we'}$ is probably smaller than σ^2_{we}, yet the *rate* of change of environmental treatment with hereditary difference, i.e., the regression, would tend to be the same.

A similar problem arises in adoptive families where the range of hereditary variance is certainly greater than usual because it includes the normal within family variance plus that due to the children coming from different biological families. Thus in the equation which begins $\sigma^2_{UT_h} = \sigma^2_{wh} + \sigma^2_{bh}$ (where UT means unrelated together) should we anticipate in the next term a new value, $r'_{wh.we}$ differing from $r_{wh.we}$ because of the greater[3] range, or suppose, as above, that the regression rate of environmental response to hereditary increment is the same, though over the greater range?

On the one hand, we are pressed toward an economy of assumptions by the anticipated dearth of further empirically derivable equations to solve for new terms. On the other, we recognize the danger of considering two terms to be identical when psychological reasons suggest it would be wiser initially to hypothesize two distinct entities. A decision on these questions, and particularly on what order of priority should be given to the eight possible r's ($r_{wh.we}$, $r_{bh.be}$, $r_{wh.be}$, $r_{bh.we}$, $r''_{wh.be}$, $r''_{bh.be}$, $r'_{wh.we}$, and $r_{wh.we'}$) is better made after the following section.

Evaluations of Origins and Possible Significances of Correlations

Although we have discussed *variances* and *covariances* systematically, through all possibilities of the model as such, our discussion of their psychological origin and meaning has so far been only for purposes of illustration. Since, among the always infinite number

[3] Earlier (Cattell, 1960) we have also split from $r'_{wh.we}$ a $r'_{bh.we}$ term. The latter seems unnecessary if we accept $r'_{wh.we}$ to hold steadily over the actual within family hereditary range due to both normal within family hereditary variance and to that component coming from between family hereditary differences. This is less confusing because if $r_{bh.we}$ enters in any other equations it does so in quite a different sense from this $r'_{bh.we}$, being due to effects of the kind discussed and exemplified above by, say, more intelligent families giving a more varied environment.

of possible models we must be assured that we have one with a reasonable chance of success, it is necessary, before adopting the MAVA model more seriously, to scrutinize more closely what we believe happens in psychological, sociological, and biological terms. In particular we should look at the forces deemed to be producing correlations.

Broadly, one has to consider three possibilities here, (1) that a genetic difference brings about a difference in environmental influences on the individual, (2) that an environmental difference brings about a genetic effect, and (3) that some third influence produces both. Since heredity and environment are by definition supplementary, and *exclusive* terms, the third possibility may seem illogical, but it must be remembered that the "environment which affects the child," as here, is still not synonymous with the universe of environment, nor are the genetic influences on the child comprehensive of all past and present sources of genetic variation. In other words, there are more influences, genetic and environmental, than those which directly account for the total variance of traits in a given population, so that a third term may lie outside of these, e.g., in an event two generations back.

For condensation, this section must arrange systematically the *results* of discussion, not indulge in discursive reflection, under the three main headings just made, considering each again under "within" and "between" family situations.

Genetic Causation of Heredity-Environment Associations[4]

Within Family

(1) *Associations by migration.* The genetic endowment causes the individual to *move* into a certain kind of environmental, ecological setting, as when schizothyme temperament causes a person

---

[4] Parenthetically, one must beware of confusing any of these true genetic effects on environment with the fallacious concept sometimes appearing in such statements as "the same environment is more stimulating of intelligence in the intelligent child because he sees more in it." This way lies madness! The only clear handling of perception ascribes part of any perception to the meaning of the situation, given by the physical definition and the species and part to the individual. The former we define here as independent environment and anything else will show up in interaction terms.

to take such occupations as forestry or research physics, where experience in social situations becomes reduced.

(2) *Associations by shaping environment.* The genetic endowment results in environment correlating because it is the means of *changing* environment (thus, in a very broad illustration, the genetic endowment of a bird causes it to experience a nest environment, denied to animals). Countless examples of individuals of given genetic make-up tending to create particular environments can be cited. For an example of a fairly subtle instance one can turn to the finding of Cattell, Blewett, and Beloff (1955) of the apparent tendency of a high I factor (tender-minded) person to create a protected, unambitious, esthetically soothing environment which cherishes his original esthetic sensitivities ($r_{we.wh} = +.10$ to $+.50$).

(3) *Associations by producing environmental response.* From the social environment (and even to some extent the physical environment) a genetic endowment may *provoke* characteristic environmental *treatment*—a relation quite different from (2). Just as physical beauty or ugliness can provoke consistently different social treatments, so also dispositional endowments will become correlated with environmental responses. For example, aggression is commonly met with aggression. The finding of Cattell, Blewett, and Beloff (1955) of strong indications of a negative $r_{we.wh}$ for the dominance factor, E, looks like such a mechanism. Here also must be included the tendency of people [excluding for the moment the subject himself, for this would be (1) above] to select their associates according to preferred characteristics.

Between Family

Examples of all three of the above, transposed into the between-family situation, will occur to the reader and require no recapitulation. For example, the upward social mobility of intelligence-endowed families as an instance of (1). One would hypothesize that correlations from (3) would be significantly smaller than within families, because society does not know each family as well as a family knows its members. Conversely, (1) and (2) should be larger, because there might be less error and happenstance in the mean reward for a whole group than in the fate of a single individual. Additionally, there will be a fourth mechanism here, as follows:

(4) *Associations by differential survival.* If a certain genetic endowment favors sheer survival, either by birth rate or death rate differentials, within a certain social environment or in conjunction with a certain family atmosphere, it cannot affect the individual, who is either alive or dead. But it *can* affect the means of families, and such selection of its members could produce $r_{bh.be}$ correlations.

Environmental Causation of Hereditary-Environment Associations

Within Family

In general, any correlation from this direction of action, upon individuals and within the family, would be expected to be zero, for the simple reason that the child is already born! His environment can no longer affect what genes he is given. However, differential death rates might theoretically operate, e.g., a more overprotective attitude to a child of weak constitution might favor his survival, producing more instances of over-protection combined with defective constitution than expected from a zero correlation.

Between Families

All four mechanisms under (1) can act here, but in the reverse direction, i.e. environment-as-cause. Environment, stimulating *migration* could act by natural selection to produce correlations, e.g., those found for lower competence and intelligence in farm-failure, city-ward migrants at time of agricultural distress. Environment can shape or at least selectively manipulate heredity, at the family level and with respect to the next generation, as when persons in a more neurosis-producing environment might avoid genetically neurosis-prone mates. [The equivalent of mechanism (1) above.] Environment can also act by selection as when parents (families) subjected to a higher education level have fewer children. This would statistically reduce the correlation of native intelligence with an intelligence-stimulating environment, for fewer cases of high intelligence with high education would exist than formerly. And it can produce differential *survival,* as in (4). One would hypothesize that correlations from (2) are trivial relative to those in (1) or (3).

COMMON CAUSE OPERATING ON BOTH HEREDITY AND ENVIRONMENT

Within Family

Only in the exceptional instance of some mutation-producer, e.g., radiation, age of mother (as in Mongolism), simultaneously affecting the heredity and environment of an individual could there be anything but a general zero correlation here.

Between Family

Many influences could simultaneously change environment while causing a hereditary selection between families. A more refined analysis would involve all four mechanisms noted above. For example, there can be deliberate selection by society, as when intelligent children are sent to college or very uncontrolled people to jail. Or again a disease may simultaneously select for environment and heredity, e.g., malaria and sickle cell anemia, or tuberculosis selecting for lower social status and leptosomatic-schizothyme constitution (Cattell, 1950). Or again, social placement agencies (see above) produce such correlations, as also do eugenic ideals in a community. [Conversely, as Fisher points out (1918), dysgenic, competitive-birth-restrictive family attitudes may become correlated with higher intelligence, and other largely genetic qualities.] One suspects that correlations from source (3) could frequently be statistically significant.

From the above it will be evident that almost any correlation obtained is likely to be the outcome not of one but of *several* influences producing it. Thus we have the situation in which we usually apply factor analysis to discover the influences back of the observed correlations. Unfortunately, with only four to six variables the possibility of employing a factor analysis to determine the number and nature of influences at work—some of which would be "doublets"—is at present remote. Interpretation and allocation of contributions to various influences will therefore have to proceed from the obtained correlations by less systematic means—by observing correlation changes with selection and manipulation and by reaching out for clues from more remote and varied fields. The hypothesis of the law of coercion to the bio-social mean is an example of such inference on correlation, for it proceeds from both MAVA results and findings in the field of misperception or projection (Cattell, 1957).

With these considerations before us, when we call for a restriction of the number of unknown r's, (as we approach the limits of available empirically based equations), elimination has to be finally carried out with regard also to (a) the unknown *variances* demanded in the solution, and (b) not only the number of equations available, but the number available which offer mutual algebraic independence. What has been called the Limited Resources Design takes as variances σ^2_{we}, $\sigma^2_{we''}$ (identical twins), σ^2_{wh}, σ^2_{bh}, σ^2_{be}, and as correlations $r_{wh.we}$, $r'_{we.bh}$, $r_{bh.be}$, $r''_{bh.be}$. The Complete Design adds one more variance, considering $\sigma^2_{we'}$ (fraternal twins) as possibly different from the environmental variance of brothers (corrected to the same age), and it adds one correlation, $r'_{we.bh}$. This combination could probably be improved upon, for $r'_{we.bh}$, as we *have seen, could well be considered the same as* $r'_{we.wh}$. Indeed, it might be better to drop one more correlation and add one more variance—namely, $\sigma^2_{bh'}$, to respect the possibility that children who are adopted are biologically selected, which we discussed at the outset. With seven unknown variances and eleven possible equations, one might then give priority to four correlations, as follows: $r_{wh.we}$, $r_{bh.be}$, $r''_{bh.be}$, and $r_{bh.we}$. This assumes that $r'_{bh.be}$ is no different from $r_{wh.we}$ and that $r_{wh.we''}$ is unlikely to be different from $r_{wh.we'}$ but that they merely act over a different hereditary and environmental range, respectively. It drops $r''_{wh.be}$, one of the placement correlations the probability of which is extremely difficult to justify, and it is only in the loss of $r_{wh.be}$ that any regret need be felt. Regarding the latter, however, it is difficult to see any possible cause of correlation other than that difference of parents, cited above (p. 105), or some oddity where extreme deviation in the offspring from the general population, as in the Siamese twins, affects the family's fortunes while the children are still young! Nevertheless, it may be desirable to seek a twelfth independent equation to permit solution for a possibly significant value for $r_{wh.be}$.

Beyond Nature-Nurture Analysis to Genetic and Environmental Mechanisms

Merely to achieve numerical values (with a given standard error) for the intra- and inter-family hereditary and environmental variances on well-known source traits, such as intelligence, ego strength, surgency, etc., and for the usual heredity-environment correlations in our culture between them, would be a substantial step from the chaos of psychologists' present speculations in this area. It would lead to much more informed theories in most aspects of personality development, and much more skilled practice, e.g., in clinical and educational psychology. However, the geneticist longs for solutions going beyond this point, namely, to some understanding in terms of Mendelian mechanisms.

Obtaining the ratios and correlations so far considered as the goal is then actually only a means to certain ends in more basic science. Those ends are dual (1) finding genetic structures and determiners, and (2) tracing the sociological determiners of environmentally produced personality change. This chapter is concerned mainly with the former, but for perspective we shall in this section glance briefly also at the latter.

The problems of inference in the former area are tremendous. Almost certainly we have to deal in most psychological source traits with polygenic determination. Only in a few highly specific bits of behavior, which might perhaps be located by intensive observation of certain families, might one hope to find potential behaviors tied to specific dominant and recessive genes. For the greater part of the broad source trait behavior which interests psychologists, we must surely look for the salvation of genetics in still greater refinements of methods of population or biometric genetics such as developed by Fisher (1930) and Wright (1960).[5] Such studies have proceeded by refined mathematical models to infer the gene structure from the form of population distributions, from changes in distribution from parent to offspring generations, and from changes in distribution with various mating restrictions.

However, the biometricians and population geneticists have

---

[5] The interested reader may also want to look at Crow (1958), Darlington (1939), Haldane (1924–1927), Malécot (1948), and Mather (1949).

worked almost exclusively with measurements of phenotypic characters which can be safely assumed to be only trivially modified by environment before the measures take place. By contrast, the psychological geneticist has first to make a separation, by the MAVA or any other available method, and to estimate what the unmodified phenotypic characters would be. Only then can he begin applying the methods of biometric genetic influence.

The strategic coordination of the nature-nurture analysis methods with the inference methods of biometric genetics perhaps awaits a methodological genius. But one can see that the designs will at any rate fall for cross-comparisons of results from several MAVA analyses carefully chosen to yield crucial relations of genetic variances for various degrees of consanguinity of parents, various measurement homogeneities of parents, etc. The single MAVA study will no longer be enough. One possible example of this strategy is as follows: (a) determination of σ^2_{wh}, σ^2_{bh}, etc., for children of random and first cousin marriages; (b) the same for parents at very similar, i.e. at high and at low levels on a trait; (c) the same for cultures in which mating is more assortative and less assortative; and (d) the same for specially selected rare instance of incestuous or bizarre mating which reproduce the inbreedings, back-crosses, etc., obtained through the freer breeding manipulations normally pursued in animal genetics.

This path toward determining the number of genes at work in given source traits, their dominance, epistacy, etc., is thus going to require two major steps. First, there has to be some original and fundamental thinking on the groups required for most potent inference in terms of new principles of biometric genetics. Secondly, research foundations need to recognize that we face an unusual type of problem in which the answer is to be obtained only by several large scale researches, carefully planned and coordinated, each requiring a considerable amount of social work to locate cases meeting very special requirements.

As mentioned earlier, although my primary concern is genetics, the reciprocal relation of genetic and learning influences makes it desirable to ask also what further analysis can be done on the environmental variance components. One of the major defects of learning theory (Cattell & Scheier, 1961), is that it has set out to explain major personality learning without stopping to ask what has to be explained. It has simply dragged in the conditioned reflex from the highly controlled animal laboratory, disregarding (a) the possibility that new phenomena *may* require new principles, and (b) the fact that the sources of human personality learning—the influences and pressures—have not been located and can probably never be subjected to manipulation. The latter points to the urgent need for a new methodology—one which locates and measures the sources of learning influence which cannot be manipulated. The MAVA method is in certain senses the only method the learning theorist has for investigating these broader personality learning principles. The measured source trait is the dependent variable and the genes and the environmental learning pressures are independent variables, yet neither of these independent variables can be manipulated as the laboratory animal experimenter is accustomed to manipulate them.

The failure to grasp this fundamental principle—including the fact that he does not even know what the distinct environmental influences in society are which he might need to manipulate—makes the present personality theories by the reflexologist peculiarly empty. Fresh from the animal cage, he fails to see that the theories which he so fluently generates about the interaction of learning and maturational influences simply cannot be tested for human personality by his methods. Instead of simply manipulating independent variables—genetic and experiential—one must now go through two stages: (1) the discovery by the MAVA method of how much of the variance of the dependent variable is due respectively to maturation (genetics) and to learning, and (2) the tracing of this variance to sources (a) within genetic mechanisms, by the biometric methods just discussed, (b) within environmental sources of experience.

In our present glance at the latter let us make first the important observation that all σ^2_{we} variance is not due to sources present only within the family and that all σ^2_{be} variance is likewise not due only to social differences between families. For example, in connection with the σ^2_{be} magnitude one thinks of the contributions from differences in family social status, localities, peer groups, income, religious and other sub-cultures, schooling, etc. The sources must reside in whatever is shared by children in one family and distinguishes them from another, up to the age at which they are tested. It is perhaps the sociologist who most needs to be warned against thinking that all between-family differences are due to positions and parameters within society. There are also those more subtle "family atmospheres," generated by interaction of personalities and partly handed down by tradition within families, such as have been studied by Baldwin (1946), Cattell (1950), Symonds (1931), and others, and part of which may be measurable by the new, objective, intrafamilial attitude measures (Sealy & Cattell, 1965). There are also such diverse environmental contributors as the gestation environment common to the sibs of one family, the common experience of certain family trauma, and historical vicissitudes, etc.

In σ^2_{we} will go typically all that distinguishes one child's experience within a family from another's, such as sib-position, birth trauma and illnesses, differing age and experience of parents at the critical (imprinting) developmental period, and differing treatment and "favoritism" by parents. Thus not all the within-family environmental variance will arise from the learning sources within the family. Children from the same family may attend different schools, attach to different peer groups, etc. Possibly a minority of influences affect both the within and the between family variances, just as, in the corresponding genetic situation, the same genes may be responsible for both within and between family genetic contributions. Consequently, the main problem is the resolution of the total environmentally determined variance into its environmental sources.

The central principle here is that conceptually distinct sources may not be statistically independent. For example, social class and religious affiliation are conceptually distinct, but in any multiple correlation expressing their effects in the United States they are correlated. It is the claim of the factor analyst using oblique simple structure that he can isolate these conceptually, functionally distinct sources. In the present case, by taking MAVA solutions for many different groups, each measured for many different environmental variance sources, he would aim to discover what influences contribute to personality learning, i.e., to recognize them as distinct contributing factors, but oblique (correlated) as in fact they would be in the community.

If we could measure every independent environmental variable the multiple correlation of this total environment with the environmental variance contribution as measured by MAVA (across a series of groups) would reach unity. But with the actually observed trait itself (genetically *and* environmentally determined) it would reach:

$$R = \sqrt{\frac{\sigma^2_{we} + \sigma^2_{be}}{\sigma^2_{bh} + \sigma^2_{wh} + \sigma^2_{we} + \sigma^2_{be}}} \tag{4}$$

$$(= \frac{\sqrt{\sigma^2_{we} + \sigma^2_{be}}}{1}, \text{ if in standard scores}).$$

An independent check on the value for σ^2_{we} and σ^2_{be} obtained by the MAVA method would thus theoretically be possible by calculating the multiple correlation, R, between *all* environmental influences (measured directly, not as effects, which we cannot directly measure), and the raw observed trait score, across a series of individuals in the community. In practice, the real problem is that we cannot hope to obtain even a major fraction of the relevant environmental influences. Thus the R value obtained would be useful only as a lower bound for the true environmental variance contribution.

However, with the multiple correlation (over individuals) of a whole set of individually obtainable values for environmental influences known to contribute to a trait we could estimate how much of an *individual's* score on a trait is due to learning, whereas in the general MAVA solution the obtained nature-nurture ratio permits us no such solution. (It may tell us in absolute terms that, say, one-fourth of the score is probably due to environment, but the regression leaves individuals in the same rank-order for their environmental as for their total observed scores.) On the other hand, for the population as a whole, a useful estimate of the *correlation* between some specific, independent variable in the environment and the purely environmentally-produced change in a trait can be made if we know the nature-nurture variance ratio and the correlation of the environmental element with the full, observed trait measurement. (We can simply take out the hereditary variance contribution from the observed trait variance assuming no correlation of the hereditary with the environmental influence or estimating it as that obtaining between heredity and total environment on that trait.)

These brief illustrations show that a rich domain of possible inference—at present unattainable in any other way—opens up as soon as we have MAVA results from a sufficient variety of subcultures and genetic groups. The major pay-off lies in the comparative analyses which then become possible. The off-handed comment is sometimes made that all one obtains in nature-nurture ratios is a quantitative statement which holds only for a particular culture and a particular racial mixture of population at a particular point in history. This is true, but this very particularity is the effective means of obtaining all kinds of valuable conclusions in *comparative* culture and *comparative* genetic studies. Here is the opportunity, for example, to relate the magnitude of environmental contribution to a trait to forms of social and political organization, or the magnitude of the obtained genetic variance to magnitude of physically measurable racial and genotypic variance in the groups. Doubtless, in time, as MAVA results on common marker factors accumulate for different cultural groups, a whole science of comparative personality-culture relations will develop here. A special instance of this which is of major interest to the clinical and developmental psychologist is the comparison of variances and correlations for groups of different ages in the same culture, or of children in different sibship positions, etc. discussed below. However, while still in the cultural realm one may note that variations in a particular ratio, namely:

$$\frac{\sigma^2_{we}}{\sigma^2_{be}}$$

would give an immediate and indisputable answer to the perennial argument between clinicians and sociologists regarding the relative importance (for a given trait and culture) of early events affecting the individual within the family and of social status and other sociological parameters broadly differentiating families.

Some Further Possible Checks and Analyses with MAVA Results

In terms of independent and dependent variables MAVA has, along with other methods, a certain family resemblance to factor analysis. This methodological class resemblance lies in its analysis of observed variances into certain abstract but theoretically important sources which cannot be directly measured. Hence, it is not surprising to find speculation on whether factor analysis itself could assist in separating genetic and environmental contributions. The present writer (1946) hypothesized that primary personality factors could be generated *either* as constitutional *or* as "environmental mold" traits. That is, *granted certain conditions of factor resolution,* a functionally unitary factor would *not* be the structure jointly produced by hereditary and environmental influences, but would be the two separate components in this structure. The hypothesis followed from the theory of *surface traits* (as correlation clusters) and *source traits* (as simple structure factors). With the methods of factor resolution somewhat crudely practiced in the 20 years following this theoretical statement, the studies published by Cattell, Blewett, and Beloff (1955); Cattell, Stice, and Kristy (1957); by Eysenck (1956) and by Eysenck and Prell (1951), have certainly produced factors which are mixtures of hereditary and environmental contributions. With the varying selection and notation conditions of the factor analytic model it is clearly possible for factors under one set of conditions to represent a final composite pattern and under other, more sensitive and refined designs, to yield the separate environmental and genetic patterns.

The approach to a U-distribution in the nature-nurture ratios found by Cattell *et al.* (1955) suggests that at this degree of refinement factor analysis is already finding some factors which are largely hereditary and others which are largely environmental, as would be required by the original theory (Cattell, 1946), if the results were blurred by some random error. Two other areas of factor analytic research have recently come up with results consistent with this theory. One is the factoring of objective motivation measurements, where patterns fall either into drives (operationally redefined as *ergs*) on the one hand, and learned *sentiments* on the other (Cattell, Radcliffe, & Sweney, 1963; Cattell, Horn, & Butcher, 1962). Another is ability measurement, where it is now claimed that what was considered a single general ability factor is actually the overlap of a genetic *fluid general ability* and an *environmentally crystallized general ability* factor (Cattell, 1963a).

Thompson (1957) and the present writer (Cattell, 1946) have

speculated that since a gene is a causal influence and usually shows pleiotropy (i.e. simultaneous effects on *several* characters), it would appear as a factor if those several characters were measured in a number of unrelated individuals and correlated. Any technically satisfactory discussion here would require more space to define the factor analytic concepts of variable density, factor order, and simple structure rotation. However, the variables in a factor analysis for this purpose should be more numerous than are the genes. Except where special foreknowledge enabled us to select just those variables appropriate for a particular gene we should need to enter literally with thousands of variables—and the biggest computers can now handle little over a hundred. What we have been obtaining as distinct "constitutional" and "environmental mold" factors in certain studies mentioned previously have been (for each of the former), almost certainly the results of a whole series of genes acting cumulatively on a particular phenotypic basis of behavior, not single gene actions.

Another speculation in this area is the possibility that one would get a single factor for all environmental influences and another for all hereditary ones. This is highly unlikely, for whereas environment might act more powerfully for one individual than another in a particular trait, it is difficult to imagine that one person gets more exposure than another to environment as a whole. Another model and method proposed by Vandenberg (1965) is what I might call *Comparative Intra-Pair Factoring*. Here one takes, on the same *n* variables, the differences for a series of identical twins (experiment A), and for a series of fraternal twins or even sibs (experiment B). A factor analysis of experiment B should yield those personality factors which we have suggested as raw material for the MAVA design. Factoring of paired differences of any two persons should yield the same factors as factoring single individuals on the same variables (but with the variance reduced).

The intra-sib variance would contain the factors due to both hereditary and environmental sources, whereas the intra-identical twin factors must be purely environmentally determined. If personality factors can indeed be readily divided into "constitutional" and "environmental mold" sources, as the present writer's theory (Cattell, 1946) required, one could get a clear separation of such factors by cancelling from the former list those which appear in the latter. But if, on the other hand, the factor model is one in which the factors obtained in ordinary R-technique are dimensions of the phenotypy—essentially surface trait patterns representing the combined shaping of hereditary and environmental influences—then one would obtain from this method the same ratios (from comparing the size of the same factor in the two series) as by the MAVA method (except for distinct within and between family values). A rough approach to this problem has been made by Freedman (1958); but real evidence on which of the above two models is more appropriate must come from the research of Vandenberg, of Loehlin, and others. It will be awaited with considerable interest as a possible cross-check on MAVA designs.

At this point, let us pause to see what the implications of the above models may be for the general use of factor analysis in psychological genetics, and for the relation of factor analytic to MAVA methods working with factors already located in ordinary R-technique. The first model—distinct hereditary and environmental mold factors (corresponding to learning influences from social institutions)—is so simple as to require no further comment. The second is more complex, for it supposes that the variables loaded by a factor are the expression of something which itself is determined partly by heredity and partly by learning. Further, it supposes that each of these influences is the sum of many smaller particular influences, each one of which can operate on more than one factor. For example, on the genetic side, a certain enzyme-determining gene may simultaneously add an increment on intelligence (U.I. 1) and on cortertia (U.I. 22). And, on the environmental side, a strict mother might simultaneously contribute to superego strength (G factor) and desurgency (F(−) factor).

With the rapid progress made in locating second and higher order factors in the last five years one can inspect certain results to see if they fit these interpretations. In certain instances, they at least fit the notion of second order factors being *either* environmental *or* hereditary. For example, the second order factors FIV (labeled Ardor) from objective tests seem largely hereditary, and the second order questionnaire factor Exvia (extraversion) may load the primaries A, F, H, and Q_2 in a pattern consistent with the hereditary component in each of the latter (Knapp, Cattell, & Scheier, 1961; Pawlik & Cattell, 1964; Gorsuch & Cattell, 1965).

Contributions to Learning and Maturation Theory From Comparative Use of the MAVA Design

Speculation about the interaction of learning and maturation has been rife, e.g., in Hunt's recent examination of Piaget and others in *Intelligence and Experience* (1961), but impotent for lack of methodology. An attempt has been made to develop concepts analogous to those of Piaget which will tie into operational analyses. The definition of "aids" offered elsewhere (Cattell, 1963b) provides a means of investigating in humans both those imprinting phenomena observed in animals and the qualitative observations made by Piaget; see, for instance, Piaget and Inhelder (1956). With animals one can withhold a learning experience and apply it at different and "unnatural" periods to compare the progress made (Hess, 1959). But with humans, the only feasible approach now is to apply the MAVA method in coordinated studies at several age levels and compare the hereditary and learning variances produced at these different levels. In this way such theories can be tested experimentally. The chief comparative designs to test interaction of learning and development include the following:

1. Comparison of abstract variances and correlations from MAVA experiments deliberately designed so that each experiment selects one type of upbringing, e.g. disciplined versus "progressive," strong father influence versus weak (or family raised by widow), upper social status education versus lower social status, religious sub-culture A versus B, and so on through those parameters which psychologists and sociologists have asserted account for within family and between family variances.

2. Comparisons of MAVA values for pairs sampled from different *types of family constellations,* e.g., large families versus small families, families from early and late marriages, pairs of sibs with a large age gap versus those only a year apart, sibs with high similarity versus low similarity of parents, the two youngest children versus the two oldest children, and so on. We may also include sex difference effects. All those MAVA designs which employ identical twins to give one of the family constellation equations imply use of same-sex pairs throughout. By using equations not involving identicals, comparisons could be made of variance ratios with and without identity of sex. (Even with same-sex pairs the effects of being

embedded in an all boy or all girl versus a boy and girl family could be investigated.)

3. Of special importance for developmental studies are comparisons from matched MAVA experiments in which only the age of the pairs differs systematically. A rich range of inferences could be drawn from systematic plots of the four variances and the intercorrelations over the whole human age range. Presumably, the *between family* environmental variance for most traits reaches a maximum at the age of dissolving the parental family ties but it may continue to show some significant effect throughout adult life. If space permitted, some analysis would be made here of the expected relations of variances at a given age, as a cumulative learning effect, to variances obtaining over earlier ages, under various learning assumptions.[6] Presumably, the relative contributions of heredity and environment with age will prove to be very different for different kinds of ability and personality traits.

4. In the above designs the obtained differences in variance contributions and correlations can be related to measured independent variables for the groups compared. This can be illustrated in the last design now to be considered—the comparison of MAVA values for persons of the same age, etc., in different cultures. (A study is now is progress in which Professors Tsuijioka in Japan, Pawlik in Austria, and Cattell in the United States are coordinating comparisons using the same objective test batteries for the same ten personality factors in the three countries.) Since measurable dimensions of cultures have now been discovered and checked (Cattell, Breul, & Hartmann, 1952; Cattell, 1961; Gorsuch & Cattell, 1965), it is now possible to relate environmental personality

[6] In this connection we must question the widespread assumption that individual differences (deviations) which manifest themselves with greater magnitude earlier are more genetically determined than those appearing later. On the most likely basic model—that each year is a random sample from a population of possible life experiences [but assuming all persons draw from the same universe and that one year's choice of adjustment paths (Cattell and Scheier, 1961) does not predetermine the next], the longer a person lives the closer his sample of environmental experience is likely to converge on the central population value for total life experiences. Consequently, both because of this more idiosyncratic character of earlier experience sampling, and because of the greater effect of early imprinting, the role of environment, relative to that of heredity, should be *greater* in the earlier years.

variances to culture patterns quantitatively, and similarly, to genetic ranges, measured as independent variables using blood groups, eye color, etc. Enough instances to yield dependable correlations would require coordinated research over many countries, but data from three or four cultures would guide in future measurement concentration.

Summary

The aim of this chapter has been:

1. To re-state briefly the essentials of the MAVA method.

2. To show how it can be developed with the additional assumption of interaction effects.

3. To ask how far one needs to proceed beyond the "limited" set of equations in order to solve for all reasonably likely heredity-environment correlations.

4. To illustrate the nature of the linear and nonlinear associations of hereditary and environmental effects in psychological and sociological terms, analyzing them as (i) genetic, (ii) environmental, and (iii) common causes of association.

5. To show how comparative use of MAVA designs can provide, via biometric and population genetic methods, an avenue to discovering Mendelian genetic mechanisms.

6. To show how such comparative use can elucidate the sources and modes of action of environmental influences, allocating fractions of the total environmentally produced effects to any one source. In both (5) and (6) the method permits conclusions about heredity and environment as independent variables which are obtainable from animal experiments, but otherwise unobtainable with humans.

7. To explore the relation of MAVA results to factor analytic models and methods of genetic research. In particular, *comparative intra-pair factoring* and the MAVA methods are examined in the light of two alternative possibilities: (*a*) that factors will appear as distinct "constitutional" and "environmental mold" source traits, and (*b*) that factor source traits will emerge as phenotypic dimensions each combining environmental and genetic effects. The two methods can be used as mutual checks.

8. To explore models involving higher order factoring. If present factor techniques yield only the phenotypic dimensions (of

composite h and e origin), then a model is indicated whereby genes and distinct environmental influences should appear as second order factors. However, because genes are extremely numerous, it may be practically impossible to locate them, due to insufficient numbers of primary factors. Conceivably factors will be obtained which are blocks of genes or cumulative gene effects on some primary factors. Instances of this may exist already in questionnaires in the second order Exvia factor and in objective tests in the second order Ardor factor (FIV).

9. To show how a basis for inferences about more complex genetic and cultural mechanisms can be provided. If the organizational difficulties of extensive comparative studies could be overcome, the MAVA method, by setting up matrices of results for various age levels, types of family constellation, cultures, racial variabilities, and types of family upbringing could provide a basis of inference for genetic and learning theory, particularly on the interaction of maturation and learning.

10. To evaluate the standard errors of the estimated abstract hereditary and environmental variance components, without which research design cannot plan for the necessary sample sizes. A beginning in this direction and was made earlier (Cattell, 1960, p. 366) as follows:

$$\sigma^2_{\sigma^2} = \frac{2k^2_1 \sigma^4_1}{n_1 - 1} \ldots + \frac{2k^2_n \sigma^4_n}{n_n - 1} \qquad (7)$$

where σ^2 is any abstract variance (e.g., σ^2_{we}), the k's are the numerical coefficients of the observed variances in the equation for the solution for the abstract variance, and the n's are the sizes of the various observed subsamples (twins, sibs, etc.). However, if, as subsequently proposed, the measured variance on any factor is corrected by subtracting from it the error estimated from the reliability coefficient for the given battery, $\sigma^2(1-r)$, then the error variance of the latter term must also be included, and the terms in (7) become:

$$\frac{2k^2_1 \sigma^4_1 (1-r)^2}{n_1 - 1}, \text{ etc.}$$

Secondly, if the standard error of the *ratio* of variances is to be determined (to allow a nature-nurture ratio), one must correct for the presence of common terms in numerator and denominator, as follows:

$$\sigma^2_R = \frac{(\sigma^2_h)^2}{(\sigma^2_e)^2}\left(\frac{\sigma^2\sigma^2_h}{\sigma^2\sigma^2_e} - \frac{2\operatorname{Covar}\sigma^2_h\sigma^2_e}{\sigma^2_h\sigma^2_e} + \frac{\sigma^2\sigma^2_k}{\sigma^2\sigma^2_e}\right) \qquad (8)$$

in which R is the ratio and σ^2_e and σ^2_h are the σ^2_{we} or σ^2_{be} above, or the combination thereof, derived from the concrete variances. This approximation[7] neglects higher powers of the ratio, and unfortunately compels us to accept our obtained $\sigma^2_{\sigma^2_e}/\sigma^2_{\sigma^2_h}$ as a substitute for the unbiased estimate required in the first two terms on the right. However, the magnitude of error introduced is small compared with the older formula which neglected the common variance in numerator and denominator. This modification narrows confidence limits to an acceptable range for a sample size of about 250 cases in each of the family constellation types, instead of the 1000 previously required. But, since the empirical variance estimates in the ten equations solution (Cattell, 1960, p. 362) from the within family sib variance and the between sib family variance must be experimentally independent the above formula still requires measurement of $10 \times 250 = 2500$ pairs of children. Fortunately, as equation (7) indicates, a shortage in one category can to some extent be made up by greater numbers in another. Thus one might use only 100 pairs of twins or sibs reared apart, and several hundred pairs in such accessible groups as sibs normally reared together.

In spite of its technical promise in solving problems otherwise unsolvable the method makes demands on funds and fortitude. But if behavioral science—and with it behavioral genetics—is to come of age, the necessary steps to more ambitious research organization must be pursued.

REFERENCES

Anastasi, A. Differentiating effect of intelligence and social status. *Eugen. Quart.*, 1959, **6**, 84–91.

Baldwin, A. H. Differences in parental behavior toward three and nine year old children. *J. Pers. Res.*, 1946, **15**, 143–165.

Burks, B. S., & Roe, A. Studies of identical twins reared apart. *Psychol. Monogr.*, 1949, **63**, No. 5. (Whole No. 300).

Burt, C., & Howard, M. The multifactorial theory of inheritance. *Brit. J. Statist. Psychol.* 1956, **9**, 115–125.

[7] The writer wishes to express gratitude to various colleagues, particularly Dr. Horace Norton, for discussion and help with this issue.

Burt, C., & Howard, M. The relative influence of heredity and environment on assessments of intelligence. *Brit. J. Statist. Psychol.*, 1957, **10**, 99–104.

Cattell, R. B. *The description and measurement of personality.* Yonkers-on-Hudson: World, 1946.

Cattell, R. B. *The culture fair intelligence test, Scales 2 and 3.* Champlain, Ill.: Institute of Personality and Ability Testing, 1949.

Cattell, R. B. *Personality: a systematic, theoretical and factorial study.* New York: McGraw-Hill, 1950.

Cattell, R. B. Research designs in psychological genetics with special reference to the multiple variance analysis method. *Amer. J. Human Genet.*, 1953, **5**, 76–93.

Cattell, R. B. *Personality and motivation structure and measurement.* Yonkers-on-Hudson: World, 1957.

Cattell, R. B. The multiple abstract variance analysis equations and solutions: for nature-nurture research on continuous variables. *Psychol. Rev.*, 1960, **67**, 353–372.

Cattell, R. B. The theory of fluid and crystallized intelligence: a critical experiment. *J. Educ. Psychol.*, 1963, **54**, 1–22. (a)

Cattell, R. B. The interaction of hereditary and environmental influences. *Brit. J. Statis. Psychol.*, 1963, **16**, 191–210. (b)

Cattell, R. B. & Scheier, I. H. *The meaning and measurement of neuroticism and anxiety.* New York: Ronald Press, 1961.

Cattell, R. B., Breul, H., & Hartman, H. P. An attempt at more refined definition of the cultural dimensions of syntality in modern nations. *Amer. soc. Rev.*, 1952, **17**, 408–421.

Cattell, R. B., Blewett, D. B., & Beloff, J. R. The inheritance of personality: a multiple variance analysis of approximate nature-nurture ratios for primary, personality factors in Q-data. *Amer. J. Hum. Genet.*, 1955, **7**, 122–146.

Cattell, R. B., Stice, G. F., & Kristy, N. A first approximation to nature-nurture ratios for eleven primary personality factors in objective tests. *J. abnorm. soc. Psychol.*, 1957, **54**, 143–159.

Cattell, R. B., Horn, J., & Butcher, H. J. The dynamic structure of attitudes in adults: a description of some established factors and of their measurement by the motivational analysis test. *Brit. J. Psychol.*, 1962, **53**, 57–69.

Cattell, R. B., Radcliffe, J. A., & Sweney, A. B. The nature and measurement of components of motivation. *Genet. Psychol. Monogr.*, 1963, **68**, 49–211.

Crow, J. F. Some possibilities for measuring selection intensity in man. *Human Biol.*, 1958, **30**, 1–13.

Darlington, C. D. *The evolution of genetic systems.* Cambridge: Cambridge Univer. Press, 1939.

Eells, K., Davis, A., Havighurst, R. J., Herrick, V. E., & Tyler, R. W. *Intelligence and cultural differences.* Chicago: Univer. of Chicago Press, 1951.

Eysenck, H. J. The inheritance of extraversion-introversion. *Acta Psychol.*, 1956, **12**, 95–110.

Eysenck, H. J., & Prell, D. B. The inheritance of neuroticism: an experimental study. *J. Mental Sci.*, 1951, **97**, 441–465.

Fisher, R. A. Correlations between relatives on the supposition of Mendelian inheritance. *Trans. Roy. Soc. Edinburgh*, 1918, **52**, 399–433.

Fisher, R. A. *The genetical theory of natural selection.* London: Oxford Univer. Press, 1930.

Freedman, D. G. Constitutional and environmental interactions in rearing of four breeds of dogs. *Science*, 1958, **127**, 585–586.

Fuller, J. L., & Thompson, W. R. *Behavior genetics.* New York: Wiley, 1960.

Gorsuch, R. L., & Cattell, R. B. The second order structure of personality in the questionnaire medium. Unpublished manuscript, 1965.

Gottesman, I. I. Heritability of personality: a demonstration. *Psychol. Monogr.*, 1963, **77**, No. 9 (Whole No. 572).

Haldane, J. B. S. A mathematical theory of natural and artificial selection. *Proc. Camb. philos. Soc.*, 1924–1927, **23**, 19–41, 158–163, 363–372, 607–615, 838–844.

Harris, D. L. Biometrical genetics in man. (This volume, p. 81.)

Hayman, B. I. Maximum likelihood estimation of genetic components of variation. *Biometrics*, 1960, **16**, 369–381.

Hess, E. H. The relationship between imprinting and motivation. In M. R. Jones (Ed.), *Nebraska symposium on motivation.* Lincoln, Nebr.: Univer. Nebraska Press, 1959. Pp. 47–77.

Higgins, J. V. Reed, E. W., & Reed, S. C. Intelligence and family size: a paradox resolved. *Eugen. Quart.*, 1962, **9**, 84–90.

Hunt, J. McV. *Intelligence and experience.* New York: Ronald Press, 1961.

Isaacs, J. T. Frequency curves and the ability of nations. *Brit. J. Statist. Psychol.*, 1962, **15**, 76–79.

Kallman, J. W. The genetic theory of schizophrenia. *Amer. J. Psychiat.*, 1946, **103**, 309–322.

Knapp, R. R., Cattell, R. B., & Scheier, I. H. Second order personality factor structure in the objective test realm. *J. consult. Psychol.*, 1961, **25**, 345–352.

Loehlin, J. C. A heredity-environment analysis of personality inventory data. (This volume, p. 163.)

Loehlin, J. C. Some methodological problems in Cattell's multiple abstract variance analysis. *Psychol. Rev.*, 1965, **72**, 156–161.

MacArthur, R. S., & Elley, W. B. The reduction of socio-economic bias in intelligence testing. *Brit. J. Educ. Psychol.*, 1963, **33**, 107–119.

Malécot, G. *Les Mathématiques de l'hérédité.* Paris: Masson, 1948.

Mather, K. *Biometrical genetics.* New York: Dover, 1949.

Newman, H. H., Freeman, F. N. & Holzinger, K. J. *Twins: a study of heredity and environment.* Chicago: Univer. of Chicago Press, 1937.

Pawlik, K., & Cattell, R. B. Third-order factors in objective personality tests. *Brit. J. Psychol.*, 1964, **55**, 1–18.

Penrose, L. S. *Outline of human genetics.* New York: Wiley, 1959.

Piaget, J., & Inhelder, B. *The child's conception of space.* London: Routledge and Kegan Paul, 1956.

Sealy, A. P., & Cattell, R. B. Can intra-familial attitudes be analyzed in the dynamic calculus by objective tests? Unpublished manuscript, 1965.

Symonds, P. M. *Diagnosing personality and conduct.* New York: Appleton Century, 1931.

Thompson, W. R. Traits, factors and genes. *Eugen. Quart.,* 1957, **4**, 8–16.

Vandenberg, S. G. Innate abilities, one or many? A new method and some results. *Acta Genet. Med. Gemell.,* 1965, **14**, 41–47.

Wright, S. Path coefficients and path regressions: alternative or complementary concepts? *Biometrics,* 1960, **16**, 189–202.

Discussion

Nichols: One specific question before we start the discussion. In following this, I didn't get whether you had taken into account that some of these empirical variances might differ with the nature of the group? Let me put it this way, is it possible that families who would put a child up for adoption have a smaller variance, than for instance families that would have identical twins, so that this may be another influence on these empirically determined values?

Cattell: As the design stands what you mention *is* another influence. If, as I suggested rather early, one designed this ideally by equating different special groups by some common social parameter, of course, we hope to get rid of it.

Vandenberg: Could you repeat your estimate of the sample size needed to reduce the variance of the variance to some reasonable value?

Cattell: Yes, we made some estimates of that, and as I have indicated here, it looks as if we would need something like 2500 pairs of children spread across ten types of family constellations.

Vandenberg: For the whole design?

Cattell: Yes. That is, taking each of the different groups in all we'd need 2500 pairs, and the testing we did last time was of about 4 hours' duration, consisting of ten personality factors and the intelligence factor, so it's a pretty formidable bit of organization.

Lasker: When you get multiple pairs in the same family group, does this provide some economy, or does it introduce new difficulties, and lack of statistical independence?

Cattell: Yes, it does reduce your degrees of freedom. The within and between variances should be experimentally independent.

Lasker: Some loss of degrees of freedom might not harm too much.

Cattell: Yes, I had the good fortune to discuss this with R. A. Fisher four years ago and his argument was in favor of doing this, because it was convenient, but reducing the degrees of freedom proportionately.

Lasker: Can you calculate the exact loss of degrees of freedom involved in using some members in these families twice?

Cattell: I think so, yes. I think it operates mainly on the degrees of freedom in the family mean in the group. So that could be done, but in practice this runs into another problem, namely that we want children of approximately the same age. In the past we have taken 11, 12, and 13 year olds, measured them, then knowing the normal rate of change with age of these personality factors, we've corrected them all to the 12.0 year old level, before we've gone into the analyses. Now if you wanted to take more children from the same family you would have difficulty in getting them anywhere near the age level you are dealing with and the correction would be greater and therefore rather suspect.

Harris: I must confess I'm not familiar enough with your previous work on this, but I did have a couple of reactions as we went through this. First let me say in general, I think that what you are trying to do is generally what we are trying to do in biometrical genetics, obtain a parametrization, or model if you prefer, for describing these observed variances in more basic, more abstract, as you call it, variances. And, as far as the genetic portion, you haven't gone as far as the sort of thing I was talking about. Maybe this is as far as you should go at this stage. You seemed to be quite concerned about this between and within correlation and I fail to see the reason for concern about that. By definition your within-family deviation will be orthogonal to your between-family variations, and it seems to me that the examples you were suggesting wouldn't lead to a correlation, as for instance when you said some families would treat the offspring more similarly or less similarly. It seems to me that the consequence of that is going to be having heterogeneous within-family variations. In some family there'll be more environmental variation and in some less, and your assumption seems to be here that these are the same so that you would have homogeneous within-family variations, but the examples you were suggesting seem to me to lead to heterogeneity of variance rather than to between and within correlations.

Cattell: Well, yes, this is right at the heart of the problem, and I would very much like to delve into this just because it gets bypassed completely in the usual discussion of ordinary analysis of variance, which is actually irrelevant here. I think you'd agree that although normally you cannot have correlation of within-group deviations with between-group deviations, but now when we're dealing with hypothetical causes and with these "abstract" contributions rather than the total variance, the possibility arises that we *do* have such a relation. Let's consider perhaps a particular case, maybe let us take the incidence of affection as an example. If we say that more conscientious or well-informed parents, or parents offering greater affection would also distribute their affection more evenly, than would families which are low in affection, the individual differences are going to be greater in the latter than they are in families high in affection, is that all right? The deviations from the mean of the sibs will be greater in the family say of low affection which neglects one sib and favors the other, than in the more emotionally stable family which tries to distribute their affection equally. This, however, would be U shaped, this would be a curvilinear relationship, and this seems to me to be the issue, whether you could get a nonlinear correlation. But certainly you could get a nonlinear correlation in this Fig. 1 on page 106. The chances of a person deviating a large amount from the mean are greater if his family mean is high, than if his family mean is low. This is producing a curvilinear relationship. Now if we get a straight linear correlation, I presume we would want to argue that the deviations are more in one direction at the lower level.

Harris: It seems that your argument is that we will have a correlation, linear or nonlinear between the within-family variance and the family mean, but the quantity you are talking about here is a correlation between the family mean and the individual values within the family. By definition of deviations this correlation mathematically has to be zero. Or grossly there is zero correlation between the between and within components, they are orthogonal. Of course, when you break the between into the H and E part and break the within into the H and E part maybe there could be some compensating relation such that the over-all relation still is zero, but where you could have some relationship within each half which cancels each other out over-all.

Cattell: The diagrams on page 106 are, of course, deviations, are they not? They are not variances, they are plotting the deviations of individuals against deviations of their families, i.e. the individual deviations from the family mean, against family deviation from the mean of all families. I have little doubt that the deviations could be so related, the question would be whether the variances are therefore related. I think I see your point here, that if we take the upper lefthand corner in the diagram, the deviations of the between environment are as great for, say, one standard deviation above as say one standard deviation below. So the high variance groups in families, as such, will be composed both of very high deviation individuals and low deviation individuals, i.e. a curvilinear relation. Now we are saying that the families which deviate highly in the positive direction will contain an excessive number of individuals deviating a good deal from the family mean, and that families which deviate a good deal in the negative direction will contain people who, as individuals, will deviate less from the family mean. We divide families into those of large variance and those of small variance and we are, shall we say, drawing a line across diagram C, then if we have a group of families which have a large variance, they will contain individuals who have either very large or very small deviations.

Harris: Yes, I think I agree with that, but that isn't the quantity which I understood you to be talking about in the r_{bw} term, or $r_{bh.we}$ term.

Cattell: What sort of plot on this thing would you feel would generate a linear association for example, an $r_{bh.we}$?

Harris: Well, let me present it this way. If we let X be the observation of the *j*th member of the *i*th family, then it is just algebraic identity that this is the over-all mean plus the deviation of the mean of the *i*th family from the over-all mean, plus the deviation of the *j*th member from the family mean. That is just an algebraic identity, and of course, what *you've* done here, you've gone on to break this into a hereditary portion and an environmental portion in the same way. But just considering it as the classification model with which we might go into an analysis of variance, you cannot get correlation between these within-family deviations and the between-family deviations.

Cattell: This, of course, could be pursued advantageously much farther. The question is whether when you pull out these abstract contributions they need behave the same way as concrete sample variances. That has been at least my line of thought so far, that these abstract deviations could behave differently from the final emergent concrete measurement variances.

Harris: Well, when you break the covariance into four parts, these four parts have to sum up to zero. You might have some positive and some negative in some way, but over-all the correlation between the within and between family variances mathematically has to be zero.

Cattell: I think I agree. When you break each of those down into their hereditary and environment components, the correlations of components must be reciprocal in direction magnitude so that the whole thing goes out. Incidentally, I am not talking about differences of sample and population parameters, but about the difference, both in sample and population of abstract and concrete variance relations.

Harris: Yes, you could get a plus two, minus one, plus one, minus two, for the four parts of this, over-all they sum up to zero, but each one of them individually might not be zero, and this is just a guess but over-all they have to sum to zero. But I'm a little vague as to what sort of forces could cause this. I don't think the ones you describe would cause this.

Loehlin: Isn't it implied with this model that if you get some within and between correlations which are positive, you must get others which are negative to exactly balance?

Cattell: Yes, that is definitely shown in my formulae. It would be worthwhile to sit down and go over them carefully to check the balance.

Harris: O.K.

Gottesman: I have a more general question. Considering the importance which we attached in genetics in general to parent-child relations in psychology, how is it that parents have not found a place in your scheme? Is there any way in which twin family study could compensate in part for the shortcomings in the twin study, and be closer to helping solve the kind of issues that you treat with the MAVA method?

Cattell: This would be desirable, but so far I have seen no way to include the child-parent difference because the parent's environment belongs to an earlier family. However, with refinement, I believe it would be possible to generate a formula for the common amount of environment, and certainly for common heredity. But there is also the practical difficulty of measuring parents and children on comparable instruments. That frightened us off in the very beginning, so we haven't explored the model further. I think there would also be problems about the amount of common environment of parents and children, in that the children have shared the environment from the beginning, but the parents haven't. When you talk about a certain within-family environmental variance, this degree of variance has applied to the children; it has only applied to the parents recently and therefore it would be exceedingly difficult to bring into the same scheme.

Gottesman: If some particular trait has a high genetic loading, then you might expect that environmental effects would be minimal on that, and you would expect higher parent-child correlations on those kinds of characteristics; so you might find something to look at more closely.

Cattell: Yes, in a rough way this relation is true, though the model needs working out. If we could overcome the practical difficulty of measurement, I think it would be very interesting to introduce this.

Gottesman: If you really believe that something like the HSPQ is a children's version of the 16 PF, then you ought to be able to substitute standard scores on HSPQ into parent's tests.

Cattell: Yes, in respect to measurement we are rapidly moving into a much better position with factored tests at all age levels. It's this difference in length of exposure to the environment that is our real problem. Incidently, I see as one useful result from this approach the examining of the nature-nurture ratio at one age, say at 5 and again at 10, and from a comparison of these results a whole new set of inferences could become possible about periods when maturation is more important and when it is less important. A whole lot of questions that we are talking about are mentioned in Hunt's recent book on "Intelligence and Experience" (1961) in which there is a lot of discussion of this interaction of maturation with the environmental effect. Unfortunately, Hunt, Piaget and other theorists develop no method for treating it positively, and the MAVA is, as far as I can see, the only way that we can get at it—by finding these variances at different ages.

Nichols: The MAVA design is I suppose an alternative to the twin study method as a means of determining nature-nurture ratio, or something of this kind. So it seems appropriate to ask the main question whether it will yield sufficiently more accurate results to make it worth the considerable extra effort. And it requires some groups that are very difficult to get, such as identical twins reared apart, and you seem doubtful whether you will ever get a large enough sample to get a stable variance on these, so the question arises whether the errors that are inherent in the twin study method are more serious or less serious than the errors that are inherent in the MAVA design.

Cattell: The difference of these two methods is one which could advantageously be brought into the limelight by the group of specialists here. One small point— one *can* get along without the twins reared apart, in the MAVA method, this is only one of several possible constellations of sibs, and indeed we have worked without them, just for the reason that you have stated, namely that it is so difficult to get samples of good size.

However, in answer to your main point, it is surely important to bring out that the MAVA is not merely more accurate. One might be unwilling to pay the price indicated if it were only a question of accuracy. Instead, it is a question of whether you get certain kinds of information *at all* from the twin method. The latter merely gives you a *within*-family variance ratio: the MAVA method doubles the information, giving you *between*-family variances too. Moreover, even the within family environmental variance is odd in the case of twins. You deal with a within-family environmental difference that is quite specific, I believe. At least I think that we should assume that it's a specific kind of environmental difference between two twins, until it is proved to the contrary. Until we find that the two values—σ^2_{we} for sibs and $\sigma^2_{we'}$ for twins will work out the same, we ought to assume that they are different. My argument in favor of the MAVA method would be that in most of the situations in clinics and schools and so on, where genetic evidence would be useful for some practical conclusion, what we are really interested in is the normal within-family variance, not this twin oddity.

Vandenberg: Well, my comment would be addressed to a different point. I think that really whether or not the two methods are getting different accuracy or different amounts of information is perhaps not of serious concern, because neither of them are going to give us genetic mechanisms. I think the twin method or the MAVA method can only lead to preliminary results, which then have to be followed through in pedigree studies, and there is a terrific difference in economy. I don't think that it is too important to know whether it is 80 or 70% as long as we don't know the exact genes, this is still an unsatisfactory answer, and even if the results might be, in some sense, more accurate, I think that the ordering of variables, as to how important hereditary components are for them, is not going to be changed very much, and that's all we want, an ordering so that we know that these are the variables which we want to investigate by other means. Secondly, I think, you are assuming that the measures which you use are close to the more basic differences that may be controlled by genes. I think that it may be more important to work on what I earlier called "purification" of the measures so that they will be closer related to real genetic variables. Perhaps this has a higher priority than to take some of the existing measures even though they may be pretty good, to see how much genetic variance there is in these.

Cattell: May I comment separately on what seem to me two very different points there. My own paper has indicated one avenue to proceed from MAVA results to genetic mechanisms but I would like to ask what is the best use we can make of these obtained ratios, or variances. (Let's speak of them as variances for the moment because the ratio is only the last stage.) As I followed Dr. Harris here, he seemed to be particularly interested in between-half-sib and between-sib genetic variances, as a means of working toward a certain Mendelian mechanism. Certainly it would seem that you would need more than one variance to do this—more than the within-family hereditary variance—and would it not be true that the MAVA method in that way constitutes a wider platform than the twin method findings for your take-off into the next stage?

Harris: Yes, I think so! First let me say that I don't really see that there is as great a difference between your method and the twin method as has been suggested—the point is, in order to get a solution to your equations, you have to have more observed variances than you have parameters in your model. In the twin study, they are restricting themselves mainly to two parameters, the numerator and the denominator of the Holzinger ratio. Now you are going at a wider set of different values. The question is, do we pour all of our resources into looking at these two quantities, or do we pour all of our resources in estimating your eight quantities? Certainly for the parameterization I was describing, if we did have a nice random mating population—just as in your procedure, in order to estimate sigma squared A, sigma squared B, etc. we have to have more values, which would be more of your genetic components of variance here, than we have terms on our MAVA. Well, if we assume that the number of the loci is quite large, we have a tremendous number of terms in the expansion I was describing. This parameterization is very good for describing covariances between relatives, but it is very difficult when we start trying to estimate the individual terms in there, because we need a tremendous number of covariances in order to estimate those. Optimistically, we might hope that the world is simple enough that these high order terms are zero, or negligibly small. To be able to verify whether this is true, we need to see from a set of observed values how well just the additive variance explains the data, how well additive variance with dominance added fits the data; and go on up on an increasing scale till we are successful, hopefully with not too many terms. Of course, if you have eight values and pick out eight terms out of this model, you can get a perfect fit, but you're not sure that if you had used more values you would not have needed more terms in this partitioning of the variance. In order to get very far with this method, you have to have more observed genetic covariances between relatives or variance components, then there are values contributing to these covariances between relatives. Then you can justify that the model includes one locus and effects and two locus interactions and that higher order effects are small. Unfortunately it takes as many terms as we have covariances to fully explain differences among observed covariances between relatives, and by your procedure of getting more of these quantities, i.e. by looking at more different types of relationships, you can come closer to relating this to the sort of parameterization that I was discussing. Whether you have enough terms can only be seen by relating the data to the model.

Cattell: Another point of difference in these methods, of course, is the omission of the correlation term and the possible interaction term from the twin analysis. If you are assuming that heredity and environment are really correlated, your twin values for the ratio of these two variances are going to be wrong, are they not?

Vandenberg: We are not assuming that there is zero correlation, we are tossing the correlation in with the hereditary variance, I guess you can do it two ways, you can toss it in with heredity or with environment, depending on the formula you would use.

Lindzey: This comment does really point to the crucial advantage which the MAVA method would have, since it does offer the possibility of dealing systematically with genotype-environment interaction. But then that also comes back to Nichols' question about the nature of the groups that would be involved. It seems to me that the failure to include the identicals reared apart introduces a major flaw in the attempt to estimate this genotype-environment interaction, because we have compelling reasons and very good data to support the notion that heredity and environment are most contaminated in the case of the identicals, in identicals reared together, so without identicals reared apart you lose, it seems to me, a major analytic hold in the attempt to get a good estimate of genotype-environment interaction.

Cattell: But if you look at the *full* MAVA design—that is, at the eleven equations, I think you'll see that you could throw out twins altogether and still get a solution. But you'd have to make certain simplifying assumptions. A further thought on Harris' point suggests that if he is right we could throw away what might be called the additional "nuisance r's." In that case the situation is actually brighter, because we could then solve, I'm sure, for the four variances and the two correlations, without having twins in the picture at all. It helps to have another equation, but I think the peculiarities due to twins, whatever happens with twins, could be avoided. If you recall, I think, Barbara Burks searched the length and breadth of the States and came up with nineteen cases, so that's a pretty hopeless group to look for.

Harris: I'd like to bring to your attention a paper which came out a few months ago in *Biometrics*, doing a similar sort of thing to what you are doing here. Here you have written out on the righthand sides the expectations of these observed covariances and you are getting solutions by equating and solving. Now you are likely on some of the covariances to have more data than on others, and so some of these are more accurate. A procedure by Hayman (1960) was described in *Biometrics* for doing this sort of thing, using weighted least squares procedure, where you account for the greater accuracy of some of your estimated observed values than others and it turns out to require an iterative procedure, but it gives a little more weight to those quantities for which you have more accurate observations, which I think would lead to an improvement in your procedure.

Cattell: Yes, I am very glad to hear of that. In my paper I wondered if there was any way we could use different size samples with advantage. Because inevitably some constellations come in quantities and others trickle in. It is possible that this type of research would be even better carried out in some other culture in which certain family relationships are more prevalent than our current ethic allows. I understand, for example, that in Japan it is quite easy to get an appreciable number of half-sibs on either side of the family, which are roughly the same age, and I think it was Newton Morton who suggested to me that it would be worthwhile to translate all the tests into Japanese just in order to be able to get these cases!

138 RAYMOND B. CATTELL

Harris: But the difficulty in that is, as I pointed out, that these parameters are dependent upon gene frequencies; if these other populations have a different genetic structure, where you are going to have a different set of values descriptive of that population, than the other population that you are interested in. In some of the other species some of these heritabilities have been similar in different populations, which is nice, but I don't think that we can always expect that to occur.

Gottesman: Do you think that all trait measures which we use in behavior genetics should be corrected for attenuation and unreliability before putting them in the various formulas, or should we play it straight the first time and then correct it and see what happens to the indices?

Cattell: Well, we have done the former, i.e. taken the reliability of the measure and allowed for that part of the variance which comes from unreliability and I rather favor that. I don't think that I can give more than an intuitive reason for doing it that way. It's tied up with the question just raised about unequal numbers in different groups. Do you see any reason for favoring the other alternative?

Gottesman: Well, when you start to correct for a scale that has low reliability you then may be giving yourself a false sense of security. I've corrected scales after looking at the uncorrected version and for the least reliable scales I get my biggest changes in the final picture. In other words, if something looks like it has high heritability without correcting for unreliability, then when I correct it may have even higher heritability or lower heritability. I don't know what kind of interpretation to make of the different heritability that would come out of it.

Cattell: Couldn't one use the argument that although you are only estimating how much of the variance is not true variance, an estimate is better than nothing? I think that an estimate is better than just letting the unreliability stay in. Another question close to this, if I may raise it here, concerns the distinction between the fluctuation of traits and the unreliability of instruments, as such. We have evidence that one of our 16 PF and HSPQ traits: surgency-desurgency (which is care-free, happy-go-lucky versus very sober, careful and exact) certainly fluctuates. Knowing the immediate test-retest reliability of this measurement on 16 PF and HSPQ and the stability coefficient from retesting people over longer intervals, we conclude that this personality trait fluctuates. For example, in correlating values on two occasions of say 1 or 2 months apart, the correlation is only about .6, i.e. people go up and down on this dimension, with alcohol for example, and with other circumstances. Then the question becomes, are you including in your estimates of the environmental variance by the present calculation some kind of fluctuation about a fixed point, and I think the answer is we are. Our estimate of the nature-nurture ratio on surgency was that about 95% is environmentally determined. Now that environment, since we tested them only once in this particular study, includes the fluctuation, and in our concept of a trait I suggest that this doesn't really belong with the environment. If people are fluctuating about a certain fixed

HEREDITARY AND ENVIRONMENTAL INFLUENCES   139

level, we want to know how much that fixed level, at that period of life, is determined by environment. So that our operations are closely tied with our concept of what we mean by environment, and perhaps we should define it at two distinct levels. (This investigation of methods was supported by Public Health Service Research Grant MH 01733 from the National Institute of Mental Health.)

An Ethological Approach to the Genetical Study of Human Behavior

Daniel Freedman

The fields of animal and human behavior-genetics are characterized by what seem to be arbitrary interests in a variety of phenotypes. In human work the exceptions that come to mind are investigations of intelligence and research with specific medical interests, e.g. schizophrenia. As for other behavior, a heterogeneous collection of phenotypes has been shown to be under partial genetic control (see Fuller & Thompson, 1960), and it is hard to see how these various traits, e.g. personal tempo, endothymia, commention, cyclothymia, or even extroversion-introversion can form building blocks for an adequate theory of personality (Freedman, 1961a). From our experience (see Table I), it seems very likely that this list will soon become inordinately long with little improvement in usefulness. What is needed is a theoretical guideline capable of specifying which phenotypes are of fundamental importance and which are not.

Such a powerful theoretical guide is available, but has yet to be used in either animal or human behavior-genetics. I refer to ethology and its underpinning of evolutional thinking. Ethology is clearly the most virile movement in animal behavior, although its introduction into American Psychology has been slow, often via Zoology. However, more and more animal studies are now appearing which are concerned with the evolutional function (survival value) of behavior.

On the other hand, the effects of evolutional thinking on studies of *human* behavior are almost nonexistent. There seems to be no reason for this. My psychologist friends believe in evolution, but somehow they prefer not to let it interfere with professional think-

TABLE I
Nancy Bayley Infant Behavior Profile (Survey of Individual Items)[a,b]

	p
1. *Social orientation:* responsiveness to persons. 　(1) Does not modify behavior to persons as different from objects. 　(9) Behavior seems to be continuously affected by awareness of persons present.	.005
2. *Object orientation:* responsiveness to toys and other objects. 　(1) Does not look at or in any way indicate interest in objects. 　(9) Reluctantly relinquishes test materials.	.02
3. *Goal directedness:* 　(1) No evidence of directed effort. 　(9) Continued absorption with task until it is solved.	.02
4. *Attention span:* tendency to persist in attending to any one object, person or activity, aside from attaining a goal. 　(1) Fleeting attention span. 　(9) Long-continued absorption in a toy, activity, or person.	.02
5. *Cooperativeness:* (Not relevant most of first year).	
6. *Activity:* inactive-vigorous. 　(1) Stays quietly in one place with practically no self-initiated movement. 　(9) Hyperactive: can't be quieted for sedentary tests.	N.S.
6.1 *Activity:* level of energy (low to high).	N.S.
6.2 *Activity:* coordination of gross muscle movements for age (smooth functioning to poor coordination).	N.S.
6.3 *Activity:* coordination of fine muscles (hands) for age (smooth functioning to poor coordination).	.04
7. *Reactivity:* the ease with which a child is stimulated to response, his sensitivity or excitability. (May be positive or negative in tone.) 　(1) Unreactive; seems to pay little heed to what goes on around him. Responds only to strong or repeated stimulation. 　(9) Very reactive; every little thing seems to stir him up; startles, reacts quickly, seems keenly sensitive to things going on around him.	.04 N.S.
8. *Tension:* 　(1) Inert, may be flaccid most of the time. 　(9) Body is predominantly taut or tense.	N.S.
9. *Fearfulness:* (e.g., reaction to the new or strange: strange people, test materials, strange surroundings).	.05

[a] Probabilities are from Mann-Whitney one-tailed tables and refer to items in which identical twin pairs exhibit significantly greater concordance than fraternal twin pairs over the first year. In no case was the opposite true.

[b] Each item was rated along a scale from "deficient" to "overendowed," with five steps usually spelled out; a nine-point scale was obtained by adding half steps. The first and last steps are supplied to clarify the items.

TABLE I—*Continued*

	p
(1) Accepts the entire situation with no evidence of fear, caution, or inhibition of actions.	
(9) Strong indications of fear of the strange, to the extent he cannot be brought to play or respond to the tests.	
10. *General emotional tone:* unhappy-happy.	N.S.
(1) Child seems unhappy throughout the period.	
(9) Radiantly happy; nothing is upsetting; animated.	
11. Endurance or behavior constancy in adequacy of response to demands of tests.	N.S.
(1) Tires easily, quickly regresses to lower levels of functioning.	
(9) Continues to respond well and with interest, even with prolonged tests at difficult levels.	
12. *Sensory areas:* preoccupation or interest displayed (none to excessive).	
(a) Sights: looking	.025
(b) Sounds: listening	N.S.
(c) Sound producing: vocal	N.S.
(d) Sound producing: banging, or other	.005
(e) Manipulating (exploring with hands)	.02
(f) Body motion	N.S.
(g) Mouthing or sucking: thumb or fingers	N.S.
(h) Mouthing or sucking: toys	N.S.

ing. Clearly, if there is any theory which can unite the field of behavior and the field of genetics it is the unifying theory of evolution.

Now to present our case in point. We have made a study of infant identical and fraternal twins, and although we used a wide variety of measures, our interest throughout was on two major aspects of behavior which seem to have strong evolutional significance: Smiling and the fear of strangers.

While previous work (Kaila, 1931; Spitz, 1946; Ahrens, 1954) had strongly suggested that smiling is deeply implanted in the biological make-up of man, we felt that the twin approach would provide direct evidence on this. Similarly, the fear of strangers, exhibited by most infants in the second half of the first year, seemed to be phylogenetically homologous to the flight response found in many other species (Freedman, 1961b), and we felt that evidence from a twin study would help to determine its biological importance in man. To anticipate, both of these behaviors were significantly more concordant in identical than in fraternal pairs.

Before going on to our data, it will be worthwhile to decide what such an increased concordance actually tells us. In the first place, I agree with Osborne and DeGeorge (1959) that heritability scores derived from twin data are largely misleading, and that it is best to use probability statements instead. If one finds greater concordance in identical twins than in fraternal twins on any trait, he has learned only that there is "some kind" of hereditary control taking place. The nature of this heredity control will most likely be as mysterious after the study as before it was undertaken. Schizophrenia, for example, has been consistently shown to have a genetic component (higher concordance in identical than in fraternal twins), but little more than this can be deduced from twin studies in this baffling area. My conclusion is that it is always necessary to complement the twin approach with other types of experimentation if one hopes to understand what is actually going into a concordance figure.

We have started to do this in our study of smiling in twins through a complementary investigation of blind infants (Freedman, 1965). As you will see, this has enabled us to say something about the role of vision in smiling, and after much more work we hope to disentangle some of the other factors which go into the finding that smiling is under "some kind" of hereditary control. I shall return to this point later.

Our Study

I should like now to present our data, some of which are the subject matter of a film (Freedman, 1963).

Mrs. Barbara Keller and I studied mental and motor abilities and personality development in twenty pairs of same-sexed infant twins. We worked independently, each seeing approximately half the twin pairs. We spent at least one morning or afternoon in the homes each month, testing, interviewing mother, and photographing; occasionally we paid an evening visit to show the accumulated films to the family. It is pertinent to say that we both became good friends of our respective families over the course of our visits and that with good rapport came a fairly accurate level of observation.

Why did we choose to study *infant* twins? Since imitation which is independent of immediate perception starts after the first year, we could rule out such mutual contagion as a factor in the results. We were also able to postpone zygosity determinations (13 blood group factors) until after the study and thereby avoid contamination from that end. Finally, we wanted to avoid the problems of retrospective histories, a major reason for most longitudinal research (Freedman & Keller, 1963).

At the end of the study we found we had an n_1 of 11 fraternal pairs and an n_2 of 9 identical pairs. The results from both Mrs. Keller's group and my own formed similar distributions on all tests and rating scales suggesting that they are readily duplicable. Within-pair differences were consistently greater among the fraternal pairs on both of our major measures, the Nancy Bayley Mental and Motor Scales ($p < .01$ for the combined scales) and the Bayley Infant Behavior Profile ($p < .001$). (p Values are all based on one-tailed tables of the Mann-Whitney non-parametric tests, Auble, 1953). Figures 1 and 2 are the separate distributions of the Mental Scale and of the Motor Scale, and it is seen that the Motor Scale more clearly differentiates fraternal from identical pairs.

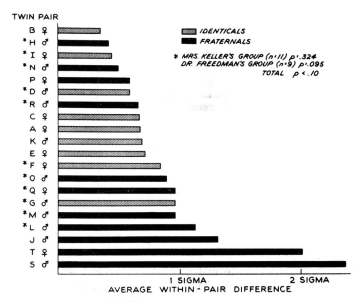

Fig. 1. The Bayley Mental Scale. Average within-pair differences in the first year, based on 8–12 monthly administrations.

We managed to eliminate possible "halo" effects by a procedure in which our films formed the basis of judgment. We had taken monthly motion pictures in which each twin of a pair was filmed separately in the same situations, and the film of each was accumulated on separate reels. At the end of the study, the filmed behavior of half of each pair was rated by a group of judges using the Bayley Infant Behavior Profile. Another group of judges rated the films of the remaining twins. Again within-pair differences among fraternal twins were significantly larger (p < .005, see Fig. 3).

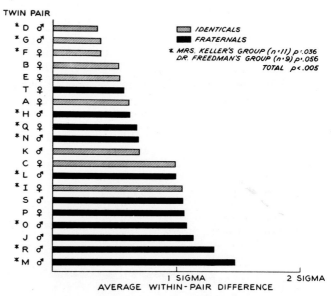

Fig. 2. The Bayley Motor Scale. Average within-pair differences in the first year, based on 8–12 monthly administrations.

Table I represents data on the Infant Behavior Profile. It can be seen that a variety of traits appear to be under genetic control, i.e. they are consistently more concordant within identical pairs than within fraternal pairs over the first year. The two items on which we have chosen to focus are *Social Orientation* (p < .005) and *Fearfulness*, (p < .05). In the first five months *Social Orientation* was scored on the basis of visual fixation of a face and social smiling, and the first part of our film (described below) traces the development of these behaviors within identical and within fraternal pairs

($p < .05$ for months 1–5, identicals more concordant). After 5 months of age, *Fearfulness* was primarily scored on the basis of fear of the investigator ($p < .02$ for months 5–12, identicals more concordant), and the second half of the film deals with this fear of the stranger.

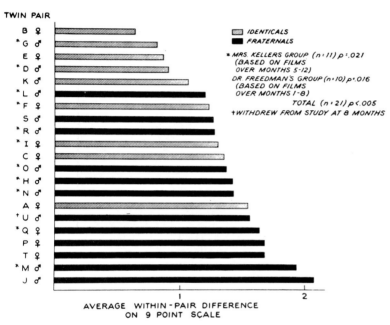

Fig. 3. Within-pair differences on the Bayley Infant Behavior Profile, based on 8 consecutive months of filmed behavior (either months 1–8, or months 5–12).

The following sections are intended to illustrate major points of the film. In order to clarify certain developmental issues, data from a non-twin and blind infants are included.

Smiling

In the first month, it was common to see smiling for which no specific external stimulation was required, and it occurred most often when the baby was sated, had his eyes closed, and was falling off to sleep (see Fig. 4). These were therefore, not social smiles. This type of smiling dropped out in most infants after the first month.

In the second month, it usually took some external stimulation to elicit smiling. This might still occur with the eyes closed but in response to touching the mouth, a voice, the tinkle of a bell, etc. This was the case with Arturo, a fraternal twin, who was a sleepyhead and rarely wide awake (Fig. 5A). His fraternal brother, Felix, was a remarkable contrast (Fig. 5B). He was wide eyed and

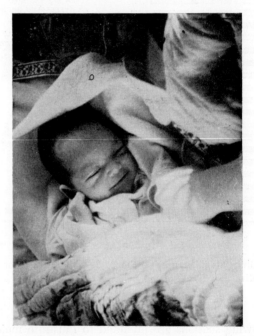

FIG. 4. Tony, singleton, 11 days of age, example of pre-social smiling.

very watchful—but he was unremittingly sober and rarely smiled. This difference in amount of smiling persisted throughout the first year.

Felix not only illustrates contrasting development, so typical in our fraternal pairs, but also his visual fixation of mother's face illustrates behavior that normally preceded social smiling. This fixation

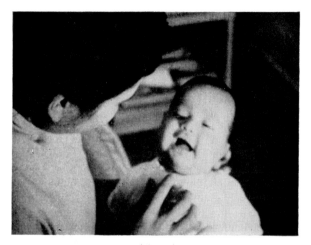

FIG. 5A

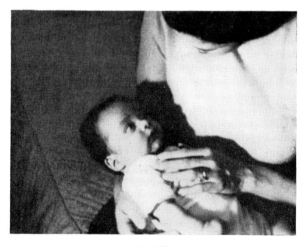

FIG. 5B

FIG. 5A. Arturo F., 2 months 4 days, smiling to touch with eyes closed. He was usually drowsy at this age.   B. Fraternal brother Felix F., 2 months 4 days, was alert but unremittingly sober.

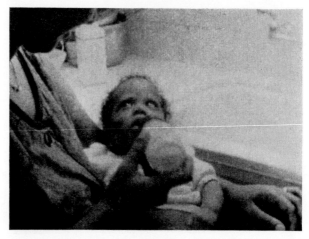

Fig. 6A

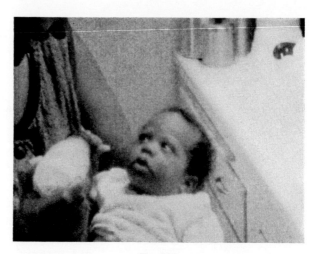

Fig. 6B

Fig. 6A. Chucky S., 2 months 0 days, staring into mother's face during feeding. B. Identical brother Marty S., 2 months 0 days, was completely concordant for this behavior.

on the face is further illustrated by Marty and Chucky (Figs. 6A, 6B), an identical pair who showed complete concordance for this behavior. Usually, about 10 days after such intense visual fixation of the adult's face had begun, the first social smile occurred.

It is of interest that in identical pairs the age at onset and the intensity of visual fixation of the adult's face was generally the same in both twins, even when there were substantial differences in birthweight. (See Price, 1950, for an explanation of such differences in birthweight.)

While genetic factors are important, we should make it clear that these events are dependent on environmental factors also. For example, Wolff (1963), found that social smiling generally appeared before the end of the first month rather than late in the second month within a small sample from the Boston Irish population. He attributed this speedy development to the high amount of stimulation given infants in this sub-population. Similarly the premature infants we have observed smile more in accord with the date of delivery, i.e., the amount of time spent in the nonuterine environment, than in accord with biological age.

As for the cause of smiling in infants, many of the major studies have stressed that visual fixation of a face or some face-like stimulus is a necessary prelude to social smiling (Kaila, 1932; Spitz, 1946; Ahrens, 1954). Consequently the deduction has been commonly made, based on ethological thinking, that a causal connection exists between a moving face-like pattern (social releaser) and smiling. However, it is clear that this conceptualization is far too simple, if only for the fact that blind infants smile. In order to understand this issue better, we are currently studying a number of infants who were blind at birth.

In the course of this study, limited because fortunately blinded infants are now rare, we find that these infants smile at the usual time, but that their early social smiles are fleeting, much like normal, eyes-closed smiling in the first month, and they seem therefore to occur at a reflex level. Our first example, Yvonne (Fig. 7), could perceive only intense light due to cataracts caused by maternal rubella in fetal life. One interesting observation made was that despite the blindness she nevertheless, while smiling, turned her eyes toward the person holding her, as in Fig. 7. This occurred with intense petting, cooing, and talking by the adult, and seems to be direct evidence that ocular fixation of the social object is initiated by central rather than peripheral stimulation. Also, although Yvonne's smiling was actually a series of rapidly succeeding smiles (reflexive firing?), the net impression was that the interaction was as joyous for her as for seeing babies.

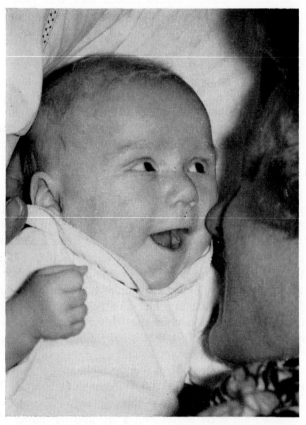

FIG. 7. Yvonne, 2 months 20 days, congenitally blind, smiling with "eyes on" mother. Smiles were fleeting at this age.

The course of early smiling was much the same in David, also a case of blindness due to fetal rubella, and his early smiles were very fleeting. However, by 6 months of age he could engage in normal prolonged social smiling (Fig. 8), as was the case with all other blind infants investigated. We also followed the development of several deaf infants and found their smiling to be quite normal.

We have concluded that since seeing infants, blind infants, and deaf infants all smile, sensory channels while important are secondary and that smiling is never a response to just a single sensory

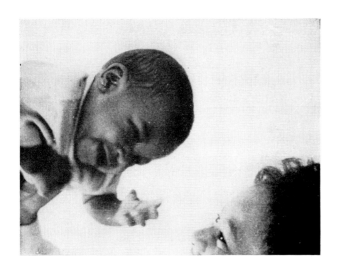

FIG. 8. David, 6 months 10 days, congenitally blind, engaged in prolonged social smiling.

stimulus. At present, smiling is probably best defined as an instinctive response of the infant to "another." (See the important paper of Goldstein, 1957, for an elaboration of this definition).

Fear of Strangers

While infants under 5 months of age will usually smile at any person, after this age they become increasingly discriminating,

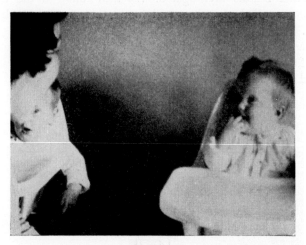

Fig. 9A

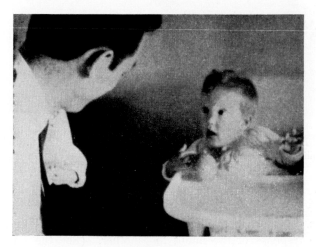

Fig. 9B

Fig. 9A and B. Lisa B., 7 months 0 days, with mother and with a male stranger.

AN ETHOLOGICAL APPROACH 155

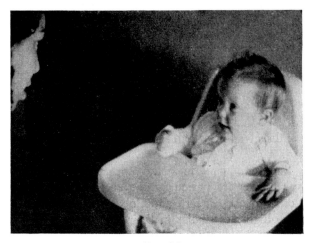

FIG. 9C

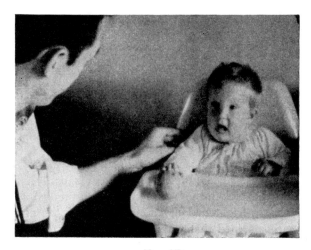

FIG. 9D

FIG. 9C and D. Identical twin Lori B., 7 months 0 days, with mother and with same stranger.

and smiles are increasingly reserved for familiar persons. Discrimination turns to wariness some time in the third or fourth trimester of the first year, and most infants begin to react with fear when with a stranger. This process is very similar to the development of the flight response in lower mammals and birds. There, too, fear of strangers often follows a period in which indiscriminate attachments are possible. (See Freedman, 1961b, for a discussion of this point).

Lori and Lisa, an identical pair, were both wary of strangers from 5 months of age through 8 months, when fear gave way to rather easy acceptance (Figs. 9A–D). As in the majority of identical pairs, the timing and intensity of their reactions were very similar over this entire period.

On the other hand, Richard and Robert, fraternal twins, (Figs. 10A–B) were highly dissimilar in their reactions. Richard was the more fearful from 5 months and on. For example he cried when Mrs. Keller arrived for the 10-month visit despite the fact that she had made a visit each month of his life. After a while he relaxed, but he cried again when the investigator donned a Halloween mask (Fig. 10A). Fraternal brother Robert (Fig. 10B), who had passed through a brief, mild period of fear of strangers at 8 months, provided a clear contrast in that he gave no indication of fright during the entire visit.

While these illustrations are typical examples, the distribution of within-pair differences in ratings of *Fearfulness* formed distinct but partially overlapping populations of identicals and fraternals. Our last example is from the overlapping cases, a pair of identical twins who were decidedly unalike in their reactions to fear-inducing situations. Susie (Fig. 11A), who was easily frightened by noises and the Jack-in-the-box in the first 3 months of life, became exceedingly frightened of strangers toward the end of the first year, and remained that way through a follow-up period of 2 years. Annie (Fig. 11B), while similar to Susie in a great many aspects of behavior, was only mildly disturbed by fear-inducing situations. There is no unequivocal explanation for such nonconcordance: Careful investigation led us to rule out as sole cause either differential treatment by parents, birth order, traumatic delivery, viability at birth, and monochorionic versus dichorionic embryogenesis. Since there was a slight stabismus, Prechtl has suggested,

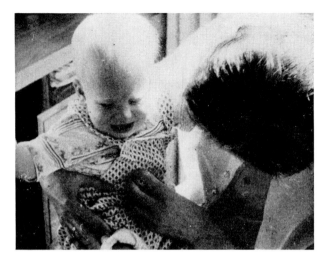

Fig. 10A

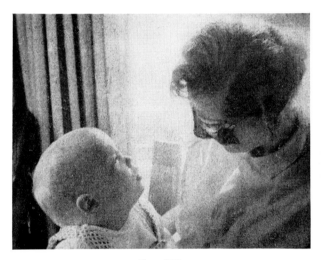

Fig. 10B

Fig. 10A. Richard C., 10 months 3 days, cries when investigator dons Halloween mask.   B. Fraternal twin Robert C., 10 months 3 days, inspects investigator with Halloween mask. His only reaction was "interest."

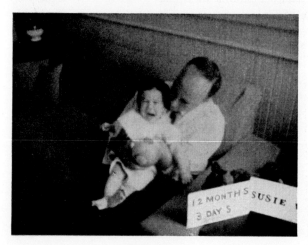

FIG. 11A

FIG. 11B

FIG. 11A. Susie W., 12 months 3 days, with investigator, rejecting both him and offered toy.   B. Identical twin Annie W., 12 months 3 days, sitting comfortably on investigator's lap.

in a personal communication, that minimal brain damage may be a factor in Susie's hyperreactivity (see Prechtl, 1963, on minimal brain damage).

Summary of Film

Heredity plays a role in the development of positive social orientation (including smiling) and in the fear of strangers. Our evidence for this is that identical twins show greater concordance than fraternal twins in these two areas of behavior over the first year of life. There seems to be no reasonable alternate explanation of these results.

The phenomena of smiling and fear of strangers are of great importance, for they are prototypes for much of later behavior. The smile is probably the first expression in human life of pleasure with another, whereas fear of strangers is normally the first expression of fear of another. As far as we know these two phenomena are universal in mankind, and they have doubtless been important in human adaptation and evolution.

Evolutionary Function of the Fear of Strangers

While the function of smiling seems to be that of a binding force between infant and parent, and later within various human diads (Goldstein, 1957), the function of the fear of strangers is not as apparent. I have argued elsewhere (Freedman, 1961b) that the fear of strangers is a homolog of the flight response found in most animals. However, I have since had to face the fact that, unlike lower animals which are highly motile when they develop the flight response, when a human infant develops its fear of strangers (at about 6–9 months) it simply does not have the motor ability to escape a predator. I have therefore been led to the hypothesis that the fear of strangers found in human infants serves primarily to intensify the bonds between the infant and those already close to him, and that it serves at the same time to prevent the dilution of these primary relationships. Let me emphasize that such speculation is open to experimentation, and to that extent has proved fruitful. For example, we have undertaken a study of infant twins in which we test their reactions to a variety of experimental "strangers." In this way we hope to find the optimal stimuli required to elicit the fear response, while maintaining a control over genotype. We have also undertaken studies with dogs in which a critical period for the formation of social bonds with humans has been found which is dependent on a rising fear of strangers. (Freedman, King, and Elliot, 1961). In a planned investigation we shall study imprinting (the development of primary attachments) in birds in which the flight response has been bred out. In this way, we can deal experimentally with the hypothesized positive relationship between strength of flight response (fear of strangers) and strength of primary attachments.

Conclusion

I hope that the above presentation is evidence that a marriage between human behavior genetics and ethology is potentially fruitful. Since it is inconceivable that an uncoordinated collection of heritabilities and concordances can be of value to a science of man, an over all conception is needed as a working guide and means of organization. This is available in evolutionary theory and in the emphasis on evolved behavior. Further, if the elucidation of function is a goal, it appears that the twin method cannot stand alone, and that various complementary studies on the units of behavior under investigation will be necessary.

REFERENCES

Ahrens, R. Beitrag zur Entwicklung der Physiognomie. Part I. *Z. fur Exp. und Ang. Psychol.*, 1954, **2**, 414–454; Part II, 599–633.

Auble, D. Extended tables for the Mann-Whitney statistic. *Bull. Inst. Educ. Res.*, 1953, **1**, 15–24.

Freedman, D. G. The differentiation of identical and fraternal twins on the basis of filmed behavior. *Proc. 2nd Inter. Congr. Genet., Rome*, 1961, **1**, 259–262. (a)

Freedman, D. G. The infant's fear of strangers and the flight response. *J. Child Psychol. Psychiat.*, 1961, **4**, 242–248. (b)

Freedman, D. G. 16mm Sound Film: "Development of the Smile and Fear of Strangers." PCR-2140. University Park Penna. Psychol. Cinema Reg., 1963.

Freedman, D. G. Smiling in blind infants and the issue of innate vs. acquired. *J. Child Psychol. Psychiat.*, 1964, **5**, 171–184.

Freedman, D. G., & Keller, B. Inheritance of behavior in infants. *Science*, 1963, **140**, 196–198.

Freedman, D. G., King, J. A., & Elliot, O. Critical period in the social development of dogs. *Science* 1961, **133**, 1016–1017.
Fuller, J. L., & Thompson, W. R. *Behavior Genetics.* New York: Wiley, 1960.
Goldstein, K. The smiling of the infant and the problem of understanding the "Other." *J. Psychol.,* 1957, **44**, 175–191.
Kaila, E. Die Reaktionen des Sauglings auf das Menschliche Gesicht. *Annales Universitatis Aboensis,* 1932, **17**, 1–114.
Osborne, R. H., & DeGeorge, F. V. *Genetic basis of morphological variation.* Cambridge: Harvard Univer. Press, 1959.
Prechtl, H. F. R. The mother-child interaction in babies with minimal brain damage. In B. Foss (Ed.), *Determinants of infant behavior,* II. London: Methuen, 1963. Pp. 53–59.
Price, B. Primary biases in twin studies. *Amer. J. Human Genet.,* 1950, **2**, 293–352.
Spitz, R. A. The smiling response. *Genet. Psychol. Monogr.,* 1946, **34**, 57–125.
Wolff, P. Observation on the early development of smiling. In B. Foss (Ed.), *Determinants of infant behavior,* II. London: Methuen, 1963. Pp. 113–138.

DISCUSSION

Gottesman: You did point out that you did the bloodtyping after the study was done?
Freedman: Yes, as a control. I will say that by 8 months our guesses were correct in all but three pairs, and it appears that differences in identicals iron out by 8 months. Monochorionic identicals are often born with considerable differences, but by 8 months the weight difference is made up, and at this time people begin having trouble telling them apart. On the other hand, fraternals are pretty well individualized by that age, so 8 months seems to be a good time for guessing. I might also mention that our parents tended to believe their twins were fraternals, and our current cultural emphasis on individuality seems to be the factor operating here. Only three parents guessed they had identicals, and they were correct. As for the obstetricians, they were right nine out of nineteen times, which is no better than chance. I think this is due to the fact that many OB books state that fraternal twins are in separate choria and identicals in a single chorion. However, in about one-third of identical pairs there are separate choria so that if you go by that method you make many mistakes. I have not yet seen an obstetrics textbook which makes the distinction. Unfortunately, Price's classic article (1950) on this subject is known primarily among geneticists. Pediatricians usually rely on the OB people, but we have had a few cases in which the pediatrician and obstetrician gave the parents opposite diagnoses.

A Heredity-Environment Analysis of Personality Inventory Data[1]

John C. Loehlin

I am going to report briefly the results of an analysis which attempts to separate the influences of heredity and environment on personality inventory data—not because I think the outcome of the study is particularly conclusive, but because I think the method used has potential value, and the results suggest some points of interest. Briefly, the method involves two factor analyses, one of variables selected to be high in genetic variance and one of variables selected to be low in such variance, followed by a comparison of the factors extracted in the two analyses.

In the present study, the variables factored were small clusters of items from the Thurstone Temperament Survey (TTS) and the Cattell Junior Personality Quiz (JPQ); whether an item was high or low in genetic variance was determined by a twin analysis, based on a separate sample. For the factor analysis, the two personality questionnaires were given to 231 eleventh and twelfth grade boys in two Lincoln, Nebraska high schools. Each questionnaire was divided in half, and item intercorrelation matrices were computed among the sets of 70, 70, 70, and 60 items—the intelligence items and the last two personality items from the JPQ were omitted. The cosine pi approximation to tetrachoric r was used for the correlations; the Thurstone items were dichotomized by including the middle category with whichever of the two extremes had the smaller frequency. Correlation clusters of three to six items each were selected from each matrix. The criterion for inclusion in

[1] I would like to express my gratitude to the Research Council of the University of Nebraska for financial assistance, to the numerous persons in the Lincoln Public Schools who made possible the gathering of data, and to the Computer Center of the University of Nebraska for providing computer time for the data analysis.

a cluster was a correlation of .30 or better with every other item, and each item was placed in only one cluster.[2] Thirty-five clusters were obtained in this manner—12, 7, 8, and 8 clusters from the four intercorrelation matrices. These thirty-five clusters were then ranked in terms of genetic variance, using data from the Michigan Twin Study (Vandenberg, 1962). I would like to thank Dr. Vandenberg for making these data available to me. I tabulated the concordances for each item of both questionnaires for the forty-five monozygotic and thirty-five dizygotic pairs in the Michigan study, and selected the fifteen clusters with the largest DZ-MZ difference and fourteen clusters with little or no difference for the two factor analyses. Scores on these clusters were obtained for each of the 231 boys in the original sample (the items being combined with unit weights), and these scores were intercorrelated to form the two matrices. Each matrix was factored by the centroid method. Four factors in each case sufficed to extract substantially all the starting communality. These factors were then rotated to oblique simple structure; the loadings (in reference vector form) are given in Tables I and II, and the angles between reference vectors in Tables III and IV. Tables V and VI present the variables loading highest on each factor. Each variable is characterized by a brief phrase, and the items composing the cluster indicated.

Every man is his own interpreter of tables of factor loadings, so only a few general comments will be made about Tables V and VI. First, it will be noticed that there are some similarities in the factors in the two sets. Factor I in each table appears to be concerned with extroversion or introversion; there is an emotionality or adjustment factor in each set, (factor III in Table V and factor II in Table VI); and a factor involving physical activities, (factor IV in Table V and factor III in Table VI). But there seems to be a difference in the flavor of corresponding factors from the two analyses: factors in the first set seem more focused on the individual himself; the ones in the second set, on his reaction to his environment. The extroversion factor in the first set stresses what the individual *brings to* group activities; the one in the second shifts

[2] More or less: the .30 line was crossed slightly four or five times, when the other correlations in the cluster appeared to justify this, and one item inadvertently got included in two clusters (fortunately, they wound up in different factorings).

TABLE I
Factor Loadings, High Heredity Clusters

Variable	Factor			
	I	II	III	IV
1	−.14	.38	−.03	−.03
2	.40	−.04	.14	.05
3	.61	.10	−.09	−.09
4	.42	.06	.10	.21
5	.40	.29	−.08	−.08
6	.09	.11	.03	.28
7	−.00	.04	.33	.36
8	.17	.00	.37	.05
9	−.00	−.07	−.52	.02
10	.42	−.10	.00	−.02
11	.16	.03	.01	−.33
12	.01	.29	.22	.12
13	−.06	−.07	.55	−.11
14	.20	.36	−.04	−.03
15	.13	−.17	−.00	.28

TABLE II
Factor Loadings, Low Heredity Clusters

Variable	Factor			
	I	II	III	IV
1	.41	−.21	−.03	.10
2	.54	−.00	−.06	−.07
3	−.05	.04	.35	−.18
4	.40	.19	.37	.19
5	.11	−.03	.55	.02
6	.53	−.04	.07	−.23
7	.56	.09	−.00	.04
8	−.10	.09	−.06	.44
9	−.42	.31	−.02	−.15
10	.11	−.14	.04	.39
11	.11	.44	.07	−.07
12	.25	−.06	.09	.31
13	−.11	.42	−.01	.10
14	.37	−.12	.33	.09

TABLE III
Cosines of Angles between Reference Vectors, High Heredity

Factor	II	III	IV
I	−.14	−.38	−.09
II		−.10	.42
III			.07

TABLE IV
Cosines of Angles between Reference Vectors, Low Heredity

Factor	II	III	IV
I	−.04	−.19	.03
II		−.03	.31
III			.25

TABLE V
Factors from High Heredity Clusters

Factor	Variable	Description	Loading
I	3	Optimistic, poised (TTS 108, 130, 133)	.61
	4	Socially outgoing (TTS 9, −32, −71, 78)	.42
	10	Has own opinions (JPQ 15, 30, 53)	.42
	2	Quick thinking (TTS 58, 68, 122)	.40
	5	Socially dominant (TTS 16, 41, 42, 43, 44, 53)	.40
II	1	Likes to take things slow (TTS −60, 85, 86)	.38
	14	Gets going easily (JPQ 80, 98, −111)	.36
	12	Adventurous, self-confident (JPQ −22, 37, 46)	.29
	5	Socially dominant (TTS 16, 41, 42, 43, 44, 53)	.29
III	13	Controls his impulses (JPQ 90, −109, 132, −133)	.55
	9	Gets angry, frightened, upset (JPQ 8, −45, −54, 65)	−.52
	8	Good social adjustment (JPQ 10, 12, −18, 47)	.37
	7	Likes to work with tools (TTS 8, 36, 83)	.33
IV	7	Likes to work with tools (TTS 8, 36, 83)	.36
	11	Intellectual interests (JPQ 13, −17, 60)	−.33
	6	Likes physical work (TTS 7, 5, 33)	.28
	15	Impatient, impulsive (JPQ 9, −113, 131)	.28

TABLE VI
FACTORS FROM LOW HEREDITY CLUSTERS

Factor	Variable	Description	Loading
I	7	Impulsive, outgoing (TTS 9, 10, 31, 52)	.56
	2	Enjoys group activity (TTS 67, −110, 128)	.54
	6	Seeks social stimulation (TTS 11, 23, 37, −55, −84)	.53
	9	Shy (JPQ −11, 25, 32)	−.42
	1	Good memory for recent events (TTS 94, 102, −124)	.41
	4	Vigorous, active (TTS 61, 65, 89)	.40
	14	Considers self fortunate (JPQ 107, 121, 130)	.37
II	11	Feels restricted by adults and rules (JPQ 117, 135, 138)	.44
	13	Nervous, suspicious, jumpy (JPQ 96, 108, 127)	.42
	9	Shy (JPQ −11, 25, 32)	.31
III	5	Enjoys team sports (TTS 64, 91, 120)	.55
	4	Vigorous, active (TTS 61, 65, 89)	.37
	3	Likes racing, boxing, betting (TTS 90, 117, 119)	.35
	14	Considers self fortunate (JPQ 107, 121, 130)	.33
IV	8	Likes school and teacher (JPQ 1, 34, −61)	.44
	10	Good behavior (JPQ 6, −69, 71)	.39
	12	Gets along well with parents (JPQ 97, 126, 129)	.31

slightly toward emphasizing the rewards and punishments the individual derives from such activities. The emotionality factor in the first table (factor III) is centered on impulses and their control; the one in the second (factor II) on response to environmental restriction and threat. The physical interest factors, IV and III, show less difference, but in the second set we may have a little more concern with the social setting of physical activity, and in the first, with the physical expression as such. Finally, compare the unmatched factors. In the first set, we have a somewhat ill-defined factor which seems to be in the temperament area (factor II in Table V), in the second, we find a factor of good socialization (factor IV).

It is not suggested that this tendency toward a greater focus on the individual in the first set of factors and a greater focus on the environment in the second set is overwhelming in these data, although it seems reasonable. However, it might be interesting to speculate for a moment on the implications of the more-or-less corresponding factors in the two analyses. This may be simply incomplete resolution of variables, in which case the overlap should tend to disappear as better methods are used on larger samples. The alternative possibility is that personality dimensions may tend, in general, to be two sided, with one aspect reflecting certain (probably complex) heredity tendencies, and the other the social institutions and interpersonal pressures which grow up around these tendencies. If this is indeed the case, it could lead to much confusion in nature-nurture work in the area of personality. For example, suppose two researchers do nature-nurture assessments on the "same" personality dimension; if one uses an "environmental" version and the other a slightly different "genetic" version their estimates could differ widely. In fact, a nature-nurture estimate obtained on a personality dimension will be as much a function of how the particular measurement technique used blends these two aspects, as it is of anything characteristic of the personality dimension itself. In short, proceed with caution in this area until the air is clearer here; or better, proceed with boldness to clear the air.

REFERENCE

Vandenberg, S. G. The hereditary abilities study: hereditary components in a psychological test battery. *Amer. J. Human Genet.*, 1962, **14**, 220–237.

DISCUSSION

Lindzey: Isn't there some incompatibility, John, between these data and interpretations and Dr. Cattell's findings and interpretation in connection with what he has called environmental mold source traits on one hand, as opposed to ergs on the other hand? Aren't you implying that there is a mapping of similar variables from one source of variation into the other, while Dr. Cattell has ordinarily looked upon the trait domain as divisible into those which are principally environmentally determined and those which are principally genetically determined?

Loehlin: Yes, I think there is a difference of opinion here. I will say, in Dr. Cattell's defense, that he has suggested such a possibility in one area, namely intelligence. He has said that you might find two factors in this area, one of native ability and one representing educational facilitation, which might be hard to distinguish, but which would be separate source traits. Now I think he has not made such a suggestion for motives in general, or for other personality factors, but I don't see that this is absolutely incompatible with his position.

Cattell: Well, I wonder how you match Tables V and VI, on what basis is this done? This is certainly a methodological matter.

Loehlin: I did it by inspection and intuition.

Lindzey: The proper answer is "using the same operation which factor analysis people conventionally use when they name factors."

Loehlin: Yes, it is essentially a factor naming process. One sees some resemblance between clusters loading on factor I in each sample, and thus assumes some relation.

Cattell: These are on the same people are they not? With different variables on the *same people* it would be possible to put the two factors in the same space.

Loehlin: Yes, I checked on the cross correlations, which were available from the computer, and there are certainly relationships. I suspect that both factor I's would have come out as a single factor if I had done a single analysis, because the cross correlations are fairly substantial between the items in factor I in each case. I'm less sure of some of these other pairs; they might have come out separately if they had been combined in a single matrix.

Cattell: Doesn't the whole argument really hinge on being able to identify and equate two factors?

Loehlin: No. The argument is that these matrices are closely related, with parallel factors in the two areas. I don't insist that the factors are identical; if anything, I insist that they are distinguishably different.

Cattell: Well, imagine a factor loading variables 1 to 10 where these were chosen odd-even for high and low hereditary determination. It would be possible to do this, of course. One can see that one marker for a factor might be much more involved with environment than another, and in that case you can get identical factors, the even ones more highly hereditary and the odds less. The implication of this seems to be that the selection of variables becomes very important when you have small numbers of variables to represent a factor. Presumably the situation would improve as you took larger and larger numbers of variables.

Loehlin: I'm not sure I see your suggestion here.

Cattell: There are a lot of complications here, this is a crossroads of a lot of assumptions, which one would have to state to get very far. At the moment I am assuming that there is a certain factor which is represented by a set of variables, say 1 to 10; before doing a factor analysis, you have made a nature-nurture ratio analysis on each of these 10 items and sorted them into two heaps, one set high and another set low. You will get different nature-nurture ratios for the factor as a whole, for set even, and set odd. The question might be how far you can extend this. I think, in the end, these items would be tails of a normal distribution of nature-nurture ratios per item. In a sense your result is an artifact of having deliberately chosen from the possible markers for a factor X, (a) those most environmentally involved, and (b) those least environmentally involved. If you were going after the measurement of this factor as such with the greatest possible efficiency, you would deliberately avoid this kind of thing, look at the distribution of nature-nurture ratios and take a random set from it. This would give a true picture of the fact. Thus you are deliberately giving yourself diplopia here, you look at the situation first through this biased glass and then through that one.

Loehlin: But I am doing this on purpose, to bring out a possible difference here. I think this is becoming a question of what one's aims are.

Cattell: Yes, you are doing this on purpose, but is it desirable to do this deliberately when your aim is finding the best possible estimate of nature-nurture ratio for the factor as a whole?

Loehlin: Well, if that is your aim, fine. But, if someone else takes a selection of items a little more from, say, the environmental end, he is going to get a different nature-nurture ratio than you do.

Cattell: Yes, this demonstrates the phenomena. It's something that you would operate against, I would suggest. Now that you have shown that this can be done, it seems to me that all correctives should be made to avoid it.

Loehlin: Yes, if that's what you're after.

Lindzey: I hope to endorse both views here, and I think we must also bring up the unpleasant expression criterion analysis.

New Techniques in Analyzing Parent-Child Test Scores for Evidence of Hereditary Components[1]

Richard E. Stafford

Two methodological chapters in this volume deal with an assumption of polygenic influences, and indeed, one is almost forced to come to this conclusion when one looks at behavioral data, which are almost always on some sort of continuum. Years ago when Mendel was rediscovered, there was a rash of studies purporting to show that all kinds of behavior were based on one, or at the most, two genes. Then we swung away from this over-simplified hypothesis and went over to the continuous data model which seems to fit a lot more of our behavioral data. Mendel himself actually found modes of inheritance which suggested continuous variables with polygenic inheritance, as well as, the discrete segregated type of inheritance—although we generally only hear about the discrete segregated type because it was with this that Mendel was most concerned. On the other hand, Galton had commented on the fact that certain traits seemed to be quite discrete, although we generally think of Galton as being father of the school of genetics which deals with traits having continuous distributions.

Unfortunately, by emphasizing their major differences, a rather bitter controversy arose. I think that East, Fisher, and others have pretty much resolved that controversy, at least for physical characteristics. But when it comes to behavioral traits, it seems to me, we are faced again with the question whether certain traits are on

[1] This research is based on a doctoral dissertation which was supported in part by the Office of Naval Research and the National Science Foundation, Dr. Harold Gulliksen, Principal Investigator; and by an Educational Testing Psychometric Fellowship at Princeton University.

a continuum or are they discrete and segregated. Just for heuristic purposes, I am going to argue that perhaps some of these traits which appear to be on a continuum, actually have an underlying dichotomy, and even though this underlying dichotomy is being masked by environmental effects and perhaps by other genetic effects, it still might be possible to tap in here somewhere and get some estimates of the genetic components.

For example, if this underlying dichotomy were hypothesized to be the effect of a single genetic unit, much more rigorous specific hypotheses could be generated from genetic theory to be tested, or in a sense, fitted to a model.

To test this hypothesis, that the trait under investigation was actually a genetic entity, a technique was devised to fit the observed continuous data to theoretical genetic models. This technique, called "dichotomic analysis," is essentially an arbitrary quartering of a bivariate distribution of parent-child scores with a successive series of artificial divisions in the continuous distributions. The frequencies observed by these arbitrary quarterings may then be compared to the theoretical expected genetic frequencies by a series of chi square goodness-of-fit tests. Should a "good fit" be found by one of the artificial divisions in the bivariate distributions of father-son scores, and also at approximately the same artificial division in the bivariate distribution of mother-daughter scores, an underlying dichotomy would be assumed.

This dichotomic analysis can also be applied to twin studies. A further extension of this technique is applicable when there is evidence that a trait has a genetic component. The possibility of it being determined by a recessive gene may be checked by making a trivariate distribution of the fathers', mothers' and children's scores.

Another technique involves correlational analysis of sex differentiated traits by fitting them to an ordered sequence of coefficients.

Since both the dichotomic analysis and correlational analysis are predicated upon fitting the data to genetic models, I would like to review briefly some population genetics as applied to families. Two of the more common models of inheritance selected for this study are the simple autosomal dominant-recessive and the X-linked dominant-recessive. In each model the simplest genetic conditions are posited: (1) there is complete penetrance, that is to say, whenever a dominant gene is present it will manifest itself; (2) there is no selective mating by the parents, at least in respect to the traits under investigation; and (3) there is no epistasis, which means there are no other genes at different loci masking the effects of the gene in the model. If these conditions are not met, then the results of the model fitting are questionable.

In the simple dominant-recessive autosomal model, we shall first assume that there are two allelic genes which we shall label "B" and "b" at a single locus on a chromosome. Since chromosomes come in pairs, four different combinations of the two genes are possible: BB, Bb, bB, and bb. It is impossible to differentiate between Bb and bB, so there are actually only three genotypes with twice as many individuals of the Bb type as there are of either the BB or the bb. If we also assume that gene B (which allows development of the trait) is dominant to gene b (which inhibits the development of the trait), then gene B will mask the effects of gene b. Individuals with either the combination of genes BB and Bb will manifest the trait equally and only those with the combination of bb will lack it. Thus, the three genotypes become only two phenotypes, BB + Bb, and bb. (Genotypes BB and bb are referred to as "homozygous" and the genotype Bb as "heterozygous.")

Since there are three genotypes possible, there are nine different possible combinations when parents mate as indicated in Fig. 1. It can be seen by referring to Fig. 1 that when a parent of type BB mates with another of type BB, all of their offspring will be alike and have the same BB combination of genes, and when a parent of type bb mates with another of type bb, all of their offspring will be alike and will also have the same combination of genes that their parents have. Similarly, when a parent of type BB mates with one of type bb, all of the offspring will be alike, that is type Bb, but they will have a different combination of genes than either of their parents. When parents that are heterozygous (type Bb) mate with other types, the resulting offspring are of different types. For example, when Bb mates with bb 50% of their offspring will be of the Bb type and 50% will be of the bb type. When a parent of type Bb mates with another Bb, three genotypes will result in the following percentages, 25% BB, 50% Bb, and 25% bb. Of course, when B is dominant over b only two phenotypes will be observed, 75% BB + Bb and 25% bb.

In addition to knowing the percentage of different types of offspring resulting from various combinations of parental matings, we also need to know the frequency of a particular gene in the population. If we discover that 64% of the population have a trait and 36% lack it, we can compute the frequency of the B and b genes in the population. We have assumed that there is only a B and a b gene at this particular locus, so we can let the percentage of B genes be "p" and the percentage of b genes be "q" where $p + q = 1$, and $0 < p < 1$. It is easier to calculate q than p because the probability

		MOTHER					
		bb	q^2	Bb	2 pq	BB	p^2
FATHER	BB p^2	Bb	p^2q^2	BB Bb	$2p^3q$	BB	p^4
	Bb 2 pq	Bb bb	$2pq^3$	BB 2 Bb bb	$4p^2q^2$	BB Bb	$2p^3q$
	bb q^2	bb	q^4	Bb bb	$2pq^3$	Bb	p^2q^2

Fig. 1. Probabilities of various genotypic combinations of parents and resulting types of offspring. (B = dominant gene; p = percent of dominant gene in population; b = recessive gene; q = percent of recessive gene in population.)

of people lacking the trait (type bb) is q^2 while the probability of people having the trait is $p^2 + 2pq$. To find q we take the square root of the percent (.36) of people lacking the trait (type bb), and find the square root to be .60; so, by subtraction, p is .40.

We can now go back to Fig. 1 and, knowing the frequency of the gene in the population, we can compute the probability of any particular parental mating combination. For example, using the hypothetical gene frequencies that we calculated, we can determine how frequently a heterozygous father (type Bb) will mate with a heterozygous mother. We see that the probability of being type Bb is $2pq$, so that the probability of both parents being type Bb is $4p^2q^2$. The chance of a type bb child resulting from this particular mating combination is 1/4. The percentage of bb children can be computed by showing that 1/4 of $4p^2q^2$ is p^2q^2 and substituting our hypothetical gene frequencies ($p = .40$, $q = .60$) we find that 5.8% of all children are of type bb and are born to parents who are both type Bb.

In this study, we are specifically interested in the similarity between parents and their children. To determine how frequently we should expect a child to resemble one parent we must derive Fig. 2 from Fig. 1. Let us assume we are studying resemblance of fathers and their sons. Fathers of the type BB + Bb have the trait and fathers of the type bb do not have the trait. We wish to know, given any gene frequency, how often we should expect their sons to be like them and how often we should expect their sons to be unlike

		CHILD	
PARENT	BB + Bb $p^2 + 2pq$	$pq^3 + p^2q^2$	$pq^3 + 4p^2q^2 + 4p^3q + p^4$
	bb q^2	$q^4 + pq^3$	$pq^3 + p^2q^2$
		bb q^2	Bb + BB $p^2 + 2pq$

Fig. 2. Expected probabilities of various combinations of parent and child phenotypes.

them. Referring back to Fig. 1 and looking across the row from father bb, we see that whenever he mates with a mother who is also bb, all of their offspring will be bb and the probability of that mating is q^4. The first entry into the lower left-hand cell of Fig. 2, where both father and son are recessive (lacking the trait), is q^4. There is still another source of sons who are recessive (type bb) from fathers who are also recessive. They come from a father of type bb mating with a heterozygous mother (type Bb). Since half of these children will be bb, and the probability of this mating is $2pq^3$, we add pq^3 to our lower left cell where both father and son lack the trait. To determine how many dominant sons (BB + Bb) there will be who have recessive fathers, we first enter the other half of the probability $2pq^3$ (father bb with mother Bb) and then add the entire probability of the mating between a father of type bb with a mother of type BB, which is p^2q^2. In a similar fashion, the frequency of recessive sons with a dominant father is obtained from the probability of the mating of a Bb father with a bb mother. This turns out to be pq^3 so this is entered into the upper left-hand cell of Fig. 2. From the probability of the mating between a Bb father with a mother also of type Bb, which is $4p^2q^2$, we see that only 1/4 of the sons are recessive (type bb) so we enter p^2q^2 into the upper left cell. This gives us exactly the same percentage in the upper left cell that we found for the lower right. By subtraction we can calculate the frequency of dominant fathers with dominant sons to be $pq^3 + 4p^2q^2 + 4p^3q + p^4$ and we enter this into the upper right-hand cell of Fig. 2. It should be kept in mind that since we are dealing with an autosomal model, the sex of the parent or child makes no difference. The same formulas could be used for mother and daughter or son. We have now shown that whenever we know the gene frequency of an autosomal trait, we can compute the frequency of parents and their offspring being alike and unlike.

Dichotomic Analysis

To demonstrate how the technique of dichotomic analysis works, let us take the hypothetical test scores of 50 fathers and their sons. We wish to know if the relationship of the sons' scores to the fathers' scores in any way fits that which we might find if the hypothetical trait has a basic hereditary component. The analysis is started by plotting the paired father-son score in a bivariate distribution with the father's score on the ordinate and the son's score on the abscissa of the graph shown in Fig. 3. The assumption is made that the trait is a manifestation of an autosomal dominant-recessive gene and that a high score is dominant to a low score (it can be shown that when a low score is dominant to a high score the order of entries is simply reversed). For our first trial we will make the hypothesis that 90% of the population have the trait, that is to say, that they are phenotypically dominant, and that the 10% who lack the trait are recessive. Therefore, the bivariate distribution is first divided into quarters with a horizontal line at the 10th percentile of the fathers' scores and a vertical line at the 10th percentile of the sons' scores. This is shown in Fig. 3 by the lines labeled "10th percentile." The expected frequencies can now be computed. The value of q is found by taking the square root of .10 (the 10th percentile must be changed into a decimal first), which equals .32, and, therefore, by subtraction, p equals .68. Substituting these values in our formulas for the four cells in Fig. 2, the expected percentages are found to be 3.2% for the lower left-hand cell, 83.2% for the upper right-hand cell, and 6.8% for each of the other cells. Since the hypothetical population consisted of 50 fathers and their sons, the above percentages are multiplied by 50 to obtain the theoretical frequencies for each cell of Matrix A in Fig. 4. These frequencies are placed on the middle

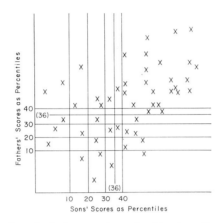

FIG. 3. Various hypothetical quarterings of a bivariate distribution of father-son scores (artificial data).

line of each cell in Matrix A. Next, an actual count is made of the number of father-son plots in each cell of the quartered bivariate distribution in Fig. 3, and these counts are entered in the top line of each cell of Matrix A of Fig. 4. The difference between each theoretical frequency and the actual frequency is then squared and a chi-square is computed according to a standard formula (Siegel, 1956). The chi square for Matrix A is 3.167. This is plotted in Fig. 5 on the abscissa against 10% on the ordinate since the hypothesis was that 10% of the population was recessive.

The process is repeated by dividing the fathers' scores at the 20th percentile with another horizontal line and likewise dividing

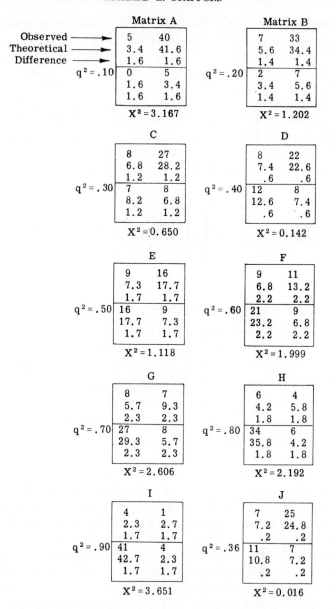

Fig. 4. Artificial data. Hypothetical dichotomies and Chi-square tests of goodness of fit (low score = recessive).

the sons' scores at the 20th percentile with another vertical line. The theoretical frequencies determined from the formulas in Fig. 2 and the actual counts made of the father-son plots in Fig. 3 are entered in Matrix B of Fig. 4. The resulting chi square of 1.202 is plotted in Fig. 5 against the hypothesis that 20% of the population is recessive. This process is repeated for each successive decile. For clarity, some deciles have been omitted from Fig. 3. The resulting graph, Fig. 5, showed that the fits with the theoretical model were particularly good at the 30th and 40th percentile. Accordingly, it was determined that the closest fit possible occurred

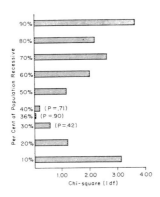

Fig. 5. Values of Chi-square for various hypothetical percentages of the population exhibiting recessiveness, fitted to a genetic model (artificial data).

at the 36th percentile, where the resultant chi-square was only .016 and that the probability value associated with this chi-square ($df = 1$) was .90. With real data, the finding that approximately 36% of the people lack the trait would be cross-validated by following exactly the same procedure with the mother-daughter paired scores to see if the best fit of their scores was near the 36th percentile. If it were, it would be further evidence that there was a real underlying dichotomy for the continuous test score distribution of the variable.

Two variables, the Symbol Comparison test of perceptual speed and the Pitch Discrimination test, a musical aptitude, gave clear evidence of fulfilling the requirement that the best fit to the genetic model was approximately the same hypothesized dichotomy for the father-son distribution of scores as it was for the mother-daughter distribution. The Letter Concepts test of inductive reasoning also showed a possible underlying dichotomy.

Incidentally, I discovered that one can use dichotomic analysis with twin data. Let me quickly explain what I did. Scores of 48 monozygous and 55 dizygous twins of both sexes, ages 13 to 18, were obtained on several mental tests from the Thurstone-Strandskov twin data (published in the Psychometric Laboratory Report #12, University of North Carolina). In Figure 6, the scores on the Seashore Pitch Discrimination test for each pair of monozygous

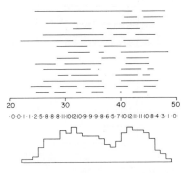

FIG. 6. Lines representing twin differences (above) and number of twin differences occurring at each specific test score (below).

twins have been plotted and joined by a line. Below this are tallied the total number of twin differences spanning each raw score and below this a histogram is constructed from the tallied numbers. The histogram for the monozygous twins shows evidence of bimodality while that of the dizygous twins shows an approximately normal curve.

Three other mental tests besides Pitch Discrimination showed bimodal histograms for the monozygous twins: Kohs Blocks test, Primary Mental Abilities test of Reasoning, and a Spelling test. These histograms with the corresponding dizygous histograms are shown in Fig. 7 with those for Pitch Discrimination.

Another way of representing this apparent bimodality is to plot the score of the first born monozygous twin against the second

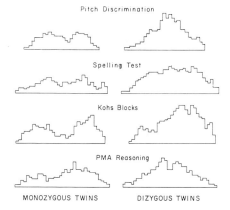

Fig. 7. Frequency of MZ and DZ twin differences occurring at each specific test score.

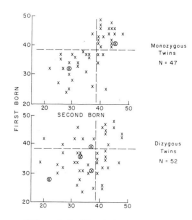

Fig. 8. Scattergram of the pitch discrimination test scores of 47 pairs of MZ and 52 pairs of DZ twins.

born on a scattergram as in Fig. 8. By drawing a line vertically and horizontally from the score at the antimode of the histogram, nearly all points should be found in the upper right and lower left cells.[2] Prediction of the number of dizygous twins having concordant and discordant scores was made by computing an approximate gene

[2] I am indebted to Dr. Ledyard Tucker for pointing out the relationship between the histogram and scattergram techniques.

frequency from the proportion of monozygous twins in the two cells with the assumption of an autosomal mode of inheritance. From a scattergram of dizygous twins divided at the same score as the monozygous twins, the actual number of dizygous twins having concordant and discordant scores was tallied and a chi square test, used for goodness-of-fit, gave the p values shown in Table I.

TABLE I

Goodness-of-Fit Values of p for Chi Square

Test	p value
Pitch Discrimination	.077
Spelling	.158
Kohs Blocks	.160
PMA Reasoning	.989

Trivariate Distribution

Another technique to analyze parents and children's test scores would be to make a trivariate distribution with the fathers' scores on the ordinate, and the mothers' scores on the abscissa, and the children's actual scores recorded as the third dimension.

This trivariate distribution would test the hypothesis that the mode of inheritance of the trait under study was a simple autosomal dominant-recessive. For example, evidence from the above dichotomic analysis of the Pitch Discrimination test scores suggests that there is an underlying discrete distribution such that 80% of the population score "high" on the test and that 20% score "low." Also the evidence suggests that scoring high is recessive to scoring low. To check this, a trivariate distribution was made of family scores on the Pitch Discrimination Test. There were 69 families where both the mother and father scored high (recessive). It would be expected that all of their children would score high, but five of their children scored low.

I'm a little disappointed here; something ran amiss, I don't think I hit the best estimate of the gene frequency, but there were several possibilities for errors, possibly in some of my assumptions, but I do not have time to go into these.

Correlational Analysis

Correlational analysis of parent-child test scores has been inadequate for investigating the presence of hereditary components because of the impossibility of assessing the effects of environment (Hogben, 1933). However, transmission of a trait determined by a gene located on the X chromosome results in some unique family correlations (Hogben, 1932; Stafford, 1961). We would expect a zero correlation between fathers' and their sons' scores, but we would expect a significant correlation between fathers' scores and their daughters' scores since the father passes only his Y chromosome to his son, and passes his X chromosome, containing the gene determining the trait, to his daughter (see Fig. 9). Since the son's

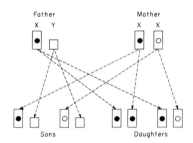

FIG. 9. The four possible phenotypes of an X-linked Trait.

X chromosome comes from his mother, the correlation between mothers' and their sons' scores should be significant and equal in magnitude to that found between the fathers' and their daughters' scores. The correlation of mothers' scores with their daughters' would be positive, although somewhat smaller in magnitude.

Presented symbolically: Mo — So ≅ Fa — Da > Mo — Da > Fa — So ≅ Mo — Fa ≅ 0.

Several tests were given to fathers and mothers and their teen-age sons or daughters. Raw scores were converted to standard scores to partial out age differences, which as Dr. Cattell pointed out in his chapter present special problems if the same test is given to adults and very young children. Males averaged significantly higher than females on two of the variables, the Identical Blocks test of spatial visualization and the Mental Arithmetic test of quantitative reasoning ability. Because males, having a single X chromosome, will manifest a trait determined by a recessive gene more often than females, it was hypothesized that the sex differences noted above were due to recessive, X-linked genes.

The family correlation coefficients corrected for attenuation due to test unreliability are shown in Table II. Although the fit to Phi

TABLE II

CORRECTED AND THEORETICAL FAMILY CORRELATIONS OF THE MENTAL ARITHMETIC TEST AND THE IDENTICAL BLOCKS TEST SCORES

	N	Pearson r's (corrected for attenuation)		Theoretical Phi		
		Mental Arithmetic	Identical Blocks	$20\%^a$	$50\%^a$	$60\%^a$
Father-Mother	99	.07	.05	.00	.00	.00
Father-Son	51	.08	.03	.00	.00	.00
Father-Daughter	63	.21	.36	.41	.58	.61
Mother-Son	50	.62	.41	.41	.58	.61
Mother-Daughter	64	.25	.22	.17	.33	.38

^{a} Based on these gene frequencies.

coefficients based on the theoretical gene frequencies should not be taken too seriously, there is evidence that these two variables have a hereditary component which is transmitted by X-linked recessive genes. Actually, the correction for attenuation was not much. The reliability of the Mental Arithmetic test is in the eighties and the reliability of the Identical Blocks test was in the high eighties. The correction for attenuation was used, of course, because the results are being compared to a genetic model, a theoretical model, and I felt that we should see what the coefficients would look like had we had a perfectly reliable test.

There were other tests with a sex difference, spelling ability, for example, but the predicted ordering of correlational coefficients did not fit the model. Then, too, with traits such as height and weight, although sex differences were significant, there was undoubtedly some homogamy entering into the picture because the father-mother correlations for height and weight were .25 and .28 respectively; so I disregarded these traits because of the assortative mating factor.

Another question arises, and I cannot give you any definite answer. What about the Mental Arithmetic test and the Identical Blocks test? Is it reasonable that out of the ten variables which I investigated that these two should be on the X chromosome? In my opinion it is not logical but the amount of covariance is relatively small, 9 to 25%, for various family members, mother being the highest, I believe. To check this covariance another way, I took mothers who scored high on both tests and examined their sons' scores. Since high scores on these two tests were hypothesized to be X-linked recessive, all sons of mothers who scored high should have also scored high. No such pattern of scores was found and I was forced to conclude that these two tests are not measuring the same trait, but that they are reasonably independent of each other, although from other studies I know a correlation between them of .3 to .4 persists. I'm still puzzled by it.

I have tried to give you rather quickly some of the techniques that I have been experimenting with, in hopes that someone else might apply them to their data. My tentative findings that spatial visualization and quantitative reasoning ability are X-linked recessive, or that pitch discrimination is autosomal recessive, or that spelling ability has an underlying dichotomy, are mentioned, not because I really believe them, but because I feel that one sometimes has to put something up to shoot at, or to put it another way, I think it is better to stick your neck way out and take a chance on being wrong than it is to be too cautious and not make any progress. So I have stuck my neck out.

In summary, these are the new techniques that I proffer for experimentation: correlational analysis with sex differentiated traits, dichotomic analysis for parent-child data and twin data, and trivariate distributions with parent-child data. I hope someone tries these out, and also perhaps works with some of the variables that I have mentioned.

References

Hogben, L. The correlation of relatives on the supposition of sex-linked transmission. *J. Genet.*, 1932, **26**, 418–432.

Hogben, L. The limits of applicability of correlation technique in human genetics. *J. Genet.*, 1933, **27**, 379–406.

Siegel, S. *Nonparametric statistics*. New York: McGraw-Hill, 1956.

Stafford, R. E. Evidence of an underlying genetic dichotomy for certain psychological traits (abstract) *Amer. Psychol.* 1959, **14**, 431.

Stafford, R. E. Sex differences in spatial visualization as evidence of sex-linked inheritance. *Perc. Mot. Skills,* 1961, **13**, 428.

Stafford, R. E. An investigation of similarities in parent-child test scores for evidence of hereditary components. Research Bulletin 63–11, 1963, Educational Testing Service.

Stafford, R. E. Spatial visualization and quantitative reasoning: one gene or two? Paper presented at Eastern Psychological Association. 1965.

Selected Areas of Development of Twins in Relation to Zygosity

Ralph L. Witherspoon

The literature on twin research reveals a recent return of interest in the role of heredity in human behavior. A partial answer to this question is likewise being sought in the role of what has been described as the "twin-situation," or the unique environmental influences due to the happenstance of being twins. During the spring of 1960, eleven Florida State University faculty members representing seven disciplines became interested in developing a 5–8-year cooperative study using twins as subjects in order to study anew some of the problems. The plan that was developed involved not less than twenty pairs of preschool age twins for at least a continuous 2-year period of study of each attribute and included: Zygosity reported by the parents is about equal. Blood type analyses will be used in the determination of zygosity before the final analysis of data is made. To date, this research has been financed by the Florida State University Research Council and the Institute of Human Development.

The availability of the Florida State University's Institute of Human Development nursery schools made possible a two-phase study: First, a study of the onset of certain early behavior in relation to developmental status in monozygotic and dizygotic twins was undertaken, including language development, response to musical stimuli, and early graphic expression; second, the influence of a controlled environmental influence was studied, as provided in the nursery school, on the development of these variables and, in addition on educational achievement, social interaction, intelligence, and personality characteristics as revealed by the Children's Apperception Test.

A study of the beginnings of graphic and musical expression was deliberately included because talents in the arts are generally believed to be strongly under the influence of heredity, while educators in these areas rightfully question such concepts. So little research has been conducted on these aspects of development that even pioneer efforts in this direction are significantly important. Current interest in the relationship of artistic expression (music, art, etc.) to personality development, adjustment, and mental health indicates the need for much developmental research in this area.

Because of the age of the twins, the two-phase problem, and the length of time the research has been in process, few analyses can be attempted at this time. Consequently, this report is of necessity descriptive and in the nature of a progress report. However, some interesting trends are indicated by an "early look" at the data. Interpretations involving zygosity were made on the basis of medical and parent reports and are subject to the limitations implied. Since only eight pairs are now in the first year of nursery school, tentative findings of the "origins" phase constitute the major part of this report.

Brief Review of Pertinent Literature

While longitudinal studies offer the most satisfactory method of determining likenesses and differences due to environment or heredity, especially in twin studies, the studies conducted to date are contradictory and inconclusive in their findings. All are open to criticism either because of inadequacy of sampling, instruments used in the collection of the data, questionable determination of zygosity, or the procedures followed. There appears, however, to be agreement among existing studies that one-egg twins are more alike in physical and mental characteristics, and there is some indication that they are more alike in personality characteristics than fraternal twins or siblings. These likenesses and differences seem to be apparent whether the twins are reared together or apart. Since, to date, consistent intra-pair likenesses have been demonstrated only in identical twins, it would seem that heredity is more important than environment in determining differences. One of the most extensive early twin studies was conducted by Newman *et al.* (1937). Subjects included 50 pairs of identical twins, 50 pairs of fraternal twins, and 19 pairs of identical twins reared separately. A time span of 10 years was involved. Identical twins were found to be more alike than fraternal twins in all traits studied except those classified under personality. Differences in mental traits increased with age. It should be noted, however, that the personality measures used may be open to considerable question in terms of present-day personality evaluation.

Previous research in language development, particularly by Templin (1957) emphasizes the need for further research on the impact of environment on speech and language development and its relation to social acceptability and the child's perception and cognition. Frequent references may be found attesting to the retardation of language development characteristic of twins. Day (1932) and Davis (1937), reported evidence which indicates that twins progressed toward an adult use of language at a relatively slower rate than singletons. There is a general consensus of opinion which attributes language retardation among twins to be due to the "twin-situation." Day (1932) points out that "if in the twin-situation each child has the other for a model in speech a larger proportion of the time, rather than adults or older children as in the case of singletons, perhaps we may expect slower progress with the poorer model." Luria and Yudovich (1959) observed that "twins did not experience the necessity of using language to communicate with each other." They further conclude that the twin-situation does not create an objective need for developing language and thus constitutes a factor that fixes the language retardation.

Traditionally, language development has been regarded as a product of the gradual maturation of intellectual structure and function. Very little thought or investigation has been given to attempting to define the precise significance of language as it is related to the development of these mental processes. Suggestions are beginning to appear in the literature, especially among publications by Russian writers, which indicate that a study of language might be the most profitable approach to the understanding of the mental development of the human organism. Templin (1957), Davis (1937), and McCarthy (1949, 1954) have contributed much to the methodology of studying language development in children.

While there is a sizable body of literature on graphic expression, music, and the development of a self concept in young children, from an extensive review of nearly two hundred studies, it would appear that there has been little systematic research with twins that relates specifically to this investigation. To date, most of the investigations in these areas have been concerned with the identification of developmental norms, and have used singletons.

Dennis (1935, 1938, 1941) reported no appreciable difference from other children in social development of twins during the first 14 months of life. However, the studies of Kent (1949), Karpman (1951), and Mowrer (1954) suggest that the social development of twins differs from that of singletons. While these studies are representative of the kinds of investigations which have been done which relate to our investigation of the social interaction of twins, their results preclude definitive statements concerning differences between twin and non-twin children because of small samples, obvious bias of clinical interpretations, and limitations inherent in the methods utilized in the collection of the data. While Slater (1951, 1953) found that seven of his one-egg pairs "despite almost identical personalities" failed to present concordant psychoneurotic histories, Shields (1953) found monozygotic twins (69%) more than twice as likely to have the same degree of adjustive difficulty as dizygotic twins (31%). The genetic contributions to "neurotic unit predisposition" was estimated to be as high as 80% by Eysenck and Prell (1951). Their findings led them to favor a classification of a neurotic personality factor as being a biological and largely a gene-specific entity. Dennis (1935, 1938, 1941), as noted previously, reported no appreciable difference from other children in social development of twins during the first 14 months of life. Burlingham (1947, 1949, 1952) concluded that in the case of identical twins (3 sets), twin interaction seemed not only to provide security under disturbing conditions but also to serve as a barrier against new contacts. Newman et al. (1937) reported a .96 correlation in achievement for unseparated identical twins, .51 for separated identicals, and .88 for fraternal twins.

Burt (1958), on the basis of an intensive study of twins, siblings, and foster children, concluded that differences in intelligence, like differences in stature, were more dependent on the action of numerous genes than on environmental conditions. These findings confirm those of Jones (1955). However, the work of Skodak and Skeels (1949) and of Newman *et al.* (1937) indicates the importance of environmental factors.

Research Plan and Some Tentative Trends

All twins are being studied longitudinally from age 6 months. At age 3, or somewhat earlier if necessary to fit the school year, the twins are admitted to nursery school for a minimum of 2 years. Gesell Developmental Schedules are administered at 3-month intervals prior to age 3, and the 1960 revision of the Stanford-Binet is given thereafter at 6-month intervals. A complete family history is made for all nursery-school children, school children, twins and non-twins, before admission to nursery school. This information, along with the results of an annual physical examination, will be used to interpret any artifacts in the final data. Periodic non-sleep electroencephalograms have been attempted but have not been very practical to date, since as C. E. Walters found, it is difficult to get the "resting" cooperation necessary. Nursery-school groupings have been arranged to study the relationship of language-enriched versus regular nursery-school experience on language development. One member of each twin-pair is enrolled in a language-enriched nursery-school program as Experimental Group B. Since there might be a tendency for the first-born of twin pairs to be more advanced in language development, both Experimental Group A and Experimental Group B consist of 10 first-born and 10 second-born twins. A similar number of non-twins in each group serve as a control group. Both experimental groups and the control group were administered the following tests at the beginning and termination of the first school year, and they will be given again at the close of the second school year: the *Templin-Darley Articulation Test*, the *Peabody Picture Vocabulary Test*, the *Templin Sound Discrimination Test*, a *vocabulary usage test*, the *Stanford-Binet Intelligence Scale, Form L-M*, and the *Children's Apperception Test*. Byrd and Witherspoon (1954) developed a normative approach for scoring personality variables from protocols of the Children's Apperception Test (CAT). Extensive longitudinal work on this method is in progress but is not ready for publication at this time. Commonly employed hearing screening tests assure that all subjects will have normal auditory perception. All groups will continue in the nursery school for 2 academic years.

Normative data indicate that the average child begins to talk —or to use oral expressive behavior—between 12 and 18 months of age. Schendel earlier found the range of onset of oral expressive language for 14 pairs of twins in this study to be from 9 months to 30 months. The mean onset was 14.6 months. The onset occurred by, or before, 18 months of age for all pairs except one. It would thus appear that the onset of oral expressive behavior in twins tends to fall within the normal age range. Schendel (1963) suggests that "it is possible that the initial attempts at 'talking' in twins are made at the normal ages but that the environmental reinforcements to these attempts are missing which then results in retardation of future growth. It is possible that onset of speech is determined by physiological readiness but that further development is influenced by environmental factors." On the basis of data collected to date using the Templin-Darley Articulation Test, the Templin Sound Discrimination Test, a test of vocabulary usage, and the Peabody Picture Vocabulary Test, it appears that articulation development in twins is about equal to or better than the published norms for their age group. However, the non-twin controls in this study appear to be even more advanced. Generally, the twins appear to be slower in developing vocabulary of recognition than the non-twin controls.

The incidence and nature of spontaneous and evoked responses to music stimuli among infant and very young twins and singletons was studied by Simons (1962) in an intensive 1-year study using 12 pairs of same-sex twins from the larger sample and a like number of singletons matched by sex and age. Subjects' ages ranged from 9 months to 31 months. Tests and observations of the forty eight subjects were made in each of the children's homes for 4 consecutive months. Each study period was approximately 40 minutes in duration. Gross and imitative responses to recorded stimuli were observed as well as responses during "free play" activities. Recorded stimuli included brief rhythm patterns, brief pitch patterns, brief musical phrases, and brief songs (unaccompanied). This study established the fact that children of this age range can and do respond to varied controlled musical stimuli.

The combined aggregate scores of twin-pairs differed from those of non-twin control-pairs, revealing that responses to music were significantly less (at the .005 level of significance) for twins than for singletons. Responses for piano stimuli were greatest to rhythmic music, less to melodic music, still less to harmonic music, and least to dissonant music. An interesting finding is that while earlier responses were observed at about the same time for twins as for non-twins, both the frequency and degree of imitative response to rhythm, pitch, phrases, and songs, decreased for the twins during the period of the study, while the same response scores of singletons increased during the same period. It is possible that some of the factors in the "twin-situation," especially those related to needs rather than developmental retardation were involved in this finding.

Mooty and Schwartz (1963) using free expression crayon drawings, on the basis of 85 test situations with three pairs of twins and judged on the Kellogg Scale, indicate that twins, regardless of zygosity, appear to be very much alike developmentally. That is, both twins of each pair tend to evidence general developmental changes in their drawings at about the same age and at about the same age as non-twins (Fig. 1). However, on measures of expan-

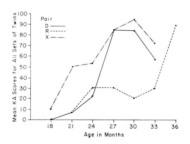

Fig. 1. Mean developmental change in graphic expression of twin pairs on Kellogg scale.

siveness, pressure, and constriction-freedom, no similar patterns were apparent. Later data will include results of the Bender Motor Gestalt Test and Goodenough Draw-A-Man Test.

Dales, on the basis of 164 Gesell Development Schedules administered to eight pairs of probably monozygotic twins and six pairs of probably dizygotic twins found no significant D. Q. differences when analyzed by zygosity groups. As mentioned earlier, zygosity determination by bloodtype testing is to be done at the end of the data collecting period. However, at all ages tested, twins as a group evidenced considerable developmental lag varying from an average of .3 months to 3 months in motor development and from 1.3 months to 6 months in language development when compared with established norms (Fig. 2). No significant intra-pair differences were found between the developmental status of first-born and second-born twins, whether the twins were monozygotic or dizygotic or as a combined group, using the Wilcoxon

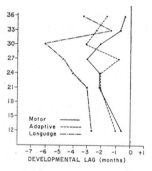

FIG. 2. Mean developmental lag in motor, adaptive, and language development of 14 pairs of twins.

Matched-Pairs Signed-Ranks Test (Fig. 3). There was a significant difference in favor of motor development when each twin was matched with self on language and motor development at 12, 15, and 21 months of age. However, the difference was not significant at ages, 24, 33, and 36 months. It must be kept in mind that the Gesell schedules are only a gross index of language development and that there is considerable over-all developmental lag in both areas as previously indicated.

Connor and J. Walters, using a 1-minute time sampling technique to appraise social interaction, indicate that twins more nearly resemble each other in their total pattern of social interaction than they resemble their singleton controls. Differences found appear to be a function of age, sex, and developmental differences rather than reported zygosity. Since this part of the research requires a group setting only a limited number of social interaction observations have been made.

Since Stanford-Binet tests of intelligence are not given prior to age 3, only sixteen reported identical-pair and eight fraternal-pair tests were available. Mean intra-pair IQ differences were 4.2 for monozygotic pairs and 6.35 for dizygotic pairs (Witherspoon, 1963). The research of Hammond and Skipper (1963) related to achievement begins at the middle of the first academic year of attendance in nursery school. Therefore, not enough data were available at this time to be included.

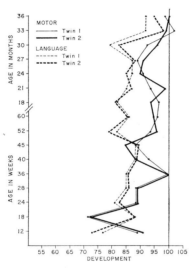

FIG. 3. Mean language and motor development of first (Twin 1) and second (Twin 2) born twins.

A highly significant observation made independently by Mooty and Schartz (1963), Schendel (1963), and Simons (1963) is the fact that the onset of early behavioral responses of twins, regardless of zygosity, appears developmentally at about the same time as for singletons. With the passage of time, however, singletons tend to make more use of and/or to perfect early graphic expression, language, and music responses while twins do not seem to. Such evidence, if supported by more extensive data and further research, would support the contention of Luria and Yudovich (1959) that a fruitful area of research lies in "a study of the child's mental processes as a product of his intercommunication with the environment, as the acquisition of common experience transmitted by speech . . ." It would, therefore, seem highly likely that the often reported speech, social, and mental retardation of twins could be due to a unique environmental circumstance, namely, the "twin-situation." This uniqueness may be even more pronounced for monozygotic twins than for fraternal twins which would account, at least partially, for the greater "alikeness" of their behavior as indicated by the results of previous research.

ACKNOWLEDGMENT

The author wishes to thank Ruth Connor, Ruth J. Dales, Sarah Lou Hammond, Mary E. Mooty, L. L. Schendel, Julia B. Schwartz, Dora Sikes Skipper, C. Etta Walters, and James Walters, all members of the faculty of Florida State University, for permission to discuss their unpublished work.

REFERENCES

Burlingham, D. T. Twins: Observations of environmental influences of their development. In *Psychoanalytic study of the child,* Vol. II. New York: Intern. Univer. Press, 1947. Pp. 61–73.

Burlingham, D. T. The relationship of twins to each other. In *Psychoanalytic study of the child,* Vol. III. New York: Intern. Univer. Press, 1949. Pp. 57–72.

Burlingham, D. T. *Twins: A Study of Three Pairs of Identical Twins.* New York: Intern. Univ. Press, 1952.

Burt, C. The inheritance of mental ability. *Amer. Psychologist,* 1958, **13,** 1–5.

Byrd, E., & Witherspoon, R. L. Responses of preschool children to the Childrens' Apperception Test. *Child Develpm.,* 1954, **25,** 35–44.

Davis, E. A. *The development of linguistic skills in twins, singletons with siblings, and only children from age five to ten years.* Minneapolis, Univer. Minnesota Press, 1937.

Day, E. J. The development of language in twins: I. A comparison of twins and single children. *Child Develpm.,* 1932, **3,** 179–199.

Dennis, W. The effect of restricted practice upon the reaching, sitting, and standing of two infants. *Pedag. Seminary & J. Genet. Psychol.,* 1935, **47,** 17–32.

Dennis, W. Infant development under conditions of restricted practice and minimum social stimulation. *Pedag. Seminary & J. Genet. Psychol.,* 1938, **53,** 149–157.

Dennis, W. Infant development under restricted practice and minimum social stimulation. *Genet. Psychol. Monogr.*, 1941, **23**, 143–191.
Eysenck, H. J., & Prell, D. B. The inheritance of neuroticism: An experimental study. *J. Ment. Sci.*, 1951, **97**, 441–465.
Hammond, S. L., & Skipper, D. S. Achievement in twins—A progress report. 1963.
Jones, H. E. Perceived differences among twins. *Eugen. Quart.*, 1955, **2**, 98–102.
Karpman, B. A psychoanalytic study of a fraternal twin. *Amer. J. Orthopsychiat.*, 1951, **21**, 735–755.
Kent, E. A study of maladjusted twins, *Smith College studies in social work.* Pp. 63–77.
Luria, A. R., & Yudovich, F. *Speech and the development of mental processes in the child.* London: Staples Press, 1959.
McCarthy, D. Organismic interpretation of infant vocalizations. *Amer. Psychol.*, **4**, 247.
McCarthy, D. Language development in children. In L. Carmichael (Ed.), *Manual of child psychology.* New York: Wiley, 1954. Pp. 492–650.
Mooty, M. E., & Schwartz, J. B. Graphic expression in twins—a progress report. 1963.
Mowrer, E. R., 1954. Some factors in the affectional adjustment of twins. *Amer. Sociol. Rev.*, 1954, **19**, 468–471.
Newman, H. H., Freeman, F. N., & Holzinger, K. J. *Twins: A study of heredity and environment.* Chicago: Univer. Chicago Press, 1937.
Schendel, L. I. Language development in twins—a progress report. 1963.
Simons, G. Comparisons of incipient music responses among very young twins and singletons. Unpublished doctoral dissertation, Florida State Univer. Graduate School, Tallahassee, 1962.
Skodak, M., & Skeels, H. M. A final follow-up study of one hundred adopted children. *J. Genet. Psychol.*, 1949, **75**, 85–125.
Shields, J. Personality differences and neurotic traits in normal twin school children: A study in psychiatric genetics. *Eugenics Rev.*, 1953, **45**, 213–246.
Slater, E. *An investigation into psychotic and neurotic twins.* London: Univer. London Press, 1951.
Slater, E. Psychiatry. In A. Sorsby (Ed.), *Clinical genetics.* St. Louis: Mosby, 1953. Pp. 332–349.
Templin, M. C. *Certain language skills in children, their development and relationships.* Minneapolis: Univer. Minnesota Press, 1957.
Witherspoon, R. L. Mental development in twins—a progress report, 1963.

A Design for the Study of the Inheritance of Normal Mental Traits[1]

Ruth Guttman

Introduction

Estimating the role of heredity in the determination of the large majority of human traits has encountered at least two major difficulties: (a) eliminating the effect of the environment and (b) virtual impossibility of conducting breeding experiments. With respect to normal mental traits, there is the additional difficulty of defining the specific behavioral characters which constitute stable and measurable human phenotypes over varied population samples, in spite of differences in cultural and educational backgrounds.

The proposed investigation is to explore a new kind of emphasis for gathering and analyzing data that may help overcome these difficulties. Past studies of human groups have concentrated largely on differences between *averages* or *proportions*. Some subpopulations are known to be taller on the average than others, or to have a higher frequency of a certain blood type than others. Most human traits are highly complex: analyses of their heritability and modes of inheritance by use of averages and proportions are often impossible. The research proposed here aims to overcome some of these difficulties and to provide a new method for examining profiles of traits in human populations.

Perhaps one of the reasons why so little is as yet known about the inherited mental potential of the individual is that studies on human abilities have hitherto been devoted largely to analyses of differences between averages of subpopulations. This has been

---

[1] The outline of this design was prepared in collaboration with Louis Guttman, Chaim Sheba, Marassa Bat-Miriam and Avinoam Adam.

true of ethnic and racial studies in the United States and elsewhere. Educators in Israel, too, have shown that there are differences in arithmetic means on tests of various mental abilities, as well as of school achievement, among the various ethnic groups.

The biological significance of such differences and the role of environment on their production remain relatively unknown. Light may be thrown on these problems by going beyond averages and studying the *interrelations,* or covariation, of such traits. Of particular interest is the question whether the same pattern of relations exists for each ethnic group. It is possible for a pattern of covariation to be invariant across ethnic groups, even though these groups may differ for each trait separately.

Cross-cultural invariance in two groups was reported by Vandenberg (1959) in a factor analytic study. Vandenberg compared the results of twenty identical tests: the first set administered in 1938 by Thurstone to American College students, the second by himself in 1954 to 92 Chinese students attending 7 American Universities.

Stability of a pattern of intercorrelations has already been shown in a previous project for scholastic achievement in 14-year-olds in Israel (Guttman & Guttman, 1963).[2] The achievements on national stipend examinations of Israeli eighth grade elementary school pupils were examined by means of a multivariate correlational analysis. 13,031 pupils out of 13,500 who completed 8th grade in 1955 took the tests. The proportion of boys and girls was almost exactly 1 : 1. Sixty-one percent of the pupils were Israeli-born. Of the immigrant children, about 50% came from Oriental countries; while 20% of the Israeli-born children had parents who immigrated from the Orient.

Table I gives the subpopulations studied according to country of origin of father, mother and pupil.

A further subdivision of the pupils with respect to the number of years they had been in Israel at the time of the test, as well as to their sex, and the type of school they attended, provided some 40 groups for whom mean scores and interrelation between scores could be calculated.

Table II shows significant differences in achievement of the different groups in the six test areas. It comes as no great surprise

[2] The research reported in this study was supported by Grant M-5159 from the U. S. National Institute of Mental Health.

that Israel-born children of European parentage generally did better on these tests. The possible causes for such differences have been discussed at length before and provide no clear information on a possible biological basis for individual differences in scholastic and other forms of achievement.

Our next step brings us, we hope, somewhat closer to such an understanding. In analyzing the intercorrelations between the scores of the six tests, we found that a definite order, namely a *simplex* pattern existed among them. That is, the correlation coeffi-

TABLE I

COUNTRY OF ORIGIN OF CHILDREN TESTED IN THE 1955 STIPEND EXAMINATIONS AND COUNTRY OF ORIGIN OF THEIR MOTHERS AND FATHERS

Pupil	Father	Mother
Israel	Israel	Israel
Israel	Iraq	Iraq
Iraq	Iraq	Iraq
Israel	Persia	Persia
Persia	Persia	Persia
Israel	Yemen	Israel
Israel	Yemen	Yemen
Yemen	Yemen	Yemen
Israel	Near East (Syria, Lebanon, Turkey)	Near East
Near East	Near East	Near East
Israel	Morocco	Morocco
Morocco	Morocco	Morocco
Israel	Other North African countries (Egypt, Tripoli, Algeria)	North Africa
North Africa	North Africa	North Africa
Israel	Southern Europe (Italy, Greece, Yugoslavia, etc.)	Southern Europe
Southern Europe	Southern Europe	Southern Europe
Israel	Israel	Europe (Central, Northern, Eastern, North America)
Israel	Europe and North America	Israel
Israel	Europe and North America	Europe and North America
{Europe {North America	{Europe {North America	{Europe {North America

TABLE II

MEAN SCORES OF DIFFERENT GROUPS OF CHILDREN ON THE 1955 EIGHTH GRADE STIPEND EXAMINATIONS

Group			Mean Score (percent)							
Country of Origin			Number Tested	Blocks	Arith- metic	Vocab- ulary	Civics	History	Bible	All subjects
Father	Mother	Pupil[a]								
Iraq	Iraq	Iraq (1–3)	181	49.31	60.17	51.37	59.67	52.74	65.26	56.47
Iraq	Iraq	Iraq (4)	470	45.71	56.73	50.76	59.00	46.67	63.11	53.66
Iraq	Iraq	Iraq (5)	163	45.26	55.28	54.82	61.37	50.26	61.15	54.69
Iraq	Iraq	Iraq (6) + Israel	178	45.94	59.56	59.63	59.67	54.63	72.11	58.59
Yemen	Yemen	Yemen (3–5)	212	34.48	49.76	51.52	53.26	46.81	69.70	50.92
Yemen	Yemen	Yemen (6)	125	40.37	50.91	53.28	57.15	61.44	72.70	55.98
Yemen	Yemen	Yemen (7+)	114	44.88	52.90	57.26	58.33	47.19	73.15	55.62
Yemen	Yemen	Israel	381	43.04	49.46	58.03	63.63	49.48	68.74	55.40
Persia	Persia	Persia (2–5)	112	38.23	48.69	45.56	52.81	49.15	77.47	51.99
Persia	Persia	Persia (6+) + Israel	112	41.74	52.59	57.01	56.63	51.33	72.48	55.30
Morocco	Morocco	Morocco (1–5)	112	49.61	51.75	58.87	58.85	48.84	52.59	53.42
N. Africa	N. Africa	N. Africa (1–5)	197	36.25	57.80	52.21	58.56	50.22	65.41	53.41
N. Africa	N. Africa	N. Africa (6+) Israel	116	48.37	51.01	58.49	60.41	52.37	69.93	56.73
S. Europe	S. Europe	S. Europe (2–5)	108	50.10	60.55	54.15	71.52	57.00	64.74	59.68
S. Europe	S. Europe	S. Europe (6)	269	49.99	61.39	61.24	64.67	58.41	69.44	60.69
S. Europe	S. Europe	S. Europe (7–8)	124	49.61	59.48	61.84	64.37	54.78	66.74	59.47
S. Europe	S. Europe	Israel	210	51.06	60.17	61.69	61.52	53.11	67.41	59.16
Eur. + N. America	Eur. + N. America	Eur. + America (1–3)	130	61.69	64.45	57.95	67.78	61.37	67.77	63.50
Eur. + N. America	Eur. + N. America	Eur. + America (4)	396	50.40	63.15	59.02	64.30	52.22	67.56	59.44

Group		Number Tested	Blocks	Arithmetic	Vocabulary	Civics	History	Bible	All subjects
Eur. + N. America	Eur. + N. America (5)	383	54.74	63.53	64.07	67.19	61.70	72.56	63.97
Eur. + N. America	Eur. + N. America (6)	482	51.44	62.00	64.76	64.22	63.11	73.67	63.20
Eur. + N. America	Eur. + N. America (7)	268	51.22	61.69	66.51	67.25	63.74	74.26	64.11
Eur. + N. America	Eur. + N. America (8)	171	56.04	63.99	68.58	68.15	61.67	63.93	63.73
Eur. + N. America	Israel	4501	57.88	67.28	72.56	79.44	66.93	76.19	69.05
Eur. + Israel	Eur. + N. America	280	55.43	64.15	71.33	69.07	63.85	74.11	66.32
Eur. + N. America	Israel	313	56.81	67.63	71.79	71.74	65.70	78.63	68.72
Eur. + Israel	Israel	524	48.39	59.48	64.68	53.19	55.52	70.15	58.57

Group	Number Tested	Blocks	Arithmetic	Vocabulary	Civics	History	Bible	All subjects
					Mean Score (percent)			
Religious schools	3449	45.64	54.56	55.15	55.22	54.87	71.55	56.17
Non-religious schools	9550	46.23	62.91	63.14	63.44	55.17	71.30	60.37
Boys—born in Israel	3843	58.66	67.03	63.47	63.15	63.24	71.76	64.55
Boys—immigrant	2598	51.50	61.10	59.30	63.32	55.11	63.69	59.00
Girls—born in Israel	3903	43.53	60.76	55.63	63.06	54.91	72.13	58.34
Girls—immigrant	2439	44.24	55.41	56.81	54.93	46.80	63.17	53.56

[a] In parentheses = Number of years in Israel

cients formed a definite pattern, in which the first item, in this case blocks, always had the highest correlation with arithmetic, the next highest with vocabulary, and so on. This pattern (see Tables III–XII) holds, regardless of the background of the child, the country of origin of his parents, whether he went to a religious or secular school, regardless of sex: the order of the intercorrelations always remains the same.

We have not attempted to explain the nature of the order found. Our hypothesis is that when we find a definite pattern that is constant over different groups, a biological basis for such a pattern may exist. Invariance of pattern may be considered a necessary, though not a sufficient, condition for an underlying biological basis of the pattern.

A somewhat similar battery of tests is administered to eighth grade pupils every year. The number of tests and their content is decided upon afresh each time. In 1959 and 1960 data were again available on ethnic background of the children and their parents. We took advantage of this fact to calculate and analyze the correlation matrices of the ten tests for over 35,000 pupils. These we subdivided into about twenty groups according to sex, area of origin, and number of years in the Israel school system. Again mean performance was highly variable among the different subcultures. A preliminary analysis indicates again cross-cultural stability of the structural pattern of the correlations between the ten test areas. This time the pattern is not essentially one-dimensional like the simplex of the six 1955 tests, but rather a three-dimensional *radex*. The configural analysis is being made possible by a new nonmetric technique. The technique is described by Guttman (1965). The results on the cross-cultural stability on these 1959/1960 radexes are being prepared for publication.

It is the knowledge and encouragement gained from this study that has led us to the design of the large scale research project which I should like to outline here.

This project is intended to probe more deeply into the problem of transmission of mental traits, not only by studying *two generations*, but also by studying simultaneously a larger number of traits some of which have a known genetic basis. It will examine the covariation among mental traits, as well as between mental, morphological, and physiological.

TABLE III
Intercorrelations of Scores on the Six Areas of the 1955 Stipend Examinations According to Type of School

	Religious Schools						Non-religious Schools					
	1	2	3	4	5	6	1	2	3	4	5	6
1. Blocks	—	.87	.60	.54	.51	.44	—	.66	.62	.51	.46	.44
2. Arithmetic	.87	—	.74	.63	.65	.60	.66	—	.73	.59	.60	.59
3. Vocabulary	.60	.74	—	.66	.72	.67	.62	.73	—	.63	.62	.66
4. Civics	.54	.63	.66	—	.86	.74	.51	.59	.63	—	.76	.72
5. History	.51	.65	.72	.86	—	.83	.46	.60	.62	.76	—	.79
6. Bible	.44	.60	.67	.74	.83	—	.44	.59	.66	.72	.79	—

TABLE IV

INTERCORRELATIONS OF SCORES ACCORDING TO SEX AND COUNTRY OF BIRTH OF PUPIL

	A. Boys—Israeli						B. Boys—Immigrant					
	1	2	3	4	5	6	1	2	3	4	5	6
1. Blocks	—	.66	.61	.44	.40	.35	—	.70	.67	.52	.47	.42
2. Arithmetic	.66	—	.74	.54	.55	.52	.70	—	.71	.60	.60	.56
3. Vocabulary	.61	.74	—	.58	.60	.54	.67	.71	—	.66	.55	.63
4. Civics	.44	.54	.58	—	.71	.37	.52	.60	.66	—	.78	.70
5. History	.40	.55	.60	.71	—	.75	.47	.60	.55	.78	—	.79
6. Bible	.35	.52	.54	.37	.75	—	.42	.56	.63	.70	.79	—

	C. Girls—Israeli						D. Girls—Immigrant					
	1	2	3	4	5	6	1	2	3	4	5	6
1. Blocks	—	.66	.61	.44	.42	.35	—	.67	.58	.49	.46	.40
2. Arithmetic	.66	—	.73	.59	.59	.51	.67	—	.73	.67	.61	.59
3. Vocabulary	.61	.73	—	.63	.61	.52	.58	.73	—	.66	.66	.62
4. Civics	.44	.59	.63	—	.76	.63	.49	.67	.66	—	.75	.76
5. History	.42	.59	.61	.76	—	.71	.46	.61	.66	.75	—	.80
6. Bible	.35	.51	.52	.63	.71	—	.40	.59	.62	.76	.80	—

TABLE V

INTERCORRELATIONS OF SCORES OF EIGHTH GRADERS, WHOSE PARENTS WERE BOTH BORN IN IRAQ[a]

A. 1–3 Years

	1	2	3	4	5	6
1. Blocks	—	.710	.639	.550	.466	.435
2. Arithmetic	.710	—	.732	.595	.600	.566
3. Vocabulary	.639	.732	—	.680	.680	.676
4. Civics	.550	.595	.680	—	.780	.769
5. History	.466	.600	.680	.780	—	.779
6. Bible	.435	.566	.676	.769	.779	—

B. 4 Years

	1	2	3	4	5	6
1. Blocks	—	.662	.639	.519	.494	.499
2. Arithmetic	.662	—	.731	.597	.602	.604
3. Vocabulary	.639	.731	—	.676	.638	.654
4. Civics	.519	.597	.676	—	.809	.785
5. History	.494	.602	.638	.809	—	.838
6. Bible	.499	.604	.654	.785	.838	—

C. 5 Years

	1	2	3	4	5	6
1. Blocks	—	.612	.611	.506	.466	.398
2. Arithmetic	.612	—	.755	.625	.701	.656
3. Vocabulary	.611	.755	—	.750	.680	.642
4. Civics	.506	.625	.750	—	.764	.626
5. History	.466	.701	.680	.764	—	.757
6. Bible	.398	.656	.642	.626	.757	—

D. Over 6 years and born in Israel

	1	2	3	4	5	6
1. Blocks	—	.638	.592	.530	.471	.378
2. Arithmetic	.638	—	.734	.608	.566	.498
3. Vocabulary	.592	.734	—	.624	.590	.576
4. Civics	.530	.608	.624	—	.733	.703
5. History	.471	.566	.590	.733	—	.735
6. Bible	.378	.498	.576	.703	.735	—

[a] A, B, C, D denote length of time pupil has been in Israel.

TABLE VI
INTERCORRELATIONS OF SCORES OF EIGHTH GRADERS, WHOSE PARENTS WERE BOTH BORN IN YEMEN

A. 3–5 Years

	1	2	3	4	5	6
1. Blocks	—	.625	.540	.568	.456	.427
2. Arithmetic	.625	—	.679	.631	.653	.606
3. Vocabulary	.540	.679	—	.588	.559	.685
4. Civics	.568	.631	.588	—	.773	.871
5. History	.456	.653	.559	.773	—	.729
6. Bible	.427	.606	.685	.871	.729	—

B. 6 Years

	1	2	3	4	5	6
1. Blocks	—	.734	.650	.629	.450	.528
2. Arithmetic	.734	—	.750	.654	.628	.648
3. Vocabulary	.650	.750	—	.685	.696	.694
4. Civics	.629	.654	.685	—	.827	.812
5. History	.450	.628	.696	.827	—	.846
6. Bible	.528	.648	.694	.812	.846	—

C. 7 or more years

	1	2	3	4	5	6
1. Blocks	—	.569	.567	.445	.422	.370
2. Arithmetic	.569	—	.695	.565	.538	.553
3. Vocabulary	.567	.695	—	.617	.536	.607
4. Civics	.445	.565	.617	—	.777	.688
5. History	.422	.538	.536	.777	—	.798
6. Bible	.370	.553	.607	.688	.798	—

D. Born in Israel

	1	2	3	4	5	6
1. Blocks	—	.628	.570	.325	.285	.322
2. Arithmetic	.628	—	.720	.467	.479	.540
3. Vocabulary	.570	.720	—	.523	.557	.560
4. Civics	.325	.467	.523	—	.784	.733
5. History	.285	.479	.557	.784	—	.783
6. Bible	.322	.540	.560	.733	.783	—

TABLE VII
Intercorrelations of Scores of Eighth Graders, Whose Parents Were Both Born in Persia

	A. 2–5 Years						B. From 6 years to Israel born					
	1	2	3	4	5	6	1	2	3	4	5	6
1. Blocks	—	.673	.563	.554	.477	.443	—	.618	.458	.422	.446	.329
2. Arithmetic	.673	—	.859	.593	.580	.658	.618	—	.628	.514	.611	.443
3. Vocabulary	.563	.859	—	.665	.640	.608	.458	.628	—	.540	.582	.539
4. Civics	.554	.593	.665	—	.799	.632	.422	.514	.540	—	.778	.660
5. History	.477	.580	.640	.799	—	.786	.446	.611	.582	.778	—	.769
6. Bible	.443	.658	.608	.632	.786	—	.329	.443	.539	.660	.769	—

TABLE VIII

INTERCORRELATIONS OF SCORES OF EIGHTH GRADERS, WHOSE PARENTS WERE BOTH BORN IN MOROCCO (A) AND OTHER NORTH AFRICAN COUNTRIES (B AND C)

A. Less than 5 years

	1	2	3	4	5	6
1. Blocks	—	.665	.597	.530	.499	.351
2. Arithmetic	.665	—	.528	.638	.635	.555
3. Vocabulary	.597	.528	—	.700	.734	.634
4. Civics	.530	.638	.700	—	.724	.672
5. History	.499	.635	.734	.724	—	.795
6. Bible	.351	.555	.634	.672	.795	—

B. 1–5 Years

	1	2	3	4	5	6
1. Blocks	—	.698	.525	.436	.446	.391
2. Arithmetic	.698	—	.690	.675	.583	.550
3. Vocabulary	.525	.690	—	.701	.720	.396
4. Civics	.436	.675	.701	—	.860	.752
5. History	.446	.583	.720	.860	—	.757
6. Bible	.391	.550	.396	.752	.757	—

C. Over 6 Years and Israel-born

	1	2	3	4	5	6
1. Blocks	—	.635	.618	.502	.518	.453
2. Arithmetic	.635	—	.707	.564	.520	.489
3. Vocabulary	.618	.707	—	.511	.555	.516
4. Civics	.502	.564	.511	—	.754	.693
5. History	.518	.520	.555	.754	—	.813
6. Bible	.453	.489	.516	.693	.813	—

TABLE IX

INTERCORRELATIONS OF SCORES OF EIGHTH GRADERS, WHOSE PARENTS WERE BOTH BORN IN SOUTHERN EUROPEAN COUNTRIES (ITALY, GREECE, BULGARIA, YUGOSLAVIA, SPAIN, EUROPEAN TURKEY)

A. 2–5 Years

	1	2	3	4	5	6
1. Blocks	—	.690	.666	.530	.504	.441
2. Arithmetic	.690	—	.744	.591	.643	.527
3. Vocabulary	.666	.744	—	.659	.667	.541
4. Civics	.530	.591	.659	—	.842	.715
5. History	.504	.643	.667	.842	—	.760
6. Bible	.441	.527	.541	.715	.760	—

B. 6 Years

	1	2	3	4	5	6
1. Blocks	—	.677	.580	.539	.482	.468
2. Arithmetic	.677	—	.705	.593	.610	.603
3. Vocabulary	.580	.705	—	.608	.642	.691
4. Civics	.539	.593	.608	—	.772	.772
5. History	.482	.610	.642	.772	—	.839
6. Bible	.468	.603	.691	.772	.839	—

C. 7–8 Years

	1	2	3	4	5	6
1. Blocks	—	.665	.625	.546	.404	.349
2. Arithmetic	.665	—	.806	.665	.674	.636
3. Vocabulary	.625	.806	—	.667	.698	.655
4. Civics	.546	.665	.667	—	.735	.712
5. History	.404	.674	.698	.735	—	.806
6. Bible	.349	.636	.655	.712	.806	—

D. Born in Israel

	1	2	3	4	5	6
1. Blocks	—	.678	.615	.376	.381	.332
2. Arithmetic	.678	—	.706	.524	.546	.550
3. Vocabulary	.615	.706	—	.603	.596	.611
4. Civics	.376	.524	.603	—	.736	.706
5. History	.381	.546	.596	.736	—	.777
6. Bible	.332	.550	.611	.706	.777	—

TABLE X

INTERCORRELATIONS OF SCORES OF EIGHTH GRADERS, WHOSE PARENTS WERE BOTH BORN IN CENTRAL, EASTERN, AND WESTERN EUROPE AND NORTH AMERICA

A. 1–3 Years

	1	2	3	4	5	6
1. Blocks	—	.706	.373	.465	.460	.289
2. Arithmetic	.706	—	.508	.460	.519	.422
3. Vocabulary	.373	.508	—	.527	.553	.507
4. Civics	.465	.460	.527	—	.726	.660
5. History	.460	.519	.553	.726	—	.797
6. Bible	.289	.422	.507	.660	.797	—

B. 4 Years

	1	2	3	4	5	6
1. Blocks	—	.689	.596	.528	.471	.406
2. Arithmetic	.689	—	.691	.626	.467	.574
3. Vocabulary	.596	.691	—	.600	.583	.483
4. Civics	.528	.626	.600	—	.777	.726
5. History	.471	.467	.583	.777	—	.793
6. Bible	.406	.574	.483	.726	.793	—

C. 5 Years

	1	2	3	4	5	6
1. Blocks	—	.694	.599	.459	.434	.432
2. Arithmetic	.694	—	.742	.552	.582	.549
3. Vocabulary	.599	.742	—	.378	.619	.570
4. Civics	.459	.552	.378	—	.771	.690
5. History	.434	.582	.619	.771	—	.758
6. Bible	.432	.549	.570	.690	.758	—

D. 6 Years

	1	2	3	4	5	6
1. Blocks	—	.667	.561	.473	.372	.374
2. Arithmetic	.667	—	.766	.582	.539	.539
3. Vocabulary	.561	.766	—	.617	.604	.616
4. Civics	.473	.582	.617	—	.765	.807
5. History	.372	.539	.604	.765	—	.799
6. Bible	.374	.539	.616	.807	.799	—

E. 7 Years

	1	2	3	4	5	6
1. Blocks	—	.656	.603	.496	.420	.448
2. Arithmetic	.656	—	.733	.572	.588	.585
3. Vocabulary	.603	.733	—	.910	.639	.574
4. Civics	.496	.572	.910	—	.808	.704
5. History	.420	.588	.639	.808	—	.765
6. Bible	.448	.585	.574	.704	.765	—

F. 8 or More Years

	1	2	3	4	5	6
1. Blocks	—	.750	.720	.458	.473	.483
2. Arithmetic	.750	—	.646	.579	.617	.611
3. Vocabulary	.720	.646	—	.575	.597	.633
4. Civics	.458	.579	.575	—	.742	.698
5. History	.473	.617	.597	.742	—	.804
6. Bible	.483	.611	.633	.698	.804	—

G. Born in Israel

	1	2	3	4	5	6
1. Blocks	—	.710	.571	.423	.375	.342
2. Arithmetic	.710	—	.712	.518	.532	.485
3. Vocabulary	.571	.712	—	.519	.541	.522
4. Civics	.423	.518	.519	—	.691	.621
5. History	.375	.532	.541	.691	—	.757
6. Bible	.342	.485	.522	.621	.757	—

TABLE XI
Intercorrelations of Scores of Eighth Graders, One Parent Born in Europe or North America, the Other Parent and the Child in Israel

	A. Father Born in Europe						B. Mother Born in Europe					
	1	2	3	4	5	6	1	2	3	4	5	6
1. Blocks	—	.625	.606	.449	.333	.212	—	.693	.628	.460	.401	.329
2. Arithmetic	.625	—	.740	.552	.521	.465	.693	—	.734	.554	.533	.471
3. Vocabulary	.606	.740	—	.513	.538	.460	.628	.734	—	.564	.564	.552
4. Civics	.449	.552	.513	—	.746	.602	.460	.554	.564	—	.744	.659
5. History	.333	.521	.538	.746	—	.754	.401	.533	.564	.744	—	.756
6. Bible	.212	.465	.460	.602	.754	—	.329	.471	.552	.659	.756	—

To aid further in the study of intergenerational transmission, two sibs of *different ages* will be used in each family, one of 18 years and the other as close to 14 years as possible. This will enable studying not only the correlations between sibs, but also parent-child correlations at different ages of the child.

TABLE XII
BOTH PARENTS AND CHILD BORN IN ISRAEL

	1	2	3	4	5	6
1. Blocks	—	.684	.608	.577	.491	.455
2. Arithmetic	.684	—	.734	.621	.594	.603
3. Vocabulary	.608	.734	—	.654	.659	.642
4. Civics	.577	.621	.654	—	.773	.680
5. History	.491	.594	.659	.773	—	.775
6. Bible	.455	.603	.642	.680	.775	—

Our purpose, then, is to focus on the *inheritance of covariation* rather than on separate traits, on a family basis.

Methodology and Procedure

THE SAMPLE

Upon reaching their eighteenth birthdays, all boys and girls who were born in Israel or are permanent residents or the children of permanent residents, are required to appear at their local Draft Board for a physical examination prior to decision whether they will be liable for military service. Our plan is to obtain our information from youths (male and female) who have both parents—and at least one sib between the ages 14–17—accessible for testing. About 25,000 eighteen-year-olds were recorded in the 1960 census of Israel. Of these, approximately 5000 families (subject, mother, father, sib), or 20,000 individuals will be chosen for study. Whenever grandparents and aunts and uncles live in the same community (which is quite common in oriental communities) note will be taken of the fact for possible subsequent analysis of an additional generation and construction of pedigrees. The countries of origin of the parents will include at least Yemen, Morocco, Iraq, Kurdistan, Algeria, Central and Western Europe, North and

South America, South Africa and marriages between members of these groups.

The 18-year-olds as well as the younger high-school age sibs will be chosen regardless of the amount of formal schooling that they or their parents had. The use of youths over 14 is suggested so that the anthropometric measurements will be most useful. Also, the same mental tests can generally be used on the parents.

THE TRAITS

The characters to be studied will be of two varieties, mental and morphological-physiological. The final list will be decided upon after pretesting in the field the traits given in Table XIII.

Of the mental traits (group B in the list), the first six are pictorial paper and pencil tests which can be administered to illiterates as well as to literates. The seventh test, using actual marbles and bags, is being used successfully in a current project conducted by us jointly with a team of the Hebrew University Dental School in villages of immigrants to Israel of North African and Near Eastern origins, among adults and children of all ages, illiterate and literate. The verbal analytical tests will be adapted from those in use by the Israel Defence Army, which are of the analogies and related types, in the light of the Army's experience with the ethnic groups involved; at least one of the tests will be administrable orally to illiterates.

The emphasis in these tests is on analytical abilities of several varieties, rather than on achievement, in order to minimize to some extent the influence of environment, especially of educational background. Environmental influences will of course remain even on the tests of analytical abilities. While such influences will certainly affect the *average* scores, they need not affect the *intercorrelations*, which is precisely the point to be investigated here. The mean scores for the adults can certainly be expected to be different from those of their children, just as they can be expected to differ from ethnic group to ethnic group. The question is: *is there any constancy of intercorrelation, within families, across ethnic groups?*

The nonverbal analytical tests will be a battery adapted from the well-known Raven's progressive matrices. The Raven's battery has been adapted for use in the Israel Defence Army. Its five parts have been found to have the simplex pattern of inter-

STUDY OF INHERITANCE 217

correlations over all the ethnic groups in the Army (Gabriel, 1954), but no study has been made yet within and between ethnic groups.

TABLE XIII
Suggested Traits for the
Multivariate Analysis

A. Morphological-Physiological and Sensory

1. Stature
2. Sitting height
3. Head length
4. Head breadth
5. Length of forearm
6. Length of leg
7. Blood groups (as many as possible)
8. Red-green colorblindness
9. Hand clapping
10. Arm folding
11. Eye color
12. Handedness
13. Relation between index and ring fingers
14. PTC (ability to taste phenylthiocarbamide)
15. Fingerprint patterns
16. Palm prints (D- line exit)
17. Relative lengths of toes

B. Mental

1. Non-verbal analytical ability
2. Non-verbal mechanical ability
3. Block counting (spatial-numerical perception)
4. Porteus mazes
5. Imbedded figures
6. Perception of Necker cube
7. Estimating number of marbles in transparent bags
8. Verbal analytical ability

The nonverbal mechanical ability test is one developed at the Israel Institute of Applied Social Research for vocational guidance among different ethnic groups and is especially designed to require no formal knowledge or training in physics, nor special acquaintance with machinery. Tests numbered 3–7 in list B are standard psychological tests often used in cross-cultural situations.

All the tests can be administered to the adults and children of the age groups to be studied. Actually, since only correlations are to be studied, it is not essential that exactly the same tests be used for the two generations, but there is some advantage in keeping the tests identical.

The final traits will be selected to fulfill a number of requirements. (1) The total time of examination should not take much more than 2 hours per individual. (2) The physical tests must not be unpleasant or painful (this is particularly important in the oriental groups who often object to blood tests; if such is the case, the taking of blood samples will need to be postponed until after the rest of the information has been obtained). (3) The answers to all questions must fall into a few, easily classifiable categories. (4) Some of the traits will serve as definite genetic markers (PTC, color blindness, blood groups). (5) Some of the anthropometric measurements are known from twin studies (see Vandenberg, 1962) to have a strong genetic component. (6) A few of the characters may be expected to be correlated.

ANALYSIS OF THE DATA

Two major sources of covariation will be studied for these data: (1) within families and (2) between traits. Each 18-year-old— henceforth called "subject"—will have, for each trait, four familial observations: on himself, his father, his mother, and his sib. Each of these four observations for any trait can be correlated with each of the four of any trait. Schematically, all the possible pairwise relationships fall into the cells represented in Table XIV.

The four sections in the main diagonal are symmetric about the diagonal (A: father-father; E: mother-mother; H: subject-subject; J: sib-sib), and will show the interdependence of the traits *within individuals*. The subject-subject section has already been studied with respect to certain achievement tests for 14-year-olds, and found to be invariant in structure across ethnic groups in Israel. We shall now be able to study the possible constancy of pattern for further traits and for different age groups.

The father-mother sections (B and B′) will reveal the assortative mating patterns for the different ethnic groups. These need not at

STUDY OF INHERITANCE 219

all be invariant as far as our hypothesis is concerned. Should different types of assortative mating (preferences in relation to different traits) be found in different ethnic groups, they will have to be taken into account in the analyses of the other sections. Statistical analysis *within* families will help overcome environmental problems of differences in means between families.

When mother traits are held constant, our hypothesis is that the

TABLE XIV
PAIRWISE DEPENDENCIES AMONG TRAITS WITHIN FAMILIES[a]

	Trait	Father 1 2 3 ... 25	Mother 1 2 3 ... 25	Subject 1 2 3 ... 25	Sib 1 2 3 ... 25
Father	1 2 3 . . . 25	A	B'	C'	D'
Mother	1 2 3 . . . 25	B	E	F'	G'
Subject	1 2 3 . . . 25	C	F	H	I'
Sib	1 2 3 . . . 25	D	G	I	J

[a] See text for explanation of symbols.

father-subject (C and C') and the father-sib section (D and D') will show a pattern which is constant over ethnic groups. Similarly, when father traits are held constant, the mother-subject (F and F') and mother-sib section (G and G') should be ethnically invariant. Also, when father and mother traits are held constant, the subject-sib section (I and I') should be ethnically invariant.

Possible sex dependence will be studied by seeing whether different patterns exist for male and female subjects (sibs).

Since the 25 traits subdivide into two types, mental and morphological-physiological, partial correlations within the sections will be studied among the mental traits holding the morphological constant.

Statistical Problems of Regression Shapes

Holding constant parental and morphological traits would be simple were we dealing throughout with numerical variables with a multi-variate normal distribution. Then the relevant product-moment coefficients of correlation could be merely inserted into Table XIV, and all required partial correlations could be calculated therefrom by matrix algebra. However, a good part of the traits on our list above are qualitative (such as eye color or PTC) and linear regression theory will not always hold. Holding constant many of the traits may have to be done by direct sub-classifications, with concomitant losses of degrees of freedom. Therefore, typologies will be sought for the "holding constant" process, in order to lose as few degrees of freedom as possible. Examples of typologies that may be relevant include scalograms (Stouffer et al., 1950), and partly ordered sets as developed for fingerprints (Bat-Miriam & Guttman, 1963). Should the traits chosen for such a study belong to different linkage groups, this should manifest itself in the typologies, since one would expect statistical independence in the absence of linkage.

Analysis of the typologies and independencies must be performed between parents and offspring—and cross culturally—in order to help distinguish between linked and otherwise associated characters.

The studies proposed here may partially answer several questions:

(a) Taking a number of heterogeneous human traits—morphological, physiological, sensory, psychological—do *all or part* of these form a profile of statistical correlations which does not depend on age, sex, or ethnic origin? If such is found, we assume that the relationship of these traits has a biological basis common to the whole population and that a study of its nature will be justified.

(b) Supposing a certain pattern is found, how can we interpret its structure? Presently we have no idea of what the structure will be like. One possibility is that we may find patterns of statistical dependence within subsets of traits, with *statistical independence holding between these subsets*. Without immediately hypothesizing causal relationships for traits *within* such subsets, their clustering will nevertheless be able to serve as a base for further research into the relationships. This would be especially true if invariant patterns were to be found around any of the genetic markers.

(c) By using a considerable number of variables and large two-generation samples, the suggested analysis should be able to distinguish between associated traits and genetic linkage.

(d) It is hoped that behavioral traits which have stable interrelations both cross-culturally, intra-familiarly, and with known genetic markers, will help define behavioral phenotypes for which modes of genetic transmission can eventually be worked out.

REFERENCES

Bat-Miriam, M., & Guttman, L. A new approach to fingerprint analysis in population studies. *Proc. 2nd Int. Congr. Human Genet.*, Rome. 1963. Pp. 1481–1483.

Gabriel, R. K. The Simplex Structure of the Progressive Matrices Test. *Brit. J. Statist. Psychol.*, 1954, **7**, 9–14.

Guttman, L. A new approach to factor analysis: the radex. In P. F. Lazarsfeld (Ed.), *Mathematical thinking in the social sciences*. Glencoe, Ill.: Free Press, 1954. Pp. 238–348.

Guttman, L. A general nonmetric technique for finding the smallest euclidian space for a configuration of points. *Psychometrika*, (in press). 1965.

Guttman, R. A test for a biological basis for correlated abilities. In E. Goldschmidt (Ed.), *The genetics of migrant and isolate populations*. Williams and Wilkins, 1963. Pp. 338–339.

Guttman, R. & Guttman, L. Cross-cultural stability of a pattern of intercorrelations between test scores: A possible test for a biological basis. *Human Biol.*, 1963, **35**, 53–60.

Stouffer, S. A., Guttman, L., Suchman, E. A., Lazarsfeld, P. F., Star, S. A., & Clausen, J. A. *Measurement and prediction.* Princeton: Princeton Univer. Press, 1950.

Vandenberg, S. G. The primary mental abilities of Chinese students: A comparative study of the stability of a factor structure. *Ann. N. Y. Acad. Sci.,* 1959, **79,** 257–304.

Vandenberg, S. G. Heritability estimates of anatomical measurements: man. In P. L. Altman & D. S. Dittman (Eds.), *Growth.* Washington, D.C.: Federation Amer. Soc. Exp. Biol., 1962. Pp. 110–114.

Discussion

Cattell: May I ask a question? On the basic assumption, I'm not quite clear, perhaps you can help me. This seems to be stating that if any structure exists then it is genetically determined, but surely this is not enough because you could get structures in social data.

Guttman: No, let me give you an analogy to show what I mean. We have all got ten fingers, but in each individual they all have different lengths, yet the fact is that the relationship of each finger to every other one forms a pattern which is more or less invariant over mankind. Now does that mean that it is determined by a simple gene mechanism?

Cattell: You have chosen something that happens to be genetic. Once you take something else that is universal over mankind, an attitude to a mother, for example, an attitude of affection of a child to a mother your hypothesis would not hold.

Guttman: Should a pattern of attitudes such as you suggest be found to be invariant over a number of different cultures, I would certainly suspect those attitudes to be based on an underlying biological process common to mankind.

Genetics and Personality Theory[1]

Riley W. Gardner

During the past several decades, our understanding of human genetics has increased at a remarkable pace. As Ham (1963) points out, however, the explosive growth in genetic knowledge has thus far influenced conceptions of human personality in all too limited ways, the outstanding work of Kallmann (e.g., 1952), Benjamin (e.g., 1962), and others notwithstanding. To take but one example, the vast expansion of knowledge concerning the human psyche that began some 70 years ago with the dramatic discoveries by Sigmund Freud has not yet been viewed through the corrective lens provided by studies of the hereditary determination of the many facets of personality organization postulated on the basis of these discoveries. Although Freud (e.g., 1937) repeatedly referred to the potential importance of heredity to the relative strengths of defense mechanisms and to other aspects of ego organization, and although Hartmann (1939), in a now classic monograph, emphasized the importance of conflict-free aspects of ego organization (thus pointing directly to innate characteristics of mental apparatuses as determinants of ego functioning), the necessary genetic studies have not yet been done.

The dearth of genetic studies of variables critical for psychoanalytic theory may be attributable to the intense interest psychoanalytic discoveries stimulated in environmental determinants of personality development. No other group of findings has provided such clear and articulated evidence of the effects of the personalities of mothers, fathers, brothers, and sisters on the development of the human psyche. The critical importance of the early mother-child interaction, normal and abnormal sex differences in sequential pat-

[1] The current research described is supported in part by Research Grant M-5517 and a Research Career Development Award from the United States Public Health Service.

terns of relationship to mother and father, and a host of related facets of personality development were first outlined in the work of psychoanalysts with clinical patients. The dearth of companion genetic studies may also be attributable to the unique methods involved in psychoanalytic treatment. Because psychoanalysis and related forms of psychotherapy are undertaken on the basis of clinical rather than research criteria, and because they are conducted with individual patients, it is rare that even small groups of subjects appropriate to the testing of genetic hypotheses have been treated by these methods.

In view of the vast amount of work that remains to be done before the contributions of heredity to the aspects of personality organization delineated by psychoanalytic methods are better understood, let me outline briefly some potential contributions of genetic studies to four aspects of personality organization: (1) temperament; (2) mechanisms of defense; (3) intellectual capacities, known to be related to certain defenses; and (4) features of cognitive organization that Klein (1954) referred to as cognitive control principles. I shall also briefly mention possible relations between dimensions of temperament and activation level—an over-all pattern of metabolic reactivity to adaptive requirements which, Duffy (1962) indicates, is in part genetically determined. In pointing to some potential contributions of genetic studies to our understanding of such aspects of personality organization, I am, of course, using psychoanalytic theory primarily as an example. The same or related kinds of information are required for our understanding of variables employed by those who would formulate the problem of personality in different terms. In some cases, for example the work of Eysenck (e.g., Eysenck & Prell, 1951), such studies have already been completed.

Psychoanalytic theory has some of its deepest roots in the theory of evolution and the assumption that human organisms are born with certain predominant temperamental factors as mainsprings of their budding personalities. The innate temperamental characteristics that have received the most attention in this body of theory are those related to sex (in a general rather than specific sense) and aggression. Studies of higher animals, such as those summarized by Diamond (1957) in his book, *Personality and Temperament*, suggest that a number of temperamental factors with important degrees of hereditary determination may be isolable, of which aggressivity, affiliation, fearfulness, and impulsivity can already be tentatively identified. An important speculative hypothesis can be generated even from these limited facts: that a *number* of human temperamental characteristics have important degrees of hereditary determination. Hopefully, genetic studies will eventually reveal the number of such hereditary temperamental attributes that must be dealt with in a fully adequate personality theory. The over-all level of organismic reactivity commonly called activation level may be related to one or more temperamental attributes, or may serve as a background variable causing similar temperament patterns to have somewhat different behavioral consequences in different individuals.

To an important degree, psychoanalytic theory rests upon the theory of conflict, including the conception of defense mechanisms as means of resolving conflicting motives. As Freud often pointed out, the emergence of particular patterns of defense mechanisms in particular individuals may be dictated primarily by heredity. Hartmann's (1939) significant contribution, referred to earlier, focuses attention also upon the contribution of hereditary characteristics of the psychic apparatus to the nature and organization of the enduring patterns of cognitive processes that emerge in development independent of conflict resolution. It seems likely to the writer that individual differences in certain defenses will prove to be attributable primarily to genetically given aspects of the physical apparatuses involved. Individual differences in other defenses may be attributable primarily to learning based on repeated interactions with the parents. If this general hypothesis proves true, anticipations about the alterability of particular defenses by psychotherapeutic procedures will achieve a higher degree of refinement. For example, the defense of repression, as Freud described it, involves both assimilation of new experiences to previously formed memories and active "countercathexis" of memories or ideas whose admission into consciousness would be anxiety-arousing. If pervasive employment of the defense of repression, which differs greatly among individuals, proves to be attributable to the possession of genetically determined memory-forming equipment susceptible to a great deal of assimilation, more accurate predictions of the alterability of repression by psychotherapeutic means may be possible.

A number of other genetic hypotheses can also be formulated concerning the emergence and patterning of defense mechanisms. Take, for example, the defense of isolation, which is known to split idea from idea in consciousness and also idea from repressed affect. Individuals who use the defense of isolation extensively are often intellectualizers who collect large funds of information which they differentiate in ever more refined ways. In keeping with these typical activities, those who employ isolation extensively often show unusual knowledge of and facility with words. Here then, is a typical link between a defense mechanism and an intellectual ability. It has long been known that over-all intellectual capacity is determined to an important degree by heredity. Although homogamous mating may have led to exaggerated estimates of the degree to which general intelligence is determined by heredity, the basic fact of hereditary determination seems clear. The recent work of Vandenberg (1965) is particularly valuable in this area since it points the way to ultimate clarification of the number of genetic factors involved in different aspects of intellectual functioning. With only these briefly outlined facts in mind, it becomes possible to pose such intriguing questions as the following: (a) Does the defense mechanism of isolation itself have important genetic determinants? (b) Since isolation seems to flower in the presence of a certain kind and degree of intellectual capacity, is it most likely to become a major defense in persons with exceptional verbal capacity? (c) Is the person who employs the defense of isolation to an unusual degree likely to be one in whom the genetically given apparatus-characteristics leading to isolation on the one hand and unusual verbal skills on the other are jointly present? These are but examples of the important questions that will ultimately be resolved only by the addition of genetic studies.

Let me turn now to an aspect of personality organization that, along with defensive and intellectual organization, has been of particular concern in my own recent research. I refer here to the studies of individual consistencies in cognitive functioning we have described as dimensions of "cognitive control" (see, e.g., Gardner et al. 1959, 1960; Gardner, 1964). We have referred to combinations of cognitive controls as aspects of "cognitive style." Our work in this area was in part stimulated by Hartmann's emphasis on conflict-free aspects of ego organization. By now, through the efforts of

Witkin and his collaborators (e.g., 1962) and members of our own research group, a number of cognitive control dimensions have been isolated and have been studied in relation to intellectual abilities, defense mechanisms, etc. The work in this area seems to be delineating some hitherto undefined aspects of cognitive organization. Questions concerning the development of cognitive controls can, of course, be answered only by combined studies of environmental and genetic determinants. Viewed in this light, many earlier findings presumably indicating effects of parent-child interactions on personality development (such as the large body of child development studies stimulated by psychoanalytic discoveries) may ultimately require reinterpretation.

Having pointed to but a few of the variables postulated by psychoanalytic theorists that will be fully understood only when extensive genetic studies are performed, let me describe some current research that may provide some direction to further work in this area. The subjects of these studies are groups of monozygotic twins, dizygotic twins, and their respective parents. The test procedures include intellectual ability tests from the battery used by Vandenberg, a battery of cognitive control tests, some of which are also being given by Vandenberg's group, a number of additional experimental procedures, clinical psychological tests, and interviews dealing with parental attitudes toward child rearing, attitudes toward similarity and difference in identical twins, etc. Major foci of the study are the relative genetic determinations of the variables referred to, including defense mechanisms as assessed on the basis of clinical and intellectual tests; within-family versus between-family similarities in cognitive style; predictive implications of similarities between individual parents and children; and a large number of related issues. It is obvious, of course, that the results of such a study will require cautious interpretation, with careful delineation of unanswered questions and areas of ambiguity. In the case of one cognitive control dimension, a proposed study of individual consistencies in Mexican monozygotic and dizygotic twins may provide important additional information. If, for example, individual consistencies in test performances of American children are not present in performances of Mexican children, we may have evidence of the effect of environment upon this control (assuming that no major genetic differences affect these performances by the two populations). If, however, similar patterns of individual consistency characterize the two populations, important questions will be answered by examination of the nature of these relationships and by comparison of monozygotic and dizygotic twin-pair similarities in the two cultures. In lieu of large and readily available samples of monozygotic twins reared separately, such intercultural studies may be of unique value.

In undertaking these studies, we are aware of the interpretive limitations imposed by the inevitable choices involved in their design, of the ambiguity surrounding interpretation of within-family similarities, and a veritable host of related problems. We also have clearly in mind Lerner's (1950) delineation of the four kinds of communality that may be responsible for correlations between performances in any two experimentally independent tests. We have undertaken these studies in the face of such interpretive obstacles because of the lack of information concerning genetic determination of several groups of variables important to one current theory of personality development and organization. Hopefully, these studies will provide some orientation points and some refined preliminary hypotheses for further explorations. Hopefully, too, our work will encourage others to begin the extensive further work necessary for ultimate understanding of the interacting genetic and environmental determinants of personality organization.

References

Benjamin, J. D. Some comments on twin research in psychiatry. In T. T. Tourlentes, S. L. Pollack, & H. E. Himwich (Eds.), *Research approaches to psychiatric problems.* New York: Grune & Stratton, 1962. Pp. 92–112.

Diamond, S. *Personality and temperament.* New York: Harper & Row, 1957.

Duffy, E. *Activation and behavior.* New York: Wiley, 1962.

Eysenck, H. J., & Prell, D. B. The inheritance of neuroticism: an experimental study. *J. Ment. Sci.,* 1951, **97,** 441–465.

Freud, S. (1937). Analysis terminable and interminable. *Collected Papers,* Vol. 5, London: Hogarth, 1950.

Gardner, R. W. The development of cognitive structures. In C. Scheerer (Ed.), *Cognition: Theory, Research, Promise.* New York: Harper, 1964. Pp. 147–171.

Gardner, R. W., Holzman, P. S., Klein, G. S., Linton, H. B., & Spence, D. P. Cognitive control: a study of individual consistencies in cognitive behavior. *Psychol. Issues,* 1959, **1,** No. 4.

Gardner, R. W., Jackson, D. N., & Messick, S. J. Personality organization in cognitive controls and intellectual abilities. *Psychol. Issues*, 1960, **2**, No. 4 (Whole No. 8).
Ham, G. C. Genes and the psyche: perspectives in human development and behavior. *Amer. J. Psychiat.*, 1963, **119**, 828–834.
Hartmann, H. (1939). *Ego psychology and he problem of adaptation.* D. Rapaport (Transl.). New York: Intern. Univer. Press, 1958.
Kallmann, F. J. Comparative twin studies on the genetic aspects of male homosexuality. *J. Nerv. Ment. Dis.*, 1952, **119**, 282–298.
Klein, G. S. Need and regulation. In M. R. Jones (Ed.), *Nebraska symposium on motivation.* Lincoln: Univer. of Nebraska Press, 1954. Pp. 224–272.
Lerner, M. I. *Population genetics and animal improvement.* Cambridge: Cambridge Univer. Press, 1950.
Vandenberg, S. G. Innate abilities, one or many? *Acta Genet. Med. Gemell.*, 1965, **14**, 41–47.
Witkin, H. A., Dyk, R. B., Faterson, H. F., Goodenough, D. R., & Karp, S. A. *Psychological differentiation.* New York: Wiley, 1962.

The National Merit Twin Study[1]

Robert C. Nichols

This chapter will be concerned with some of the methodological problems and some early results of the National Merit Twin Study which is still in progress and when finished will be reported in detail elsewhere. Analyses completed so far have been primarily concerned with the relative effects of heredity and environment on ability measures, which will therefore be the major focus of this chapter.

Of course, both heredity and environment are necessary for any behavior to occur. If heredity had not produced a behaving organism there would be no behavior, and if environment had not provided the appropriate situation for the development of the organism there would also be no behavior. Both are necessary and completely interdependent influences, and it is not possible to say that one is more important than the other. If, in an experimental situation, either heredity or environment were completely controlled, variation in any behavior could be produced by variation of the other. However, this is not the question we are asking. We are interested in accounting for variance in human behavior as it is observed in the natural social setting, and it is a perfectly legitimate question to ask how much of this variance is due to variation in heredity and how much is due to variation in the environment. The question "If everyone had the same heredity, how much would the variance in behavior be reduced?" is a convenient simplification of the question we are asking. It is a simplification because there are undoubtedly many instances in human behavior of a phenomenon which has been demonstrated in animals that the effect of environ-

[1] This study is part of the research program of the National Merit Scholarship Corporation and is financed by grants from the National Science Foundation, the Carnegie Corporation of New York and the Ford Foundation.

ment on behavior varies with the genetic characteristics of the animal.

ESTIMATION OF HERITABILITY

Twins raised together in the same family form a natural experiment in which the two kinds of twins vary in hereditary similarity, but are presumably about equal in similarity of environment. The influence of heredity can be inferred from the greater similarity of identical twins than of fraternal twins. The estimation of the proportion of the variance in the trait which can be attributed to heredity poses some additional problems, but with certain assumptions it is possible to estimate the proportion of variance of a trait which is due to variation in heredity.

Figure 1 shows a schematic representation of the sources of variance in twin data. The left-hand vertical line represents the total variance of a trait in identical twins and the right-hand line the total variance in fraternal twins. The possible sources of varia-

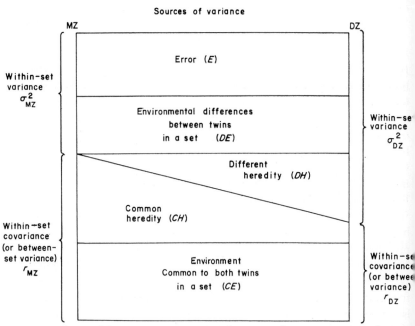

FIG. 1. Schematic representation of sources of variance in twin data.

tion are listed between the two lines. Both hereditary and environmental variance is divided into that common to twins of a set and that which is different for the two twins of a set.

The particular power of the twin method lies in the fact that the difference between the intraclass correlations for identical and fraternal twins is equal to the proportion of the total variance due to hereditary differences between fraternal twins. Since fraternal twins have on the average half their genes in common, this is half the hereditary variance in the trait. This fact can be used to construct heritability coefficients from twin correlations which give estimates of the proportion of the variance in a trait attributable to heredity. The coefficient h^2 proposed by Holzinger (1929) is the ratio of half the hereditary variance to the variance within sets of fraternal twins.

$$h^2 = \frac{\sigma_{DZ}^2 - \sigma_{MZ}^2}{\sigma_{DZ}^2} = \frac{r_{MZ} - r_{DZ}}{1 - r_{DZ}} = \frac{DH}{DH + DE + E}$$

Since error of measurement enters into the within-set variance the correlations are usually corrected for attenuation. Another coefficient which we have developed for use in our twin study is called

$$HR = \frac{2(r_{MZ} - r_{DZ})}{r_{MZ}} = \frac{2DH}{CH_{MZ} + CE}$$

HR. This is the ratio of the hereditary variance to variance due to heredity and environment common to both twins of a set. If one is willing to assume that the major environmental influences on a trait, at least those which might be measured or manipulated, are common to both twins of a set, this ratio is the proportion due to heredity of the variance attributable to heredity and major environmental variables. This ratio also offers the advantage of not including error variance and thus not requiring correction for unreliability of the measuring instruments.

The schematic representation in Fig. 1 and the logic behind the heritability coefficients make certain assumptions which may or may not hold in a given study or with respect to a given trait being investigated. The four major assumptions are the following: (a) that the similarity of environmental influence is the same for fraternal and identical twins; (b) that for the trait in question there is no correlation between parents due to assortive mating (although if the correlation between parents is known it can be corrected for);

(c) that hereditary variance in a continuous trait being studied shows no dominance or interaction effects; (d) that hereditary and environmental influences are not correlated (although small correlations make little difference).

Relatively minor violations of these assumptions may not be serious, since the heritability coefficients are subject to large sampling fluctuations and even with large samples can give only rough approximations of the heritability of a trait. Since the heritability coefficients depend on ratios between correlations, they are much more stable with high correlations than when the correlations are relatively low. Thus, in any given sample of twins we may be able to get stable estimates of the heritability of intelligence measures where the correlations tend to be high, but much less reliable estimates of the heritability of personality traits where the correlations tend to be low.

Sampling Procedures

As a part of National Merit's annual talent search, a 3-hour scholastic aptitude test is administered to about 600,000 high school juniors each spring. In 1962, we included an item on the test asking if the student was a twin. From those checking they were twins, the 1507 pairs who met all of the following criteria were selected: (a) same sex, (b) attended same high school, (c) same last name, and (d) same home address. This procedure missed many sets of twins, probably those with the greatest differences in ability, those with differences in educational or health histories, and those who were separated. Assuming approximately one twin birth in every 100 of which two-thirds are fraternal, about 4000 sets of like-sex twins would be expected from a total sample of 600,000, and less than half that many were obtained. Yet this procedure yielded a large number of twins of the same age for whom test scores were available. The exclusion of twins with markedly different experiences is desirable for most research objectives.

Twin Diagnosis

The first problem, and a crucial one for our study, was to develop an adequate method for the diagnosis of zygosity. The older studies of twins have almost always diagnosed twins as identical or fraternal on the basis of similarity of appearance. Recently, however, blood typing has made these diagnoses much more precise, since blood can be typed with great accuracy for a number of independent, categorical, genetically determined characteristics which have high penetrance. Identical twins are of course alike on all blood groups and it is possible to calculate the probability of getting any combination of blood groups alike by chance in fraternal twins (Smith & Penrose, 1955). If all available blood groups are used an average accuracy of around 98% can be obtained. Since we had a large sample of twins which was geographically dispersed, we were very interested in finding an easy method of diagnosis which had sufficient accuracy to allow us to use our large sample. If we had to rely on blood typing the sample size would be greatly reduced. Most investigators have tended to assume that diagnoses based on physical similarity are of questionable accuracy, but nevertheless we decided to attempt the construction of a diagnostic index based on physical similarity.

We sent all of our twins a four-page questionnaire asking questions about similarity of physical appearance, and replies were received from both twins of 79% of the sets and from at least one twin of 82%. By contacting twins who were in college through the college health services, we were able to obtain usable blood samples on both twins of 124 sets. The blood was typed for 22 anti-sera, and all analyses were run twice using anti-sera from different suppliers. On the basis of the blood analyses, 42 sets were diagnosed fraternal and 82 were diagnosed identical. All 42 of the fraternal sets and an equal number of the identical sets were set aside for cross-validation. We used the remaining 40 identical sets and some fraternal sets diagnosed by marked differences in hair and eye color to develop a preliminary diagnostic index on the basis of the questionnaire physical similarity data. The index which was developed is applied on two levels. The first-level diagnoses are based on a single marked dissimilarity in height, weight, eye or hair color or on marked similarity as indicated by mistaken identity. The second-level diagnoses are based on combinations of several less striking indications of similarity or difference. Table I shows how well this preliminary index worked on the blood-diagnosed cross-validation cases. About half of the cases were diagnosed at the first level and about half at the second level. The over-all accuracy of diagnosis was 93%. The index was revised on the basis of the experience with the cross-validation sample and all cases were diagnosed using the revised index. As a further indication of the validity of the index diagnoses, intraclass correlations for the NMSQT (National Merit Scholarship Qualifying Test) composite score were calculated for identical and fraternal twins diagnosed by various methods. Less accurate diagnoses would be expected to lead to a higher intraclass correlation for fraternal twins and a lower correlation for identical twins. As can be seen from Table II, the correlations for identical

TABLE I
Cross-Validation of Preliminary Diagnostic Index

	Blood diagnosis	
Index diagnosis	MZ(N = 42)	DZ(N = 42)
Level one		
MZ	15	2
DZ	1	21
Undiagnosed	26	19
Level one and level two		
MZ	38	4
DZ	1	35
Undiagnosed	3	3
Index with intuitive diagnosis of uncertain cases		
MZ	40	4
DZ	2	38

TABLE II
Intraclass Correlations of the NMSQT Composite Score for MZ and DZ Twins Diagnosed by Various Methods

	MZ		DZ	
Method of Diagnosis	Correlation	N	Correlation	N
Blood, hair, and eye (index development and cross-validation cases)	.88	82	.68	199
Index level one	.89	92	.65	137
Index level two	.87	513	.59	146
Undiagnosed by index, Diagnosed intuitively	.87	50	.77	20

twins were about the same for all methods of diagnosis, and the correlations for fraternal twins actually decreased (but not significantly so) from the blood diagnoses to the less certain index diagnoses. Those twins who could not be diagnosed by the index—cases which gave no clear indication either of similarity or dissimilarity—were apparently mainly identical, but they were not included in any further analyses.

On the basis of experience with the development of the index, we have concluded that it is not difficult to diagnose twins, and that diagnoses almost as accurate as blood diagnoses can be made very simply. The feasibility of diagnosing twins by mail opens up many new possibilities for large-scale twin studies. All extensive testing programs should include an item inquiring about twin status, and diagnosis by the physical similarity index would provide large samples of twins on which test scores were already available.

Inheritance of General and Specific Ability

The first analysis we have done with the twin data is a study of twin similarity on the National Merit Scholarship Qualifying Test, the three-hour aptitude test by which the twin sample was first identified. In this study, we were concerned with the heritability of general intelligence, which has been well researched, as well as the heritability of specific abilities, which has been little studied.

Most previous studies of the inheritance of ability have been concerned with general intelligence, although it is now generally recognized that intelligence is not a unitary trait. There are a number of different abilities, which tend to be positively correlated. There are two major explanations for this structure of intelligence. The first explanation, growing out of Spearman's theory of general and specific abilities, is that the general level of ability is determined mainly by heredity and that the ability to perform particular tasks is either facilitated by or is not facilitated by environmental experience producing specific abilities. The second explanation, growing out of Thompson's sampling theory, is that many very specific abilities are hereditarily determined and that the phenomenon of general intelligence arises from the use of the same specific abilities in many different tasks and from the generalized effects of environmental influences. Intelligence, like all behavior, is determined by some complex interaction of heredity and environment, but the question remains whether the herediatry component is mainly general—or mainly specific.

The NMSQT consists of five subtests measuring achievement in different academic areas. A factor analysis of these five subtests indicated that there is one general factor and that once the influence of this general factor is taken out, the residual variances of the five subtests are relatively independent. We wanted to study the heritability of general ability and of the specific abilities measured separately by each subtest. General ability is well represented by the composite score (the sum of the standard scores for the five substests). In order to get a measure of the specific factor measured by each subtest, we computed multiple regression equations for predicting each subtest score from the other four subtests, and then computed residual scores for each subtest which were independent of the other four subtests. The reliabilities of the subtests vary around .90 and the multiple correlations of the subtests with the other four vary around .75, so that there is a reliable portion of each subtest which does not overlap with the others. However, the residual scores contain a large proportion of error so that we would not expect large correlations.

The heritability of the composite score will show the inheritance of general intelligence, and the heritability of the residual scores will show the inheritance of certain fairly specific abilities which are independent of general intelligence. These heritability coefficients are shown in Table III.

As we would expect, the identical twins were much more alike than the fraternal twins with respect to the composite score. Both heritability coefficients indicate that about 70% of the variance in the composite score can be attributed to differences in heredity, and this is about the same value that has been obtained by other investigators, some of whom have used different methods of estimating heritability.

The twin correlations for general ability show a most encouraging consistency with the results of other twin studies. Table IV gives a summary of results of studies using a variety of tests conducted in different countries, in different languages, over a period of 35 years. The consistency of the findings is striking.

Now turning to the heritability of the residual scores, we can see

TABLE III
INTRACLASS CORRELATIONS AND HERITABILITY RATIOS FOR NMSQT SUBTEST, COMPOSITE AND RESIDUAL SCORES

Test	Intraclass correlation		t^a	HR	h^{2b}	Intraclass correlation		t^a	HR	h^{2b}
	MZ $N=315$	DZ $N=209$				MZ $N=372$	DZ $N=273$			
Composite	.87	.62	6.62^c	.73	.72	.88	.65	7.76^c	.65	.74
Subtests										
English Usage	.71	.64	1.46	.22	.27	.77	.49	6.16^c	.99	.67
Mathematics Usage	.74	.42	5.69^c	1.16	.80	.70	.47	4.48^c	.87	.65
Social Studies Reading	.76	.50	5.08^c	.92	.67	.79	.52	6.01^c	.92	.60
Natural Science Reading	.69	.52	2.96^c	.60	.51	.66	.48	4.48^c	.68	.50
Word Usage (Vocabulary)	.85	.64	5.32^c	.60	.69	.86	.64	6.78^c	.62	.75
Residual Subtests										
English Usage	.40	.42	−.25	−.10		.48	.25	3.37^c	.96	
Mathematics Usage	.48	.21	3.47^c	1.14		.43	.23	2.74^c	.92	
Social Studies Reading	.33	.16	2.09^d	1.06		.29	.17	1.60	.83	
Natural Science Reading	.27	.32	−.61	−.36		.31	.10	2.77^c	1.37	
Word Usage (Vocabulary)	.55	.37	2.50^d	.65		.55	.26	4.48^c	1.07	

[a] t-test of significance of difference between MZ and DZ correlations.
[b] h^2 was computed with correlations corrected for attenuation.
[c] $p < .01$
[d] $p < .05$

the instability of the heritability coefficient when dealing with low correlations. We have not reported Holzinger's h^2 for the residuals, since correcting the residual scores for attenuation would take us pretty far from reality. All residual subtests showed significant differences between fraternal and identical correlations for at least one test and all residual subtests showed significant heritability in

TABLE IV

SUMMARY OF RESULTS OF STUDIES OF THE INTELLECTUAL RESEMBLANCE OF TWINS

Study	MZ twin sets		DZ twin sets		Test
	Intraclass correlation	N	Intraclass correlation	N	
Holzinger (1929)	.88	50	.63	52	Binet IQ
Newman, Freeman, Holzinger (1937)	.91	50	.64	50	Binet IQ
Newman, Freeman, Holzinger (1937)	.92	50	.62	50	Otis IQ
Blewett (1954)	.75	26	.39	26	PMA Composite
Husen (1959)	.90	215	.70	416	Swedish Military Induction Test
Husen (1960)	.89	134	.62	180	Reading Achievement
Husen (1960)	.87	134	.52	181	Arithmetic Achievement
Erlenmeyer-Kimling and Jarvik (1963)	.87	14[a]	.53	11[b]	Various Intelligence Measures
Nichols (present study)	.87	687	.63	482	NMSQT Composite

[a] Median of 14 studies.
[b] Median of 11 studies.

the total sample of both sexes combined. The question now is whether we should attach any significance to the sex differences in the heritability coefficients. I am inclined to think that we should not, and that at least until the study is replicated (we are collecting another sample of twins with NMSQT scores next year) we should attribute the differences in the various heritability coefficients of the residual subtests to the unreliability of ratios between low correlations. If we are willing to assume that the sex differences and the differences between subtests in heritability are due to sampling fluctuations, the average heritability coefficient of the ten shown may be a good indication of the heritability of specific abilities such as those measured by the residual subtests. The average heritability of the residual subtsets is .75, about the same as that obtained for the composite score. This result supports the point of view that very specific abilities are independently inherited.

One of the assumptions of twin studies which has been most frequently questioned is that the similarity of environmental influences is the same for fraternal and identical twins. If identical twins have had more similar experiences, this might account for their greater behavioral similarity without any hereditary influence. In order to see if this could account in part for the differences shown in Table III, we asked the twins to report all periods of separation and any differences in experiences and illnesses. We repeated the analyses shown in Table III after excluding from the sample all sets of twins who reported separation for 1 month or more, those who reported that one twin had a major illness or disability not shared by the other, and those who reported a major environmental experience (such as marriage or special training) by only one twin. These criteria led to the exclusion of 18% of the identical twins and 25% of the fraternal twins. The correlations tended to be slightly higher for all scores when twins with different experiences were excluded, indicating that the differences in experience tend to produce differences in ability (or perhaps to reflect differences in ability rather than causing them), but correlations tended to be affected about equally for both fraternal and identical twins, so that the conclusions concerning the effects of heredity were not affected.

FUTURE PLANS

In the National Merit Twin Study we have also collected a great deal of personality data on our sample of twins, but results of these data are not yet available. The following briefly outlines the sort of data that have been collected and the kinds of analysis which are planned.

All twins who were diagnosed by the physical similarity index were sent a packet of questionnaire materials which required about 3 hours to answer, and complete returns were obtained from 845 sets (about 72%). These materials included the California Psychological Inventory, the Holland Vocational Preference Inventory, an experimental behavior inventory, and a number of self-rating and life history items. In addition the twins' mothers completed a questionnaire concerning the early life and development of the twins with particular attention to differences between the twins in early environmental experiences. We also obtained teacher and peer ratings of personality differences between the twins for most of the sample. We have just completed a followup of the twins 1 year after the initial personality data were obtained and at the end of the freshman year for those twins attending college. In this followup some of the original measures were repeated and particular attention was given to differences in experiences of the twins during the intervening year.

In analysis of these data we shall be concerned with the heritabilities of the various personality measures and shall try to find a rough ordering of personality traits from those with a large hereditary component to those dependent mainly on environment. We shall attempt to relate personality differences between twins of a set to differences in environmental experiences, and we shall attempt to get some idea of the dimensions of hereditary influences by factor analyses of correlation matrices divided into hereditary and environmental components and by separate factor analyses of items with high and low heritabilities.

REFERENCES

Blewett, D. B. An experimental study of the inheritance of intelligence. *J. Mental Sci.*, 1954, **100**, 922–923.

Erlenmeyer-Kimling, L., & Jarvik, L. F. Genetics and intelligence: a review. *Science*, 1963, **142**, 1477–1479.

Holzinger, K. J. The relative effect of nature and nurture influences on twin differences. *J. Educ. Psychol.*, 1929, **20**, 241–248.

Husen, T. *Psychological twin research.* Vol. I. *A methodological study.* Stockholm: Almqvist & Wiksell, 1959.

Husen, T. Abilities of twins. *Scand. J. Psychol.*, 1960, **1**, 125–135.

Newman, H. H., Freeman, F. N., & Holzinger, K. J. *Twins: a study of heredity and environment.* Chicago: Univ. Chicago Press, 1937.

Smith, S. M., & Penrose, L. S. Monozygotic and dizygotic twin diagnosis. *Ann. Human Genet.*, 1955, **19**, 273–289.

DISCUSSION

Cattell: Well while we are on that topic, how do you feel about the reliability of mail questionnaires? Personally I have had a rather disappointing experience, where the same questionnaire administered verbally in a family group situation, was mailed to an equivalent group, the reliabilities dropped appreciably; and I presume that the individual is apt to be under suggestion from someone administering the questionnaire in the homes. This may be quite a serious thing, I feel, and I wonder what your experience is?

Nichols: The usual individual may be influenced, but we have had very good luck with it, perhaps because we have been working with very bright students. We also sent this questionnaire to a random sample of people who took the National Merit exam and got a lower response rate from them. In checking them for errors we also found that they made many more errors, e.g. marking their sex incorrectly.

Cattell: My circulation was to housewives and it appears that they usually asked their husbands, "Well, my dear, am I talkative?"

Nichols: We have tried to check the accuracy of reporting some objective things, and we found that if you ask a straightforward unambiguous question "Did you or did you not do this? What was your grade point average last semester?," you get very good reports. We've checked grade point averages against transcripts from college and found remarkable accuracy in reporting of these things. Now when it comes to attitude items and self-ratings you do run into kibitzers, etc., however, I do feel that the responses are valid.

Gottesman: Was it your intent to mail out things like the CPI (California Psychological Inventory) and have them filled out and mailed back to you? I wonder (*a*) if this is ethical and (*b*) if the Psychological Corporation and other companies that publish these tests for restricted use would be happy with that?

Nichols: I would answer yes to both of your questions. Gough has been really intrigued with our use of the CPI, and has supported the idea.

Gottesman: So there are a lot of CPI's floating around the United States right now.

Nichols: Yes, we sent some to Harvard as a matter of fact.

Suggestions from Animal Studies for Human Behavior Genetics[1]

John L. Fuller

Since my work deals with animals I do not have to make the compromises that people working with human beings have to accept. Animal laboratory problems seem trivial in comparison with those faced by human behavior geneticists. If my task were simply to give a review of contemporary animal behavior genetics, I would have no emotional response. The request that I provide suggestions for human behavior genetics based on animals bothers me, because I'm not sure that any of my suggestions would be truly helpful.

There are three kinds of problems that concern geneticists working on animal behavior: (a) formal genetics including identification and enumeration of genes, and their linkage, dominance and epistatic relationships; (b) physiological genetics which deals with the processes by which little packets of DNA (deoxyrebonucleic acid) affect behavioral phenotypes; and (c) the population genetics problems, the attempt to separate observed behavioral variance into environmental and hereditary components. The behavioral aspects of biological evolution also belong in this category. Animal studies in behavior genetics are usually pure science without immediate application, though in agricultural schools there are projects which are motivated by economic considerations. But no matter how academically oriented we may be, we can generate principles of wide generality and in doing so develop models which will be applied to human problems. Dobzhansky's (1962) recent

[1] Original research described in this paper was supported in part by the Ford Foundation, the Foundations Fund for Research in Psychiatry and Public Health Service grant MH-01775.

book, *Mankind Evolving*, is based as much upon research with *Drosophila* as upon human biology.

Physiological Behavior Genetics

Let us look initially at the physiological genetics of behavior for suggestions from the field of animal genetics. Is there any possibility, for example, that we can find homologs in animals for inherited neurological or behavioral disorders? For biological phenotypes such homologs (or analogs) are well established. Animals have inherited hemophilia, ataxias, muscular dystrophies, various anemias, and differing susceptibility to degenerative and infectious diseases. Animals with specified genotypes are used widely by cancer biologists, pharmacologists, pathologists, and others in experimental studies, the results of which hopefully can be applied across species. Drugs which are effective against mouse leukemia are subjected to clinical testing in man. At the cell and tissue level, the parallels between human and subhuman syndromes are sometimes quite precise, but I do not think there is any prospect that the animal behaviorist is going to find as good equivalents for the human phenotypes of special interest to psychologists.

The problems which might be studied in animals and man at a comparable molecular level lie outside the main stream of behavior genetics at the present time. Phenylketonuria is an inherited disorder produced by a gene whose chemical action is known. The mental retardation of homozygotes for this gene is severe. In the mouse, the *dilute* gene has a similar metabolic action, and dilute mice could be used as models for research into the mechanisms of gene action upon intelligence. However, interest here is in genetic influences upon behavior variation which is distributed continuously and quasi-normally. Now it is certainly possible, and even easy, to demonstrate that strains of animals which are genotypically unlike almost always show quantitative behavioral differences. Whether such strain differences provide a useful model for explaining individual differences in man is a question. The farther one goes from metabolic similarities, the less precise is the homology between species.

Differences in alcohol preference shown by inbred strains of mice provide an example of the physiological genetics approach.

Rodgers and McClearn (1962) have proposed that the preference difference is produced by, or at least correlated with, a difference in the alcohol dehydrogenase activity of the liver. No gene for the enzyme has been isolated, but strain differences are fairly large. Furthermore, *in vivo* oxidation rates for ethyl alcohol are higher in a preferring strain than in an avoiding strain (Schlesinger, 1964). The evidence is fairly strong, and yet McClearn has not concluded that this enzyme is the key to preference behavior. Even if it is, we must find the way in which a chemical process controls behavior. This may not be too difficult in the present case. Possibly, when alcohol is readily oxidized it serves primarily as a source of energy and is primarily reinforcing. If it is not readily oxidized it may produce discomfort which would be negatively reinforcing.

Experiments based upon this hypothesis have been successful in modifying preference. Mice allowed access to both water and alcohol for one 5-minute period per day take less alcohol than those given the same choice for five 1-minute periods. We account for this by assuming that forcing all fluid consumption into one short period per day overtaxes the oxidative systems and leads to increased aversion to alcohol. The difference between high- and low-preference strains remains under all the conditions studied, although the absolute preference level can be moved up and down by manipulating the drive state. The proportion of alcohol ingested is not an invariant phenotype under all conditions. It changes with many factors but genetic determinants are detectable amid the changes.

The alcohol-preference phenotype can also serve as an example for research on mode-of-inheritance. Recently I took four strains of mice, ranging from DBA/2J, which take almost no alcohol if given a choice, to C57BL/6J, which have a fairly high preference. For reasons concerned with homogeneity of variance, I measured their preference as the concentration of alcohol above which one-half the fluid intake was observed when six concentrations were simultaneously present. Hybrids were obtained between all pairings of the four strains, and their alcohol preferences were determined (Fuller, 1964).

When mice of intermediate-preference strains were bred to the high-preference strain, the hybrids had high preference. The same intermediate animals bred to a low-preference strain produced hybrid offspring with low preference. If one tries to apply a classical term such as "dominance" to the results one encounters difficulty. Low preference is dominant in one cross, high in another. Instead I have analyzed the preference scores in terms of general and specific combining action of the contributions from each parental genotype. The general combining action of a strain is based upon the mean performance of all its hybrid offspring. Deviations from the value expected from the general action of both parents provide estimates of specific combining action. In this study, about two-thirds of the variance was contributed by specific combining action and one-third from general.

The experiment makes two points. First, it illustrates the dangers of generalizing upon mode-of-inheritance from breeding experiments between a single pair of "high" and "low" lines. Until recently, most of the behavior genetics literature in this area was based on single crosses.[2] Second, the preponderance of specific combining action over general implies that dominant and epistatic effects are more important quantitatively than additive effects. If the additivity hypothesis does not work well in controlled studies with animals, one should be hesitant about accepting it in human psychogenetics. Nevertheless, the additivity concept is implicit in most of the models commonly used.

Physique and Behavior

We shall now turn to some other findings from animal research, in particular the subject of correlations between behavioral and physical characters. Several years ago at Bar Harbor, Scott and Fuller (1965) compared behavioral and physical measurements in five pure breeds of medium-sized dogs. We also made Mendelian crosses between two of these breeds, the cocker spaniel and basenji, which differed in behavior and were of diverse origin. A correlation matrix which included weights and body dimensions along with results of thirty psychological tests was factor-analysed by Brace (1961). The physical and behavioral measures were completely separate; no mixed factors turned up in the analysis.

[2] This point has been expressed strongly by Bruell (1964) since my original remarks.

At least in this species which is so phenotypically variable the idea of a somatotype which is correlated with behavior finds no support.

What about correlations between behavior and physiological traits which represent a more dynamic state than do body dimensions? A few decades ago, strains selectively bred for activity or timidity by Hall, Tryon, Heron, and others were compared on various endocrine gland characteristics (see Fuller & Thompson, 1960, for references). More recently, an emotionally reactive strain of rats has been found to be slightly hypothyroid (Feuer & Broadhurst, 1962). The results of such comparisons have not been perfectly consistent, but low thyroid activity does seem to be frequently associated with inactivity or timidity. Many studies of this type are based upon comparison of a pair of strains differing on a behavioral character. If they turn out to be different on a physiological trait too, a hypothesis can be stated whichever way the second difference falls. There is a need for broadly based correlational studies based upon many independently derived true-breeding strains.

The Choice of Phenotypes

Animal research has indicated that genetic differences in the *response* to hormones may be more important as causes of behavioral variation than the difference in amount of hormones produced. The best example is the work by Young and his associates at Kansas (Young, 1957) on sexual behavior in guinea pigs. They found that sexual activity can be restored by androgens in castrated males, but only to the level characteristic of the precastrate state. A male with low sex-drive could not be converted to high-drive status by supplying more hormone. Once the amount of androgen is over a threshold, the critical factor for motivating sexual behavior is not chemical but somatic response.

There are important genetic differences in sexual behavior among animal strains, and Young's work suggests that the genes produce these effects, not by regulating the factory which makes the hormones, but by modifying the target tissues on which they act. This idea in turn suggests to me that looking for correlations between behavioral traits which are dynamic and physiological traits under steady-state conditions may be a lost cause. Perhaps one should expose individuals to disequilibrating forces and observe tolerance limits, dose-response curves, or rate of re-equilibration. Now this is harder than analysing blood or urine obtained under steady-state conditions, but it may be much more profitable. Recent experiments in our laboratory employing behavioral measures and psychoactive drugs have shown the value of this approach. Such experiments are not impossible with human subjects, though the recruitment of subjects is not easy. Recently I visited an operant conditioning laboratory in which people were employed as professional subjects for experimentation on a full-time basis. Perhaps a few pairs of twins could be recruited to work in such a capacity. Paying for their services would be as legitimate a research expense as any other. Naturally there are more restrictions when using human subjects, but this should not be a serious handicap.

In closing this section I wish to make one further comment regarding the choice of phenotypic variables. I have discussed this with Dr. Guttman and we agree that behavior geneticists may make a mistake in selecting phenotypic characters on the basis of their importance in psychology. Dr. Guttman has included in her battery, traits such as hand clasping and arm folding, which are trivial in a sense, but which may yield neater genetic results than more important personality factors. Also, at the present state of knowledge, no soundly proved genetic facts should be called trivial.

The Genotypic Approach

One can, to an extent, circumvent the problems of choosing a phenotype by starting at the other end and determining the behavioral consequences of having alternative genotypes, AA, $A'A$, and $A'A'$ for example. Perhaps one should include as many measurements as possible in such a study to increase the probabilities of finding a correlation between genotype and behavior. Would it be worth while to look for behavioral correlates of the numerous genetic polymorphisms which are being discovered in man? There is considerable evidence that some of these, such as the blood groups, are significant factors in natural selection (Reed, 1961). Although psychological effects of such genes are certainly conceivable, the work on animals indicates that the contributions of individual loci (which do not produce gross morphological effects) to behavioral variation is small and extremely difficult to detect. With animals, at least in mice, we have a further advantage in our ability to place the variants at one locus on an otherwise genetically homogeneous background. The results of the single locus studies, including some unpublished ones of our own, are relatively meager. Eventually, as we obtain more biochemical mutations with which to work, the picture may change. At present, the animal studies suggest that the genotypic approach to human behavior genetics is not likely to be productive.

POPULATION GENETICS

Concerning the contribution of animal studies to population behavior genetics I shall add a few words about the Jackson Laboratory dog experiment. Scott and Fuller (1964) gave a large variety of tests to five breeds of dogs. Scott found that the breeds differed significantly on 49 of the 50 variables used by Brace in the factor analysis which was mentioned earlier. On a more restricted and only partially overlapping domain, emotional reactivity, I obtained 30 measures at each of three ages. Among the 90 possible breed comparisons only four failed to reach to .05 level of significance. There was a slight, questionably significant, increase in the intra class correlation (the proportion of variance attributable to breed) as the animals grew older. Thus, the effects of genetic differences did not wash out with age, but possibly became slightly stronger. It would be misleading to leave the impression that members of a breed conform to a standard and that the behavioral boundaries between breeds are sharp. Overlapping was the rule, though breed differences characteristically were responsible for from 10 to 60% of total variance. We came to think of breed in behavioral terms as a population characterized by a range of variation, rather than as a type as described in breed standards.

GENES AND EARLY EXPERIENCE

The concluding remarks concern the relationship between characteristics influenced by changes in genes and those affected by changes in environment. For the past few years I have been investigating the effects of early experience (isolation) upon two breeds of dogs—beagles and wirehaired fox terriers. These breeds are quite different in behavior, but they respond similarly to experiential isolation. Upon emergence, the dogs are less social with people and with other dogs. They contact objects less and are either hyper- or hypoactive, depending upon circumstances. Some forms of behavior are more affected by isolation than others as we have shown by comparisons with pet-reared puppies of the same age. The point of the experiment is that the measurements on which the beagles and terriers are most unlike are the ones which are most readily modified by the isolation procedure. We do not find a class of traits determined by environment and another controlled by heredity. Instead some characters are relatively insensitive to either genetic or experiential modification, while others are readily modified by either manipulation.

Finally, it is clear that these somewhat sketchy remarks do not provide a guide to quick success in human behavior genetics. Animal workers have problems too. We have the advantage of possessing experimental control over the lives and breeding histories of our subjects. Bridging the gap between a gene and observed behavior is much the same in mouse and man.

REFERENCES

Brace, C. L. Physique, physiology and behavior. Unpublished doctoral dissertation, Harvard University, 1961.

Bruell, J. H. Inheritance of behavioral and physiological characters of mice and the problem of heterosis. *Amer. Zoologist*, 1964, **4**, 125–138.

Dobzhansky, T. *Mankind evolving*. New Haven: Yale Univer. Press, 1962.

Feuer, G., & Broadhurst, P. L. Thyroid function in rats selectively bred for emotional elimination II. Differences in thyroid activity. *J. Endocrin*, 1962, **24**, 253–262.

Fuller, J. L. Measurement of alcohol preference in genetic experiments. *J. Comp. Physiol. Psychol.*, 1964, **57**, 85–88.

Fuller, J. L., & Thompson, W. R. *Behavior genetics*. New York: Wiley, 1960.

Reed, T. E. Polymorphisms and natural selection in blood groups. In B. S. Blumberg (Ed.), *Proc. of the Conference on Genetic Polymorphisms*. New York: Grune & Stratton, 1961. Pp. 80–101.

Rodgers, D. A., & McClearn, G. E. Alcohol preference of mice. In E. L. Bliss (Ed.), *Roots of behavior*. New York: Hoeber, 1962. Pp. 68–95.

Schlesinger, K. Genetic and biochemical determinants of alcohol preference and alcohol metabolism in mice. Unpublished doctoral dissertation. Univer. of California at Berkeley, 1964.

Scott, J. P., & Fuller, J. L. *Genetics and the social behavior of the dog.* Chicago: Univer. Chicago Press, 1965.

Young, W. C. Genetic and psychological determinants of sexual behavior patterns. In H. Hoagland (Ed.), *Hormones, Brain Functions, and Behavior.* New York: Academic Press, 1957. Pp. 75–98.

A Longitudinal Study of Children with a High Risk for Schizophrenia: A Preliminary Report[1]

Sarnoff A. Mednick and Fini Schulsinger

In the past 14 months in Copenhagen, Denmark, we have examined 207 children whose mothers are severely and chronically schizophrenic. We have also examined 104 control children carefully matched with the 207 for age, sex, father's occupational status, rural-urban residence, years of formal education, and institutional versus family rearing conditions. The examination included biochemical, psycho-physiological, social, cognitive, and personality measures.

Our long range plan is to follow these children until some appreciable number become schizophrenic. We will then be in a position (1) to look back at the results of our current examinations and determine the childhood characteristics which distinguish those that succumbed, (2) investigate the interaction of these characteristics with life circumstances including conditions surrounding the onset of psychosis, (3) observe characteristics which differentiate these children and plot them as they change from the time of the first examination to frank psychosis.

We study high-risk children because the prevalence rate of the disorder in the general population is only approximately .29% (Lemkau & Crocetti, 1958). In view of the extensive testing involved the project would not be possible if subjects were selected

[1] This report is the essence of a paper presented to the XV International Congress of Applied Psychology, 1964, at Ljubljana, Yugoslavia. This research is supported by the U. S. Public Health Service, National Association for Mental Health, Scottish Rite Committee for Research on Schizophrenia, and the Human Ecology Fund. This research is being conducted at the Psychological Institute, Kommunehospital, Copenhagen, Denmark.

without bias. It would be necessary to examine 10,000 individuals in order to include perhaps 29 who would subsequently become schizophrenic. To maximize the number of cases of schizophrenia which will occur in our sample we study individuals whose mothers have been schizophrenic. The psychiatric genetics literature suggests that 16% to 21% of individuals with one schizophrenic parent will become schizophrenic (Fuller & Thompson, 1960).

The purpose of this paper is to report the background, method and preliminary results of this project. We shall begin by briefly relating the theoretical orientation which prompted the choice of variables.

Schizophrenia: A Learned Disorder of Thought

This theory has appeared in a number of publications (Mednick, 1958; 1961; 1962); the presentation will be brief and discursive.

In this discussion of the theory we are going to suggest that the origin of the schizophrenic disorder lies in the interaction of certain predisposing factors (perhaps genetically determined) and unkind environments. In the context of behavior theory, we shall first try to reconstruct how an individual may experience an acute breakdown; then we shall try to explain the transition from the acute state to the chronic state. Following this, we shall discuss the theory in the light of research on the reactive-process continuum.

The theory begins with the hypothesis that there are three characteristics which predispose an individual to schizophrenia: (1) excessively strong reactions to mild stress, (2) excessively slow recovery from autonomic imbalance, (3) excessive stimulus generalization reactiveness.

These hypotheses were originally suggested by the ease with which schizophrenics form defensive conditioned responses and the frequent finding that schizophrenics over-generalize (Mednick, 1958). Much research on the arousal state in schizophrenia lends background support to these hypotheses. There is certainly strong evidence for Arieti's description of the first phase of schizophrenia as a "period of intense anxiety and panic." There is a body of psychophysiological literature which points to a low threshold of emotional arousal in schizophrenia. This includes studies of heart rate, GSR (galvanic skin resistance), EMG (electromyography), the startle reflex, etc. A review of the literature on psychophysiological research in schizophrenia by Silber (1962) pointed up the equivocality of results and failings of method in this area. However, there is very consistent evidence that the schizophrenic takes longer to recover from autonomic imbalance than does the normal. The more recent and methodologically sophisticated work in EMG and GSR is especially supportive (Ax et al., 1961; Williams, 1953). We are highly cognizant of the research that conflicts with this interpretation. We do not pretend to explain it all. However, frequently, where the schizophrenic is found to be underaroused, psychophysiologically, the experimenter has been testing very chronic patients. In some research where schizophrenics have not shown elevated psychophysiological responsiveness the experimenter has pointed to the hyper-aroused basal condition of the schizophrenic as placing a ceiling on responsiveness (Williams, 1953). We are also aware of the questions recently raised concerning the hypothesis of greater stimulus generalization responsiveness. The results of these experiments, again, can frequently be better understood by careful study of the chronicity of the subjects.

Explanation of Acute Breakdown

We shall try to explain how an acute schizophrenic episode might ensue in an individual with these three predispositions. In this instance we shall deal with an incipient *reactive* schizophrenic. We shall then use the theory to explain chronicity and the difference between the reactive and process schizophrenic.

In reactive schizophrenia we frequently find a precipitating event or series of events. Let us closely examine the effect of such a stress. A normal individual will show an arousal response to stress. Because of his heightened sensitivity, the preschizophrenic will show a larger response and will recover at a slower rate than the normal. If at this point the individual can get away from the troublesome situation and allow his anxiety to subside to its resting level we would expect no further untoward consequence to this specific upset. However, if he continues to operate in society (for example if he is an individual from a lower socioeconomic group and must stick with his daily job) there are certain dangerous consequences which might ensue. For one thing, the increase in anxiety or arousal level will cause an attendant increase in the level and breadth of stimulus and associative generalization. This increase in generalization will affect *all* of the individual's habits including those related to learned anxiety reactions. In other words, there will be a broadening of generalization gradients around all stimuli the individual fears. The increase in the breadth of generalization will cause many new stimuli to become potential anxiety arousers; in addition, those stimuli which previously elicited anxiety reactions now will elicit stronger anxiety reactions. Because of the increase in the breadth of the gradient of stimulus generalization, the possibility that he will encounter an anxiety arousing stimulus has increased. If he continues in society and does encounter such stimuli this will increase his arousal level. This, in turn, will cause his generalization responsiveness to increase. This will increase the probability of his encountering fear-arousing stimuli by increasing the *number* of such stimuli and the *magnitude* of his response to such stimuli. This will cause an additional increase in his anxiety level which will, in turn, again increase his generalization responsiveness, and so forth. This process has been named *reciprocal augmentation*. An empirical investigation testing aspects of the reciprocal augmentation hypothesis has been strongly supportive to the hypothesis (Mednick & Wild, 1962). Perhaps the symptoms we observe in the acute patient can be understood in terms of excessive generalization responsiveness. As generalization mounts, the individual will experience atypical thought sequences and tangential intrusions into his stream of ideation. These will be due to associative generalization. His perception will be distorted through stimulus generalization. Noises in the plumbing will sound like speech; shadows will resemble people. Because of the uncontrolled nature of some of his thought and the irreality of some of his perceptions he will begin to fear that he is "going crazy." This will only lead to further anxiety. His psychophysiological state will be producing internal stimuli which will partially cue ongoing thought processes. These thoughts will therefore tend to dwell on previous periods of anxious discomfort and create further anxiety-producing stimulation. Eventually some idea may be hit upon which will tend to reduce slightly the torturing fears. This solution (perhaps involving the conviction that he is being subjected to special scrutiny by the FBI) may consist of a "rational" explanation of his difficulties in thinking or of his generalization-produced fears of police or of his employer. This "rational explanation" may be enough to cause the admitting psychiatrist to diagnose him as a paranoid schizophrenic. For surely by this time his behavior will have become atypical or annoying enough to others to bring about this hospitalization.

The crux of this explanation of the acute process lies in the reciprocal augmentation of generalization and anxiety. One question which may be asked is "Why doesn't everybody become schizophrenic after a stress experience?" The answer lies in the three predisposing characteristics of the preschizophrenic.

1. Low threshold for anxiety arousal.
2. Slow recovery rate.
3. High generalization reactiveness.

An individual may have a low threshold for anxiety arousal but recover quite quickly. Such an individual will not suffer a reciprocal augmentation process since the building up of anxiety would be prevented by his fast recovery. He would only be threatened by such a process in conditions in which stresses followed one another at a very rapid pace. Such circumstances may be found under severe military combat conditions. Indeed, there is good evidence that combat experience tends to produce acute psychotic breakdowns (Benedict, 1958).

Chronicity in Schizophrenia

We have discussed the essentials of the explanation of the onset of schizophrenia in a reactive schizophrenic. Now we shall attempt to follow him into chronicity. Meyer and Kraeplin both described schizophrenia as an evasion of life. We believe that this evasion or avoidance is learned. We believe that the learning takes place gradually over a period of time and that *it is this avoidance learning which is schizophrenia*. This learning is especially prevalent during periods of high anxiety. During such periods thought with anxiety-producing content dominates the thinking of the patient. Due to generalization, remote, tangential associations interfere with the anxiety-producing stream of thought. Thus, an anxiety-producing thought will be momentarily replaced with one which is irrelevant, neutral, or perhaps even relatively comfortable. As a result, the associative transition will, at times, be accompanied by a momentary reduction of anxiety. This will be due to the momentary removal of anxiety-producing thought from ideation. This drop in anxiety level will reinforce the preceding associative transition. This will increase the probability of the occurrence of this associative transition in the context of anxiety-provoking thoughts. As this process occurs over and over again for a long period of time, it will provide the patient with a repertoire of thought responses which have the power to reduce anxiety. After several acute breaks the patient will have acquired a built-in anxiety regulator which will remove him from any stimulation with a hint of threat. If at this point the patient responds with the anxiety provoking thought, "I am going crazy," he may defend against it by making an immediate associative transition to an irrelevant, tangential thought or by making use of a well-learned rationale such as "the radiators are broadcasting to me." This avoidant thinking will be continually self-reinforcing since it will enable him to escape anxiety-producing stimuli. It will assume the status of a conditioned response which will be elicited automatically upon the presentation of anxiety-producing stimulation. This will make it extremely difficult to reach his awareness with any stimulation which is even mildly anxiety-arousing. As more of these avoidant associates are learned the patient will experience less and less anxiety and may very well be described as having "flat affect." What is especially damning is the effectiveness of this means of reducing anxiety and the fact that it is automatically self-sustaining through reinforcement. Whenever an anxiety-producing stimulus presents itself, the individual's anxiety-level will momentarily rise. An avoidant associative transition will remove him from the threatening stimulus and result in a reduction in anxiety which will once again reinforce the learned avoidant thinking mechanism. At this point it is difficult to see how this avoidant habit can be extinguished. Most avoidant responses are quite resistant to extinction because of their built-in self-reinforcement. Extinction can be hastened by physically preventing the avoidant response in the presence of the danger signal and omitting the punishment. It is difficult to imagine how an avoidant thought might be physically prevented. One aspect of this explanation should be emphasized. Schizophrenia is here seen as a learned disorder of thought and emotion. The acute phase of the illness in the reactive patient is not in and of itself schizophrenia. It is the learning of the avoidant thoughts which is the *essence* of the schizophrenic disorder. Chronicity is determined by the degree to which the patient's thinking is dominated by avoidant thought sequences, and we believe that chronicity may best be measured in this manner. We have attempted to do such measurement in terms of the relative percentage of associations which are avoidant under conditions of mild arousal. This measure related well to the length of hospitalization.

Reactive and Process Schizophrenia

As an example of how the theory may be seen as being relevant to other aspects of research on schizophrenia, let us now turn to the question of reactive versus process or good premorbid versus poor premorbid schizophrenia. In the light of the above explanation, the difference between the reactive and process patient is simply one of level to which the patients' thoughts are dominated by avoidant associates at the time of hospitalization. For purposes of this discussion the process patient is seen as initially having the same three predisposing characteristics as the reactive patient.

The reactive schizophrenic comes to the hospital with a relatively healthy premorbid history and learns the bulk of the avoidant thought sequences at the time of his acute break. The process patient, on the other hand, has suffered repeated trauma and stress early in life and has acquired a highly developed set of avoidant associative thought patterns *before hospitalization*. Basing the following speculations on some of the work of the Duke group, we would guess that in those cases where the reactive and process schizophrenics both have the same 3 predisposing characteristics, differences between them may be explained in terms of their family situation. Briefly, there is good evidence that the process patient is born into a family with a strident, carping, and punishing mother (Farina, 1960). A highly sensitive infant, with such a mother, is exposed to continual overstimulation of autonomic functioning. With techniques appropriate to his age, the child soon learns many ways of avoiding or withdrawing from the noxious stimulation that the mother presents. These avoidant responses first learned in reaction to the mother will generalize to teachers and playmates and in adolescence to girls. Throughout childhood and adolescence we would expect the process preschizophrenic to manifest the classical withdrawn pattern. During this time his thinking is gradually becoming dominated by avoidant associative sequences. We find that he typically enters a mental hospital with very inadequate precipitating stresses or else he simply drifts over the line of environmental tolerance.

On the other hand, the reactive patient with the same predisposing characteristics is probably presented with a warm, supporting maternal environment. His premorbid history is typically healthy. His acute break results from a combination of a number of severe stresses and his predispositions. He begins to learn most of his schizophrenia at the time of his first break.

Role of the Theory in Research

We chose to organize this investigation around a theory of schizophrenia. It could have been done differently. Other investigators may have chosen a shotgun approach. Our personal preference and scientific strategy is always to try to work within some organizing conceptual framework. The role of the theory is not only to help organize but also to delimit and motivate. It is true that the theory helped us to select the measures that will be described. However, most of these measures has been shown in previous research to discriminate schizophrenics from normals and could have been chosen on this basis alone. Each of our variables has relevance to an area of research on schizophrenia and will contribute useful data aside from whatever reflection it casts on the theory.

CHECK ON PSYCHIATRIC GENETIC LITERATURE

We are studying children with schizophrenic mothers because it has been reported that the probability of schizophrenia is high in such samples. In view of the reliance we place on this assumption it seemed prudent to check these reports on a sample more or less identical to our sample. Dr. Niels Reisby (a young Danish psychiatrist) and Dr. Fini Schulsinger conducted this investigation in 1963. We shall summarize some of the relevant findings.

First, they found the names of 98 women patients in four mental hospitals. From the hospital records, the Folke-register (an up-to-date record of the current address of every individual in Denmark), and the Human Genetics Institute they prepared a list of their children. Again, at the Human Genetics Institute (where every psychiatric hospitalization is registered) they checked to see the percent of these children that had been hospitalized for schizophrenia. This figure proved to be 12.7%. (They used Strömgren's modification of the Weinberg correction. The average age of the children was 34.8 years.) The average age at first admission for those children who succumbed was 24.6 years. We can consider 12.7% to be a sizable under-estimation of the prevalence of schizophrenia in this sample because well over 50% of the children live on small farms in isolated areas of Denmark, but only 27% of those who were hospitalized for schizophrenia were from such farm areas. In terms of clinical experience and from surveys done in equivalent areas, it is known that the farm people shelter many harmless "trouble-free" schizophrenics and avoid hospitalization if at all possible. They are currently doing a follow-up to obtain a reduction of the number of unidentified children; this will probably increase the estimate of risk. In any case, the figure, 12.7%, is not far from Kallmann's (1938) figure of 16%.

Of great interest in this study is the distribution of ages of the children at first admission (Table I). In view of the fact that the

TABLE I

CHILDREN WITH SCHIZOPHRENIC MOTHERS[a]

Children's age of first hospitalization for mental illness

Age	Percent
Below 20	38
21–25	14
26–30	10
31–35	24
36–40	10
41 and over	5

[a] The total number of cases is 21. The total number of mothers is 98. The mean age of the mothers at the time of investigation was 61.5 years; the children were 34.8 years.

mothers were selected on the same basis as the mothers of the proposed long-term study, this table should enable us to estimate the ages of breakdown of the children in the long-term study. Of course, since we propose to re-examine the children periodically we should be able to diagnose the illness well in advance of the first hospitalization.

While not relevant to the aim under examination, it is of interest that by reviewing the hospital records Reisby and Schulsinger were able to divide the schizophrenic group into process and reactive patients. In the case of the process patients, the mothers had been with the children until the age of 13.0 years, before separation. (The separation usually coincided with the hospitalization of the mother.) In the case of the reactive patients, the mothers had been with the children until the age of 7.4 years.

With respect to the major aim of this study, it is clear that the results are consonant with Kallmann's (1938). They suggest that our sample of younger children of schizophrenic mothers will yield schizophrenics in numbers sufficient for purposes of our research. We have two additional pieces of information. We can maximize our rate of hospitalization for schizophrenia by concentrating on an urban sample. In addition to schizophrenia, the children of these mothers manifested a number of other varieties of socially deviant behaviors, e.g., suicide, mental deficiency, and psychopathy. While the resulting Ns for each of these deviancies may not reach levels appropriate for statistical analysis, by having followed such individuals from childhood we may be able to formulate hypotheses concerning etiology. It will be of special interest to relate differential pathological outcome within sibling groups to differential individual characteristics and environmental conditions.

Selection of Site for Study

The senior author first attempted this study in Michigan in 1960 (Mednick, 1962). The attempt had to be abandoned because the hospital records proved inadequate and the high residential mobility rate made it improbable that we could follow our subjects for any appreciable amount of time. Fifty percent of the entire Detroit population moves every 4 years. When the mobility patterns for individuals in the age range of greatest danger of incidence of schizophrenia were examined, a 47% change of address every year was noted. This figure is even greater for the low-income non-homeowner. The group of greatest interest is also the most mobile (U. S. Bureau of the Census, 1956; Detroit Area Study, 1956). If 10 of the subjects should move to California one would be hard pressed to locate them there. If one were able to do so it would cost up to $5000 to bring them with a responsible adult for one day of testing. On the basis of these data and other considerations the decision was made not to attempt this study. Anything less than certainty of 90% follow-up success was unacceptable.

Such certainty seemed unattainable until an address by the internationally eminent Danish research worker in psychiatric genetics, Professor Erik Strömgren, directed the attention of the senior author to opportunities for research in Scandinavia. As an example, Fremming (1951) carried out a 60-year follow-up of 5500 individuals. He was successful in locating 92% of his sample. Now, Myschetzky, in Denmark, is attempting (with great success) to locate this same sample. At this point his effort represents a 78-year follow-up.

Clearly the study just abandoned as impractical was readily feasible in Scandinavia. Further investigation and a visit to Scandinavia in the summer of 1961 revealed that the epidemiological aspects of schizophrenia are about what we experience here in the United States (Lemkau & Crocetti, 1958). A tour of the mental hospital system of Denmark in the summer of 1961 suggested that the clinical picture which emerges is not essentially different from what we observe in U.S. hospitals. The remarkable and complete national psychiatric register at the Copenhagen University Institute for Human Genetics would greatly facilitate various aspects of the study. The Folke-register (which maintains an up-to-date record of the current home address of every individual living in Denmark) completely explained the excellent record of the longitudinal research mentioned above.

What further recommended Denmark as a site for the proposed research was the compact size of the country (16,619 square miles, somewhat smaller than Massachusetts and New Hampshire combined). The population is concentrated in greater Copenhagen. These factors all tend to minimize the distance that subjects must be transported. The major loss of subjects in a longitudinal study in Denmark will be through external emigration. If any Ss from our sample do emigrate an attempt could be made to locate them. The low rate of emigration of Danes suggests that in any case this will not be a serious problem in our research (Statistiske Meddelelser, 1961).

Finally Dr. Schulsinger (Director of the Department of Psychiatry, Kommunehospitalet) became actively interested in the study and was able to solve the subtle and difficult organizational problems which the project presented.

Method

SAMPLE PROCUREMENT

Our procedure entailed several steps. First, we sent social workers to state hospitals in the vicinity of Copenhagen. (We wanted most of our children to be urban residents so as to make it more likely that they would be hospitalized upon becoming psychotic.) The social workers obtained lists of resident female patients who had children between the ages of 10 and 18. These lists were checked with the University Institute for Human Genetics and the Folke-register. The mothers' hospital records were examined by two psychiatrists who certified chronic schizophrenia. (After almost perfect reliability of diagnosis was established only one psychiatrist was used.) Social workers then went to the homes of the children and enlisted their cooperation and the cooperation of the responsible adults. The children were then scheduled for testing.

After some children of schizophrenic mothers (Experimental group) had been tested we began forming *pairs* of matched Experimental (E) subjects. When some of the Experimental subjects begin to become schizophrenic we shall already have made an unbiased selection of comparison E subjects who did not become schizophrenic. The pairs of E subjects were matched on sex, age, father's occupation (the best measure of social class in Denmark; Svalastoga, 1959), rural-urban residence, years of education, and upbringing in a children's home versus family life. Next, a single Control (C) subject was selected who was matched on these same variables "individual for individual" with each E pair. We wrote to a school attended by one or both of the E subjects and asked for the names of children, at that school, meeting all the matching criteria. The school typically would send through some names. The name of the E child was removed; the other names were then sent to the Institute for Human Genetics, which returned a report on each child concerning *any* history of psychiatric hospitalization or civil disturbance for the child, his parents or his grandparents on either side. For example, on Child A we may have learned that his grandfather on his mother's side had been hospitalized for two months in 1936 for severe depression. Child B's record might be clean. We would choose Child B, of course, as our control subject. A social worker would visit Child B's home and engage the cooperation of the child and his guardians.

In the case that an E child was in a children's home, he would be paired with another child in a children's home. C children were also sampled from children's homes. These were checked as usual with the Institute for Human Genetics. In most of these C cases the parents had divorced, died, or were alcoholics or criminals. These controls from children's homes will be examined carefully in the data analysis. In any case, they represent some degree of control for the broken home aspects of being a child of a schizophrenic mother. Having summarized our subject-procuring methods, we shall now back-track and present some of the stages in greater detail, discuss problems which arose and our solutions.

SELECTION OF SCHIZOPHRENIC MOTHERS

We chose to study children whose *mothers* are *process* schizophrenics for a number of reasons.

1. Paternity is sometimes questionable in normal middle class families. It might be difficult to be quite sure of our major independent variables with alleged fathers, especially if they are schizophrenic.
2. Schizophrenic women have more children than schizophrenic men (Goldfarb & Erlenmeyer-Kimling, 1962) giving one the luxury of a larger subject pool. In a pilot study in Nykøbing, Sjaelland, we found schizophrenic women had about five times as many children as schizophrenic men.
3. Psychodevelopmentally, mothers presumably play a greater role in shaping children (Sanua, 1961). Studying mothers has permitted us to carry out some research on the effects of being reared by a schizophrenic mother.

4. The offspring of process schizophrenic mothers yield a higher rate of schizophrenia (Schulz, 1939, 1940; Lewis, 1957).

There are aspects of family dynamics that will be biased because we chose mothers. For example, Rosenthal (1962) has shown that in a family with a schizophrenic the family members of the same sex are more likely to be concordant for schizophrenia. We shall expect more of our Experimental girls to become schizophrenic.

DIAGNOSIS OF MOTHER

Two experienced psychiatrists trained together and then independently tested their reliability in making judgments from hospital records on the form prepared to record the mothers' symptoms. Their agreement as to diagnosis on 20 test cases was found to be 100%. They merely had to judge whether the mothers were typi-

TABLE II
CHARACTERISTICS OF THE EXPERIMENTAL AND CONTROL SAMPLES

	Control	Experimental
Number of cases	104	207
Number of boys	59	121
Number of girls	45	86
Mean age[a]	15.1	15.1
Mean social class[b]	2.3	2.2
Mean years education	7.3	7.0
Percent of group in children's homes[c] (5 years or more)	14%	16%
Mean number of years in children's homes[c] (5 years or more)	8.5	9.4
Percent of group with rural residence[d]	22%	26%

[a] Defined as age to the nearest whole year.

[b] Determinations based on Svalastoga (1959).

[c] We only considered experience in children's homes of five years or greater duration. Many of the Experimental children had been to children's homes for brief periods while their mother was hospitalized. These experiences were seen as quite different from the experience of children who actually had to make a children's home their home until they could go out and earn their own living.

[d] A rural residence was defined as living in a town with a population of 2500 persons or less.

cal schizophrenics. They were instructed to discard any questionable cases. Following this reliability check only one psychiatrist checked each record.

CHARACTERISTICS OF THE SAMPLE

An initial letter was sent to the family informing them of the project and the fact that they have been included in the sample. The project has received copious newspaper, radio, and TV coverage, so that in most cases the family was already somewhat prepared. Both C and E families were told that we were interested in the effect of "nervous breakdown" on the members of the family. They were also informed of our need for normal families. We did not identify to which group we thought they belonged, but this was usually obvious to them.

Table II presents the identifying characteristics of the E and C groups.

Procedure

Until testing was complete none of the researchers was informed regarding whether the children tested that day were C or E. (The interviewing psychiatrist could usually guess from the interview material.) The social worker scheduled all visits. The procedure was identical for C and E subjects.

THE SUBJECT'S DAY

Testing began at 8 A.M. First S had his height and weight taken. He was then escorted to the psychophysiological laboratory where he underwent mild stress, conditioning, and stimulus generalization procedures. He then received a full Wechsler Intelligence Scale for Children and an abbreviated MMPI. Then came lunch. After lunch he took two Word Association Tests (single word response and continual association) and completed an adjective check list describing himself. He then returned for a second psychophysiology session where he underwent mild stress, semantic conditioning, and mediated generalization procedures. Finally, S was interviewed by a psychiatrist and given an honorarium. A sub-sample also provided urine samples.

PSYCHOPHYSIOLOGY PROCEDURE

Morning Session

Recording was done on an Offner Type R Dynograph. We made use of a device designed in Dr. Ax's laboratory to reverse polarity of the electrodes every 1.2 seconds. The psychophysiology laboratory was in the basement of a hospital, in a room separated by two tight-fitting doors from a lightly traveled corridor. The humidity and temperature of the laboratory were relatively constant but recorded before and after each session. After washing and alcohol sponging at points of electrode placement, S reclined on a hospital bed and was asked to relax. Respiration, heart rate, GSR, and EMG electrodes were attached.

Respiration—An Offner respiration transducer was fastened just above the waistline and below the diaphragm.

Heart rate—Electrodes were attached with rubber straps to the ankles and to the left arm just below the elbow. Electrode paste was vigorously rubbed in.

EMG electrodes—EMG electrodes were fastened with elastic tape 2 cm above the eyebrows and 5 cm on either side of the noseline. Electrode paste was vigorously rubbed in.

GSR electrodes—GSR electrodes consisted of zinc, 7 mm in diameter, embedded in a plastic cup. Small sponges saturated in zinc sulfate solution were inserted into the plastic cup.

When the transducers were attached, recording was started and was continuous until the conclusion of generalization testing. S was permitted to relax for 15 minutes. Earphones were then attached.

CONDITIONING AND GENERALIZATION PROCEDURE

Approximately 30 seconds after the tape recorder was started, S heard instructions informing him of the procedure to follow. At the end of the instructions there was a silence of 70 seconds followed by 8 desensitization presentations of the CS (conditioned stimulus—1000-cps tone). Nine seconds after the final "desensitization trial," conditioning trials began. Unconditioned stimulus (UCS) was an irritating noise of 96 db (where 0 db = .0002 dynes/cm^2) presented for 4½ seconds following ½ second after the onset of CS (54 db). There were 14 partial reinforcement trials (9 CS-

UCS pairings, and 5 interspersed presentations of the CS alone). Trials were separated by intervals which varied from 17 to 77 seconds.

Following the final conditioning trial, there was an interval of 3 minutes, following which conditioning and stimulus generalization testing began. Generalization stimuli were tones of 1311 cps (GS_1) and 1967 cps (GS_2). There were 9 trials, 3 each of CS, GS_1, and GS_2. Duration of CS, GS_1, and GS_2 was 2 seconds. The order of the CS, GS_1, and GS_2 was counterbalanced.

The final conditioning and stimulus generalization testing trial marked the end of the morning session, which took approximately 50 minutes. S was disconnected from the apparatus and escorted from the laboratory.

Afternoon Psychophysiology Session

Word Association Norms

Before beginning this pilot study we obtained and tabulated word association norms for the Kent-Rosanoff List from 145 Danish school children between the ages of 9 and 16 years. This was in preparation for the semantic conditioning and mediated generalization procedures of the afternoon psychophysiology session and the Word Association Test.

In this procedure a psychophysiological response was conditioned to the word "light" ("lys" in Danish). In mediated generalization testing we presented the stimuli "lamp" ("lampe") in Danish), "dark" ("mørk" in Danish), and "fruit" ("frugt in Danish). These words as stimuli vary in the probability that they will elicit "light" as an associate (61%, 20%, 0%, respectively in the Danish norms). "Lamp" or "dark" elicit the implicit association "light" which elicits the conditioned psychophysiological response. The strength of this response has been shown to vary with the strength of association between the generalization test stimulus and the original conditioned stimulus (M. T. Mednick, 1957; Mednick & Wild, 1962).

The conditioning and generalization test procedures in the afternoon were otherwise equivalent to the morning session.

Wechsler Intelligence Scale for Children—A Danish translation of the WISC in common use in Denmark was used. Danish norms will be available in the future. All subtests were administered.

MMPI—A Danish translation was used. It should be noted that norms do not exist for Danish scales. We intend to attempt to develop a preschizophrenia scale. The test was shortened to 304 items by removing items deemed inappropriate or offensive to children.

Word Association Tests (WAT)—A Danish version of the Kent-Rosanoff Test was administered. The word "mutton" was omitted since it is rather unusual in Danish. The words "sun" and "star" were added to make a total of 101 words. The words were read to S with the instruction to respond to each word with the first word that came to mind. Response latency was recorded.

Subjects were also presented with 30 other words and asked to associate continuously to each of them for one minute. Responses and response latencies were recorded. We attempted to avoid associative chaining by having the stimulus word on a card before S during the entire minute period.

Adjective Check List (ACL)—A specially constructed list of 241 adjectives was administered to each S with instructions to check those items which he would use to describe himself. The interviewing psychiatrist, WISC, and WAT administrators also used the ACL to describe each S.

Psychiatric Interview

Each S underwent a psychiatric interview to assess his current diagnostic status. The interview was largely precoded and highly structured. The questions concerned S's mental status, social history, and attitudes. There are 94 coding items in the interview. Two coders independently coding 20 cases agreed in 100% of the 20 cases on 69 of these coding items, 95% on 9 of these items, 90% on 10 of these items, and 85% on 6 of these items. We considered the coding reliable for these items. Six other items were discarded since they could not reach 85% agreement.

The psychiatric interview took 30–40 minutes.

Measures Obtained Outside of Testing Day

PARENTAL INTERVIEW

The social workers interviewed the individual responsible for the child on questions concerning the child's current behavior, social development, school behavior, parental behavior, and parents' socioeconomic status. The interview was highly structured. Working with 30 cases and 88 coding items, two coders achieved 100% agreement on 61 items, 95% agreement on 11 items, 90% agreement on 13 items, and 85% on 3 items. We considered the coding reliable.

SCHOOL REPORT

A questionnaire was mailed to each S's school. Items concerned S's relationship to the teacher and classmates as well as academic achievement. Of 311 forms sent out, 310 have been received to this date. Items were in the "true, not true" form. The questionnaire was filled out by the teacher most familiar with S.

Results

We shall report a group of initial analyses comparing the Experimental and Control groups on selected measures. The psychophysiological data, WISC and Word Association Test have been analyzed by analysis of variance and covariance. The remainder of the measures have been analyzed by chi square tests.

PSYCHOPHYSIOLOGY

Description of Measures and Scores

The GSR data of the morning psychophysiology session have been subjected to some analysis. The other channels have as yet not been scored. The afternoon session has not yet been punched in IBM cards. (It should be mentioned that it takes approximately 8 hours of an *experienced* scorer's time to score one GSR record. Thus, scoring the GSR alone took over 2500 man-hours. Each S's morning session takes 15 fully punched IBM cards.)

The following variables were measured from the GSR records:

1. *Latency of response.*
2. *Time to response peak.*
3. *Time from the highest point of the response to full recovery from the response.*

4. *Time to half-recovery.* If the response was a 10,000-ohm drop in resistance, we measured the number of mm to recovery of 5000 ohms.
5. *Basal level.* This was measured in ohms resistance by referral to prepared conversion sheets based on calibrations. This was the pre-response level.
6. *Amplitude of response.* This was measured in ohms resistance change. We intend to transform these scores into Lacey's Lability Scores (Lacey & Lacey, 1962). This will serve the function of allowing us to evaluate responsiveness while removing its correlation from basal resistance level. This is essential for interpreting individual records. However, for group comparisons we have achieved *precisely* the same purpose by analyses of covariance using basal level as our control variable.
7. *Amount of recovery.* This was measured in ohms resistance.
8. *Angle of recovery.* This was measured in degrees of deviation from the horizontal that the record showed in the 10 mm immediately following response peak. While this variable has been scored for the entire sample for every response it has not yet been subjected to analysis.

Scoring the GSR record was a real chore. Each scorer's work was checked and often double-spot-checked. Scorers were almost only employed on a part-time basis in order to reduce fatigue-induced errors. Regular scorer-conferences were held in order to discuss scoring problems and eliminate differences between scorers. Our checking procedures leave us confident in the reliability of the final scores.

STATISTICAL METHOD

A selected group of GSR measures have been subjected to two-way factorial analyses of variance and covariance. Our two ways have been groups (E vs. C) and age (Group 1: 8 to 13.2 years; Group 2: 13.3 to 16.6 years; and Group 3: 16.7 to 22 years). Lacey and Lacey (1962) report that in their study of the psychophysiological behavior of 20 boys and 17 girls, "age was a more pervasive and powerful factor than sex." This was our reason for including age rather than sex as the second "way" in these initial analyses of variance. Since our C and E groups differed in N we computed our Fs for unequal Ns. The age groups (1, 2, and 3) were established so as to achieve proportionality of numbers of subjects across the two ways (Table III).

TABLE III
ANALYSIS OF GSR DATA[a]

Number of subjects by group and age

Age Groups	Group		Total
	Control	Experimental	
1	34	68	102
2	33	66	99
3	37	74	111
	104	208	312

[a] Cell mean scores were inserted for two E-Ss in Age group 1, one E-S in Age group 2, and one C-S in Age group 3 in order to achieve proportionality.

PRELIMINARY ANALYSES

Our psychophysiological laboratory was not temperature or humidity controlled. Since certain GSR measurements might be related to these variables we recorded temperature and relative humidity before each session. The mean temperatures for E and C groups were 23.3 and 23.6°C. The mean temperatures by Age groups (across E and C groups) were 23.8, 22.5, and 23.8°C. As explained above, the reliability of these differences were analyzed by a 2-way factorial analysis of variance. The E-C difference was not significant; the Age groups differences were significant ($F = 8.91$, 2305 df, $p < .01$).

The mean relative humidities for the sessions of the E and C groups, respectively, were 42.7% and 38.9%; this difference was highly significant ($F = 36.30$, 1305 df, $p < .01$). The mean relative humidities by Age group were 43.3%, 41.6%, and 47.1% for Age groups 1, 2, and 3. These differences were also significant ($F = 7.01$, 2305 df, $p < .01$). We shall take account of humidity and temperature in analyses reported below.

LATENCY OF GSR RESPONSE

Here we chose to analyze the latency of response to the first Desensitization Tone and the first two presentations of the CS-UCS (stress stimulus). As can be seen in Table IV the latency of the GSR was consistently and significantly shorter for the E group. No significant differences were observed as a function of age. The reader may note that the mean latencies are apparently shorter to the Desensitization Tone than to the CS-UCS. This is probably because only the more reactive individuals would respond to the Desensitization Tone while everyone (even the "sluggish" responder) responded to the CS-UCS.

TABLE IV
LATENCY OF GSR[a]

Points of Measurement	Experimental	Control	F	df	p
Desensitization Tone I	1.51	2.09	37.97	1305	<.01
CS-UCS$_1$	1.85	2.23	10.76	1305	<.01
CS-UCS$_2$	1.46	1.75	5.08	1305	<.05

	Age Groups				
	1	2	3		
Desensitization Tone I	1.70	1.63	1.75		n.s.
CS-UCS$_1$	2.02	1.99	1.94		n.s.
CS-UCS$_2$	1.46	1.56	1.66		n.s.

[a] Comparison of E-C groups and age groups on mean latency in seconds at three points of measurement.

PEAK RESPONSE TIME

The mean Peak Response Times for the E and C groups to CS-UCS$_1$ were 11 and 9 seconds, respectively. Clearly the E group took longer to reach the peak of its response. However, as we shall see below the E group showed a larger GSR to CS-UCS$_1$. It would take longer for this group to reach the peak of its response even if the rate of approach to this peak were the same for both groups. Consequently, an analysis of covariance was made with "Size of response to CS-UCS$_1$" being the control variable. The difference between the E and C groups proved not to be significant in this case; nor were there significant differences between the Age groups. The correlation between "Peak Response Time" and "Size of Response" for CS-UCS$_1$ was .92.

BASAL LEVEL

Figure 1 depicts changes in basal level of skin resistance for chosen points across the experimental session for E and C groups and Age groups. Table V reports these in terms of mean level and significance of the observed differences. First we should point out that the analysis of basal level during the Third Rest Period showed the groups to differ significantly even when covariance controls for temperature and humidity were employed. The correlation be-

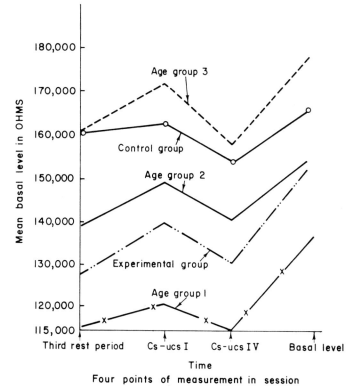

FIG. 1. Basal levels in ohms across the experimental session.

tween temperature and basal level was —.12, between humidity and basal level was —.15. This is the typical level of correlation between these variables. These are the only analyses which took into account the temperature *and* humidity differences. While these analyses suggest that the influence of these two variables might not be of great importance, we shall take these variables into account in further analyses before dismissing them completely.

TABLE V

BASAL LEVELS IN OHMS RESISTANCE[a]

Points of Measurement	Experimental	Control	F	df	p
Third rest period[b]	128,340	161,700	9.26	1305	<.01
Third rest period[c]	128,340	161,700	7.52	1305	<.01
CS-UCS$_1$	140,860	163,180	4.28	1305	<.01
CS-UCS$_4$	131,150	154,410	6.19	1305	<.05
Final basal level	152,920	166,500	—	—	n.s.

	Age groups					
	1	2	3			
Third rest period[b]	116,880	139,000	160,620	7.79	2305	<.01
Third rest period[c]	116,880	139,000	160,620	7.93	2305	<.01
CS-UCS$_1$	121,280	149,060	172,470	8.65	2305	<.01
CS-UCS$_4$	115,010	141,460	158,570	8.39	2305	<.01
Final basal level	137,060	154,690	178,640	4.23	2305	<.05

[a] Comparison of *E-C* groups and age groups on mean basal level at four points of measurement.
[b] Covariance control for temperature.
[c] Covariance control for humidity.

There are several interesting features of Fig. 1 deserving comment. The *E* group is consistently at a lower basal level than the *C* group. The *C* group shows little change in basal level across the experimental session while the *E* group shows a rapid increase in basal level. The Final Basal Levels are not significantly different. It is possible that the *E* group responded more vigorously to the electrode hook-up conditions than did the *C* group. Consequently, the *E* group had greater distance to go to recover its pre-experimental basal level.

The differences in basal level among age groups were even greater than the E-C differences. This difference has been observed in previous research (Lacey & Lacey, 1962).

AMPLITUDE OF RESPONSE

As is typical for GSR research, we observed sizable correlations between the amplitude of a response (in ohms) and the preceding basal level (in ohms). In view of the significant differences obtained in basal level, we used this variable as a covariance control in conducting our analyses of variance on the amplitude of response data. In each case we used as a control the basal level immediately preceding the response under analysis.

The results of the analysis of response amplitude is presented in Table VI. The largest difference and certainly the most reliable (in terms of the size of the F) came on the first exposure of the subject to the UCS. In general across all trials, the E group responded with greater drops in resistance. On the trials testing for extent of conditioning, the E and C groups differed significantly on only one of the three trials analyzed. The difference in amplitude of response on this trial (the final test trial with the CS *only*) was not large, though apparently consistent. It is clear that the E group was considerably more reactive to the UCS (stress) stimulus than was the C group.

The Age group differences are inconsistent and not especially statistically reliable.

RECOVERY

This initial analysis of the recovery of the GSR was not completely satisfactory. There are several analyses that must be completed before we will be able to draw any firm conclusions. The analyses of this run will be reported. First, we analyzed differences in amount of recovery expressed in ohms. All of these analyses yielded significant Fs. The E group recovered significantly more ohms than the C group. However, since they had responded more, the results were ambiguous. For example, to CS-UCS$_1$ the average amount of recovery was 13,890 ohms for the E group and 9570 ohms for the C group. However, the E group's response was 23,080 ohms while the C group's response was 11,070 ohms. Thus the E group had recovered only 60% of its response while the C

group recovered 87% of its response. This suggests a data analysis using a ratio of amount recovered to amount responded. However, this would tend to exagerate the slowness of recovery of the group giving the larger response. There is only a finite amount of time in

TABLE VI

AMPLITUDE OF RESPONSE[a,b]

Points of Measurement	Experimental	Control	F	df	p
Instructions on	9,340	6,520	7.00	1305	<.01
$CS\text{-}UCS_1$	23,080	11,070	37.98	1305	<.01
$CS\text{-}UCS_2$	12,020	7,870	13.56	1305	<.01
$CS\text{-}UCS_3$	13,020	5,590	—	—	n.s.
$CS\text{-}UCS_4$	9,366	6,726	9.95	1305	<.01
$CS\text{-}UCS_5$	9,170	5,690	17.02	1305	<.01
CS_1	2,824	3,363	—	—	n.s.
CS_4	2,970	2,360	—	—	n.s.
CS_5	3,588	2,396	5.54	1305	<.05

	Age groups					
	1	2	3			
Instructions on	4,760	8,930	11,270	—	—	n.s.
$CS\text{-}UCS_1$	15,210	22,370	19,690	—	—	n.s.
$CS\text{-}UCS_2$	7,740	12,950	11,240	4.04	2305	<.05
$CS\text{-}UCS_3$	6,340	9,270	15,550	—	—	n.s.
$CS\text{-}UCS_4$	7,218	9,635	8,626	—	—	n.s.
$CS\text{-}UCS_5$	6,550	10,030	7,510	4.47	2305	<.05
CS_1	3,329	2,565	3,096	—	—	n.s.
CS_4	3,010	3,710	1,700	4.52	2305	<.05
CS_5	4,004	3,269	2,375	4.52	2305	<.05

[a] Comparison of E-C groups and age groups on mean size of response in ohms of GSR at nine points of measurement.

[b] All of the analyses in this table were conducted with the immediately preceding basal level as a covariance control.

which recovery can occur. This time ends with the advent of the next stimulus. A small response has a better chance of fully recovering than does a large response. (Clearly in future testing we will have to allow some large standard amount of time following a stress stimulus to evaluate this variable adequately.)

Given our current intertrial intervals (17–77 seconds) a recovery measure based on amount of recovery in ohms is a biased measure. We also scored our results in terms of the number of mm (amount of time) it took for the individual to return to his prestimulus basal level. Here again the measure is biased by size of response and intertrial interval. A measure which is less biased by intertrial interval is the number of mm (amount of time) it took for recovery half the way (in ohms) back to basal level. (How long does it take to recover 5000 ohms of a 10,000 ohm response?) Since we only ask for recovery of half of the response, the length of the intertrial interval should not be as important. Unfortunately, in this run, we only included a single analysis of this nature—time to half recovery of the response to $CS-UCS_7$. The E group showed "half-recovery" in 8.30 seconds while the C group "half-recovered" in 6.26 seconds. This difference was significant ($F = 6.51$, 1305 df, $p < .05$). However, when a covariance control for response amplitude was applied the F dropped to 3.01 which is not significant. It is not possible to draw a firm conclusion on the basis of this one analysis; if anything, it suggests that the E and C groups do not differ, significantly, in *rate* of recovery. We must assess more of the points of measurement using the half-recovery measure with a size-of-response covariance control before we can reach any firm conclusions.

In analyzing differences in recovery as a function of age we are beset with the same problems as those encountered with the E-C comparisons. We shall report the one analysis of half-recovery of $CS-UCS_7$ with size of response as a covariance control. After responding to this stimulus Age Groups 1, 2, and 3 showed half recovery in 6.10, 7.68, and 8.96 seconds, respectively. These differences are significant ($F = 4.31$, 2305 df, $p < .05$). Younger subjects recover more quickly from the effects of a stress stimulus.

GENERALIZATION

Figure 2 presents the generalization gradients for the E and C groups. This figure summarizes the data for the first two test trials of each stimulus. As can be seen the E group showed considerably more generalization responsiveness than the C group. Not only are the gradients more elevated but they are also flatter. Table VII presents the same data in terms of mean generalization response amplitude and includes the results of analyses of variance with the immediately preceding basal level as the covariance control. The differences apparent in Fig. 2 are statistically reliable.

Differences in generalization level may be due only to heightened responsiveness to the CS. Relative generalization is a measure

TABLE VII
AMPLITUDE OF GENERALIZATION RESPONSES[a,b]

Points of Measurement	Experimental	Control	F	df	p
First test					
CS (1000 cps)	3100	2300	—	1305	n.s.
GS$_1$ (1311 cps)	3140	1830	5.96	1305	< .05
GS$_2$ (1967 cps)	1982	625	9.68	1305	< .01
Second test					
CS (1000 cps)	2466	1349	5.26	1305	< .05
GS$_1$ (1311 cps)	2120	820	9.50	1305	< .01
GS$_2$ (1967 cps)	2740	587	2.47	1305	< .05

	Age Groups					
	1	2	3			
First test						
CS (1000 cps)	2510	2960	3020	—	—	n.s.
GS$_1$ (1311 cps)	2520	2930	2670	—	—	n.s.
GS$_2$ (1967 cps)	1479	2163	1011	—	—	n.s.
Second test						
CS (1000cps)	2015	2656	1665	—	—	n.s.
GS$_1$ (1311 cps)	1540	1480	2000	—	—	n.s.
GS$_2$ (1967 cps)	1853	3426	925	—	—	n.s.

[a] Comparison of E-C groups and age groups on mean response amplitude in ohms of GSR at six points of measurement.

[b] All of the analyses of response amplitude in this table were conducted with the immediately preceding basal level as a covariance control.

which expresses amplitude of response to the CS. Computed on the basis of group means the relative generalization percentages for the E group were 92% for GS$_1$ and 87% for GS$_2$; these figures for the C group were 67% and 38%. The E group also shows more relative generalization than the C group.

As can be seen from Table VII the age groups did not differ significantly in generalization response amplitude.

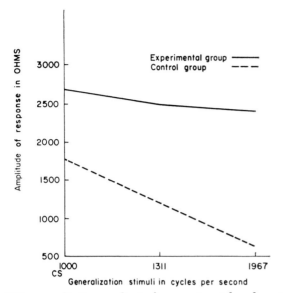

FIG. 2. GSR generalization gradients for experimental and control groups. Summary of first two test trials for each stimulus.

CONCLUSIONS CONCERNING GSR RESULTS

1. We predicted differences in response amplitude, speed of recovery, and generalization responsiveness. The two predictions concerning response amplitude and generalization were strongly supported. While the recovery data are not yet fully analyzed, the analyses completed up to the time of this report do not support the hypothesis.

2. Consistent and highly significant differences were observed for latency of the GSR response. This difference was greatest for the First Desensitization Tone. This difference was not predicted but is not dissonant with the general theoretical position. The short latency suggests that the E group is characterized by a volatile autonomic nervous system which is quickly and easily aroused.

3. The E group is at a considerably lower level of basal resistance than is the C group at the point at which we computed our first F (third section of the Rest Period). They show a large amount of recovery over the course of the psychophysiology session. It seems likely that their final basal level at the end of the

session is not higher than their presession level. If that is so, then it follows that their initial response to the experimental situation must have been much greater than any other response we observed and measured. Unfortunately the major part of this response occurred before or during the time the electrodes were attached so that we have no record of it except by inference. We may be missing the most differentiating responses in the research. It may be of value to consider attaching at least the GSR electrodes before introducing the subject to the psychophysiology laboratory, either that, or consider the possibility of the use of telemetry.

We can conclude that a group with a relatively high risk of becoming schizophrenic manifests behavior suggesting that, compared with a well-matched control group, their autonomic nervous system is at a higher state of alertness or arousal, is more labile in its response to both mild and stress stimulation, and responds more vigorously to both mild and stress stimulation. This group also exhibits more generalization of the responsiveness of its autonomic nervous system.

WORD ASSOCIATION TEST

The following variables were measured from the Word Association Test.

1. *Number of stimulus repetitions by the testee.* This is the number of instances the testee repeats back the stimulus word.

2. *Number of failures to respond.*

3. *Number of idiosyncratic responses.* It is impossible for the principal investigator to make a clinical judgment regarding the deviancy of a Danish word association. For this reason a Danish psychologist went through every response and judged whether it could be said to be a completely personal association to the stimulus word. A second judge went through a sample of 40 records making the same judgment. Agreement on a subject's score was almost perfect ($r = .98$) although this was sometimes based on slight variations in the contributing items.

4. *Number of fragmented responses.* These include responses which were letters, unrecognizable sounds or not whole words.

5. *Number of multi-word responses.* These include responses of two or more words. The instructions ask for single word responses.

6. *Sum of deviant responses.* To get an over-all score for deviancy, all the scores on the first five items above were simply summed for each individual.

7. *Number of most common responses.* The most common response to each stimulus word was determined by reference to the Danish Word Association Norms. The number of such common responses was summed for each S.

8. *Number of latencies greater than 6 seconds.*

9. *Number of latencies greater than 15 seconds.*

All of the nine variables were analyzed by analysis of variance.

TABLE VIII
WORD ASSOCIATION TEST[a]

Measure	Mean Experimental	Control	F	df	p
Number of repetitions of stimulus word	.54	.20	—	—	n.s.
Number of failures to respond	1.50	2.67	—	—	n.s.
Number of "idiosyncratic" responses	14.49	10.14	6.16	1305	<.05
Number of fragmented responses	.68	.30	4.57	1305	<.05
Number of multi-word responses	2.52	1.92	—	—	n.s.
Sum of deviant responses	19.70	15.30	4.73	1305	<.05
Number of most common responses	23.19	24.66	—	—	n.s.
Number of latencies greater than 6 sec.	14.15	15.30	—	—	n.s.
Number of latencies greater than 15 sec.	3.39	1.98	6.44	1305	<.05

[a] Comparison of *E-C* groups.

Table VIII presents the data and statistical analyses of the WAT measures. As can be seen the differences are significant on four of the nine measures. The *E* group gave more idiosyncratic and fragmented responses and had more response latencies greater than 15 seconds. Of interest is the fact that the *C* group had more failures to respond and more latencies greater than 6 seconds. (Neither difference is statistically significant.)

The fact that we find significant differences between the groups suggests that already some of the E group are beginning to develop deviant associative processes. The low magnitude of the group differences reflects the fact that no members of either of the groups was seriously mentally ill at the time of the examination. It will be of considerable interest to observe the shifts in these scores in future tests with these Ss.

The largest difference between the E and C groups comes on the number of idiosyncratic responses. In general, these are responses that would require some explanation in order for an observer to understand the stimulus-response relationship. It will be important to prepare a distribution of scores on this variable and select out the most deviant individuals for more intensive study. For example, do they tend to be "pre-process" subjects on the basis of the social data? Do they respond in a deviant manner on the GSR measures, or the intellectual measures? Table VIII also suggests that we can improve the discriminatory power of "Sum of deviant responses" by omitting "Number of failures to respond" from this score.

Wechsler Intelligence Scale for Children

A form of the WISC in common use in Denmark was employed. This form consists of a translation of the WISC as we know it. Norms are now being gathered in Denmark; however, in the meantime we have analyzed the results in terms of raw scores (Table IX). Also included in Table IX is an analysis in terms of Verbal, Performance, and Full I.Q. based on U.S. norms. This is included for reference purposes only.

It can be seen that the differences observed between the E-C groups are not dramatic. However, there is a significant difference in Full I.Q., Verbal I.Q., and in the Arithmetic, and Coding subtests. In addition, on every subtest the C group outscores the E group.

Lane and Albee and their students (Schaffner) have made an intensive study of the premorbid intellectual performance of individuals who are now schizophrenic. They find that in lower socioeconomic groups the preschizophrenic has a lower I.Q. than the non-preschizophrenic. In middle and upper socioeconomic groups this relationship does not hold. Instead, the preschizophrenic is at or above his classmates in I.Q. However, he is below his sibs in I.Q. At both socioeconomic levels the I.Q. level for the preschizophrenic seems to decline in adolescence.

TABLE IX
WECHSLER INTELLIGENCE SCALE FOR CHILDREN[a]

Measure	Mean raw score		F	df	p
	Experimental	Control			
Verbal scales					
Information	17.80	18.56	—	—	n.s.
Comprehension	14.07	15.07	—	—	n.s.
Arithmetic	11.34	12.08	7.40	1305	<.01
Similarities	14.13	14.49	—	—	n.s.
Vocabulary	48.84	52.04	3.55	1305	n.s.
Digit span	9.70	9.91	—	—	n.s.
Performance scales					
Picture completion	13.12	13.31	—	—	n.s.
Picture arrangement	30.11	30.55	—	—	n.s.
Block design	32.20	35.03	—	—	n.s.
Object assembly	24.72	25.74	3.72	1305	n.s.
Coding	48.23	52.24	6.42	1305	<.05
Mazes	16.12	16.49	—	—	n.s.
Intelligence Quotients based on U.S. norms					
Performance	100.04	102.48	2.25	1305	n.s.
Verbal	101.26	104.98	4.07	1305	<.05
Full	100.70	104.30	4.57	1305	<.05

[a] Comparison of E-C groups.

The findings reported in Table IX add some support to the Lane and Albee results. Some sizable fraction of the E group will become schizophrenic. Their scores may be pulling down the average of the E group to produce the differences noted. Table IX also provides more detailed information concerning the specific area of deficit of the preschizophrenics, Arithmetic and Coding. Both of these subtests share the characteristic of requiring more intense concentration than most other subtests. Lubin, Gieseking, and Williams (1962) observed the intelligence test scores taken on entrance into the army of a group of veterans hospitalized for schizophrenia. Tests similar to the Arithmetic and Coding subtests of the WISC were the best tests to differentiate these veterans from those that did not later become schizophrenic.

In future analyses we plan to use socioeconomic class as a second way in the analysis of variance and look for an interaction between this and *E-C* groups as is suggested by the Lane-Albee findings. It is also planned to isolate the sib groups in the *E* group and study carefully the sib in each group with the lowest intellectual performance.

In follow-up studies of these Ss it would be of interest to test for the decrement in I.Q. in adolescence which has been reported in the retrospective studies. It will be possible to evaluate this index as a prognostic measure. We can do this both longitudinally and cross-sectionally.

School Report

The school report form consists of precoded items of a yes-no nature dealing with the subject's school behavior. Table X presents the content of each item, the percent "yes" ascribed to each group and an indication as to whether the difference is statistically significant. Interestingly enough when asked if the pupil, in their judgment, is likely to suffer from a serious mental illness in the future, the teachers said "yes" for 19% of the *E* group Ss. This is not far from our best estimate of the morbidity risk in the *E* group. It will be of great interest to see whether these data agree (in the selection of candidates for breakdown) with our other sources of data. If so, this is very encouraging from the point of view of our future interest in preventive work. It means that teachers' judgments might be used as an *initial* screening device.

Table X reveals the *E* group subjects as having some history of having come to the attention of the school psychologist and having been placed in special classes for one reason or another. They are also described as more nervous and easily upset, rejected by their peers, tending to play alone, passive, withdrawn, and subdued in activity. This is a rather classic picture of the adolescent preschizophrenic. However, in addition to this picture formed by the statistically significantly differentiating items there is a less ob-

TABLE X
School Report Form[a,b]

	Percent "Yes"	
Item content	Experimental	Control
*1. Was pupil left back?	10	3
*2. Was pupil seen by psychologist?	13	1
3. Is pupil's level of performance lower than his potential?	26	16
4. Is pupil quiet and disinterested in class?	50	37
*5. Is pupil passive, taking no initiative?	52	36
*6. Is pupil a loner, avoiding his classmates' activities?	33	21
*7. Is pupil rejected by his classmates?	18	8
8. If pupil is isolated is he content with his isolation?	6	3
9. Is pupil shy, reserved, silent?	19	11
10. Is pupil awkward and ill at ease with classmates?	10	5
11. Is pupil uncomfortable and awkward with teacher?	14	10
12. Is pupil unduly upset when mildly criticized by teacher?	24	22
13. Is pupil unresponsive to praise and encouragement?	16	15
*14. Does pupil get upset for almost no reason?	23	12
*15. Does pupil become withdrawn when excited?	22	8
16. Does pupil disturb class with chatting?	17	10
17. Is pupil domineering and aggressive, creating conflicts?	9	4
18. Is pupil a disciplinary problem?	9	4
19. Is pupil abnormally active?	12	9
*20. Is pupil normally active?	63	80
*21. Is the pupil abnormally subdued?	24	15
22. Is the pupil abnormally emotionally sensitive?	22	18
23. Is pupil of average emotional sensitivity?	70	79
24. Is pupil of below average emotional sensitivity?	9	7
*25. Is pupil nervous?	37	14
*26. Do you think the pupil will have a serious psychiatric illness in the future?	19	4
*27. Did pupil attend any special classes (for retarded, for poor readers, etc.)?	24	12

[a] Item content, percentage agreement for each group.

[b] Comparisons of the E and C groups were made for each item by means of chi square. The items which significantly differentiate the E and C groups at least at the .05 level are asterisked. It should be pointed out that many of the other items almost reached this level.

vious contradictory trend in the data. Note Items 16, 17, 18, and 19. These items show that some small subgroup in the E group is also described as behaving aggressively and more actively than the C group. Thus, we find a picture rather typical for data on schizophrenics—extreme variability.

The analysis described did not take into consideration age or sex. It is a rather imprecise picture since many of the questions undoubtedly have sex and age interactions. These analyses will be reported in future reports.

PARENT INTERVIEW

The social worker conducted the Parent Interview with S's guardian or the adult with the greatest responsibility for raising the child (most often the father). The interview consists of 88 coded items many of which contain several parts. For purposes of this interim report we have selected 20 questions drawn from 15 items. A briefly worded form of the question and the percentage of parents saying "yes" is presented in Table XI.

The parents' descriptions tend to support the School Report. As children, E group Ss are described as both more passive and more domineering and aggressive. The mothers (schizophrenic) of E group Ss tend to have scolded S to an unusual degree; they are reported to have a poor relationship with S. There are reported to be many quarrels in the home of the E group; the father tends to win these quarrels less often than does the father in the C group. The E group S is reported to fear the father more than the C group S but to prefer to confide in him in preference to the mother.

This brief, superficial glimpse of the Parent Interview data adds to the picture we have of the distinguishing characteristics of the E group. The information concerning the mother-child relationship and the quarreling in the home are reminiscent of the findings of the Rodnick-Garmezy group.

One note: the picture given by the above analysis of the Parental Interview is exceedingly coarse-grained. For example, the significant chi square relating to the relatively aggressive and domineering behavior on the part of the E group has its main contribution from the relatively strong showing of the female E-group Ss. They are more aggressive and domineering than is any other subgroup.

TABLE XI
Parent Interview (Selected Questions)[a]

	Question[b,c]	Percent saying "yes"	
		Experimental	Control
*1.	As a baby S was passive.	13	5
*2.	As a baby S was energetic.	87	95
3.	As a baby S was actively interested in new toys.	91	87
4.	As a baby S was emotionally sensitive.	25	21
5.	As a child S was friendly and interested in strangers.	53	62
*6.	S was aggressive and dominant with other children.	20	10
7.	S was shy with other children.	10	8
*8.	S is shy or afraid of current father figure.	31	5
9.	S has a quiet temperament.	71	80
10.	S is friendly.	98	99
*11.	S and mother do not get along.	20	5
*12.	Mother scolded child quite often.	31	17
*13.	Child confided in mother only sometimes or not at all.	59	10
*14.	When S had a choice he preferred to confide in the father.	45	6
*15.	Family harmony was poor (many quarrels).	35	7
16.	If there were frequent quarrels the mother was usually the winner.	33	18
*17.	If there were frequent quarrels the father was usually the winner.	41	92
*18.	If there were frequent quarrels there was usually a compromise.	26	0
*19.	Does the child have friends at school?	94	99
20.	S's hobbies and interests are mainly active rather than passive.	86	93

[a] "Parent" refers to the child's guardian or the individual with the greatest responsibility for raising the child.

[b] The wording of the questions as listed does not have a one-to-one correspondence with the form in which the questions were asked. The wording has been altered for purposes of brevity, clarity, and in order to fit it into this form of presentation.

[c] An asterisk next to a pair of C-E "yes" percentages indicates that the frequencies differ significantly by chi square test.

OTHER ANALYSES

Unfortunately the Psychiatric Interview, Adjective Check List, and MMPI are not yet analyzed. From some preliminary analysis of the MMPI, it looks as though the E group is being extremely cautious in its responses. This is especially true on items with intimations of abnormality. On these items a higher proportion of the C group gives responses admitting more pathology. The analysis of the Psychiatric Interview will give us the opportunity of assessing the reliability of judgments of the Ss across the Parental Interview, School Report, and Psychiatric Interview.

Summary of Results

From this superficial pass at the data we can delineate certain *group* characteristics that distinguish the children with schizophrenic mothers.

Since they all have mothers who are schizophrenic, in general, their home life has not been harmonious but has been marked by frequent parental quarrels. The mother has apparently been relatively dominant in the home. However, her influence has not been benign. The child apparently sees her as scolding and unreliable or at least not worthy of his confidences.

This difficult environment has been imposed upon (or perhaps has been responsible for producing) a child whose autonomic nervous system is highly labile, reacting to threat abnormally quickly and with abnormal amplitude. To make things still more difficult, his reactions are not specific but over-generalized. This serves to broaden the range of stimuli that are adequate to provoke this sensitive autonomic nervous system.

In school, his teachers recognize his tendency to get upset easily. He seems to react to excitement by withdrawing. He handles peer relations and classroom challenges by passivity. (Perhaps this mode of reaction is learned since it is likely to be followed by the reduction of his anticipatory fear.) Despite the use of passivity and withdrawal he is still approachable and is performing relatively adequately. He still *shows* his "nervousness" enough for his teacher to remark on it. However, having begun to learn avoidance behavior it is difficult for him to stop since this takes him away from the very social situations in which he might learn more direct manners of dealing with his anticipatory anxiety. Since he withdraws, his peers reject him. And the circle gets closer and more difficult to break out of.

While, in general, he performs adequately he has already learned to effect momentary withdrawal responses whenever pressures build up. On tasks which require continuous concentration and effort (Arithmetic, Coding) his performance will begin to slip.

He is a loner much of the time. He does not share associations with his peer group as much as does his schoolmate. In addition, he is beginning to learn to escape from autonomic arousal by drifting off into idiosyncratic thoughts.

Granted, this interpretative summary is heavily biased toward the theoretical orientation of this research project. But while this is true it must be asserted that the interpretation has treated the facts with some respect. It seems reasonable to conclude that the evidence to date does not warrant the abandonment of the whole of the theory.

This description summarized the bulk of the factors that differentiate the E group from the C group. However, there is an implicit assumption that there is a single subgroup of the E group that is responsible for all these findings. There is absolutely no evidence that this is so, as yet. Further, it implicitly assumes that it is this subgroup that will become schizophrenic. Again, there is no direct evidence. One could present an argument for the assertion that these findings have relevance to etiology in schizophrenia if one would be willing to accept the meaningfulness of the variables used to match the E and C groups.

REFERENCES

Ax, A. F., Beckett, P. G. S., Cohen, B. D., Frohman, C. E., Tourney, G., & Gottlieb, J. S. Psychophysiological patterns in chronic schizophrenia. Paper read at Society of Biological Psychiatry, 1961.

Benedict, P. K. Special aspects of schizophrenia. In L. Bellak (Ed.), *Schizophrenia: A review of the syndrome.* New York: Logos, 1958.

Detroit Area Study. *A social profile of Detroit, 1956.* A report of the Detroit Area Study of The University of Michigan.

Farina, A. Patterns of role dominance and conflict in parents of schizophrenic patients. *J. abnorm. soc. Psychol.,* 1960, **61,** 31–38.

Fremming, K. H. *The expectation of mental infirmity in a sample of the Danish population.* Papers on Eugenics, No. 7. London: The Eugenics Society, 1951.

Fuller, J. L., & Thompson, W R. *Behavior genetics.* New York: Wiley, 1960.

Goldfarb, C., & Erlenmeyer-Kimling, L. Changing mating and fertility patterns in schizophrenia. In F. J. Kallmann (Ed.), *Expanding goals of genetics in psychiatry.* New York: Grune & Stratton, 1962.

Kallmann, F. J. *The genetics of schizophrenia.* New York: Augustin, 1938.

Lacey, J. I., & Lacey, B. C. The law of initial value in the longitudinal study of autonomic constitution: Reproducibility of autonomic responses and response patterns over a four-year interval. *Annals N. Y. Acad. Sci.*, 1962, 98, 1257–1290, 1322–1326.

Lemkau, P. V., & Crocetti, G. M. Vital statistics of schizophrenia. In L. Bellak (Ed.), *Schizophrenia.* New York: Logos, 1958.

Lewis, A. J. The offspring of parents both mentally ill. *Acta Genet.*, 1957, 7, 309–322.

Lubin, A., Gieseking, C. J., & Williams, H. L. Direct measurement of cognitive deficit in schizophrenia. *J. consult. Psychol.*, 1962, 26, 139–143.

Mednick, M. T. Mediated generalization and the incubation effect as a function of manifest anxiety. *J. abnorm. soc. Psychol.*, 1957, 55, 315–321.

Mednick, S. A. A learning theory approach to research in schizophrenia. *Psychol. Bull.*, 1958, 55, 316–327.

Mednick, S. A., & Higgins, J. (Eds.), *Current research in schizophrenia.* Ann Arbor: Edwards, 1961.

Mednick, S. A. Schizophrenia: a learned thought disorder. In G. Nielsen (Ed.), *Clinical psychology.* Proceedings of the XIV International Congress of Applied Psychology. Copenhagen: Munksgaard, 1962.

Mednick, S. A., & Wild, C. Reciprocal augmentation of generalization and anxiety. *J. exp. Psychol.*, 1962, 63, 621–626.

Rosenthal, D. Familial concordance by sex with respect to schizophrenia. *Psychol. Bull.*, 1962, 59, 401–421.

Sanua, V. D. Sociocultural factors in families of schizophrenics. *Psychiatry*, 1961, 24, 246–265.

Schulz, B. Empirische untersuchungen uber die Beidseitigen Belastung mit endogenen Psychosen. *Z. Neurol. Psychiat.*, 1939, 165, 97–108.

Schulz, B. Kinder schizophener Elternpaare. *Z. Neurol. Psychiat.*, 1940, 168, 332–381.

Silber, D. Studies relating to arousal in schizophrenia. Unpublished paper, University of Michigan, 1962.

Sobel, D. E. Children of schizophrenic parents: Preliminary observations on early development. *Amer. J. Psychiat.*, 1961, 118, 512–517.

Statistiske Meddelelser, *Befolkningens Bevaegelser, 1959.* Det Statistiske Departement, København, 1961.

Strömgren, E. Trends in psychiatric genetics. In F. J. Kallmann (Ed.), *Expanding goals of genetics in psychiatry.* New York: Grune & Stratton, 1962.

Svalastoga, K. *Prestige, class, and mobility.* Copenhagen: Gyldendal, 1959.

U. S. Bureau of the Census. U. S. Census of Population: 1950, Vol. IV, Special Reports Part 4, Chapter B, Population Mobility—States and State Economic Areas. U. S. Government Printing Office, Washington, D.C., 1956.

Whitehorn, J. C., & Zipf, G. K. Schizophrenic language. *Arch. Neurol. Psychiat.*, 1943, **49**, 831–851.

Williams, M. Psychophysiological responsiveness to psychological stress in early chronic schizophrenic reactions. *Psychosom. Med.*, 1953, **15**, 456–462.

General Discussion

Gardner Lindzey (Chairman)

Vandenberg: I have been accumulating a few points that I would like to introduce and I will try to keep this very short. As a result, it will look as if I am jumping from one topic to another. I could connect them in a more elegant way but I would have to spend a lot more time than I think that I should. A few of these questions have also been suggested by previous speakers. They could be very interesting topics on which we should perhaps focus more directly in future work.

Picking them off in random order, the first question is whether or not those skills which are earlier manifested in a child's development are going to show higher heritability. We really haven't any evidence on this, but I would like to mention that Stewart Hunter in his 1959 Ph.D. thesis at Michigan found what seemed to be a remarkable correlation between the time of eruption of teeth and the heritability index; that is, the earlier a tooth erupts the more clear-cut the evidence in the twin data for genetic control. This might also be true in human skills that don't depend much on specific training, for example, walking, or, at least certain aspects of motor development. Perhaps walking is already too complex. Complex skills are probably too complicated to expect such a simple relationship. The next idea which I would like you to think about is that maybe we are looking at the whole thing in the wrong way; maybe abilities or traits are not hereditary at all. Of course they must be governed to some extent by genetic mechanisms, but the overlay of experience etc. may be so great that we never get to the genetics. Perhaps we should look for deficiencies, marked absences in certain skills, whether cognitive or personality, which most people have, but which are occasionally grossly deficient. This may be a better approach. Then, relating to my first question, there's the question of whether a more trainable or changeable variable would show greater or lesser evidence for hereditary influence. This might be the most interesting place to start. On one hand you can say theoretically, if it's possible to reduce environmental variance you should push up hereditary variance; but it may not be all that simple. If there is a third component, the interaction, that is really the most important, you could get all kinds of unpredictable results. The next quasi-random thought I'd like to throw out is that perhaps the variables on which there has been the strongest selection pressure might show the greatest genetic determination. We aren't going to find it because almost everybody has been brought up to a certain level. In this connection there are two interesting papers by Post (1962a, 1962b), a review of relaxation of selection pressures on anomalous color-vision and on aberations of acuity in vision. Although these are just reviews of published material, the evidence he was able to collect fits together. There seems to be a somewhat higher incidence of anomalous color-vision and of various acuity aberations in civilized societies than in primitive societies. You can readily hypothesize that there is, has been, and indeed continues to be relaxation of selection pressure for these things because they are no longer important in our culture. This kind of model might be interesting to use on some more behavioral phenomena. I would like to close with a few remarks about the differences in heritability which were reported. First of all I feel that we should be very much aware of the fact that estimates of heritability are going to differ between populations. In addition I would like to emphasize that the degree to which various methods measure hereditary components may differ for very subtle reasons. I was very much impressed with the fact that two tests which appear very similar and have high correlation in populations give very different estimates of heritability. This need not be a question of the validity or reliability of these tests; it could lead to somewhat more experimental work. That concludes what I wanted to get in before the general discussion.

Sutton: I should like to mention some lessons I have learned in dealing with biochemical traits by the twin study method. There is really no new information here, but I think many of us are aware of how important the nature of the variables in the population really is. One of these has to do with the excretion of β-aminoisobuteric acid, one of the amino acids in urine, and the one which has been shown to be clearly under genetic control. In a typical Caucasian population perhaps 5% of the people excrete very much larger amounts than do the rest of the population. High excretors occur within certain families. If we look into racial groups, such as Orientals, we find that the frequency of high excretors goes up to 50%. Here demonstration of the genetic control is a fairly simple matter. Whether it is a simple Mendelian trait or not is subject to debate but certainly it is genetic. I was therefore somewhat surprised when we analyzed the data by the Holzinger method to find no evidence of heritability. Further reflection made it quite obvious why there was no evidence of heritability since out of our 80 pairs of twins only perhaps 2 or 3 pairs would have fallen in the high excretor range, and we were dealing with a trait in which most of the variability was simple laboratory measurement. Now we cannot really conclude that we are dealing with a trait that is not heritable; we must just make sure that we are dealing with one in which there is biological variability. Had our twins come from a Chinese population I am quite confident that the gene frequencies would have been such as to give a very high heritability. Another trait which we studied was lysine excretion, where the evidence for a genetic factor was less clear, although an occasional Caucasian excretes consistently large levels of lysine. We have found that among Orientals the frequency of high lysine excretors is high, but unlike the β-amino-isobuteric acid this excretion is readily manipulated by diet. On a typical western diet they excrete high lysine; but if you cut down protein, lysine excretion drops out. This is particularly of interest because of possible selective factors. Lysine is an essential amino acid and it is the one which is most likely to be absent in the diet. Again we found no clear-cut evidence of lysine heritability, although the racial studies indicate quite a bit of genetic variation, simply because we failed to focus attention on the environment for this population. The third trait which I shall mention is PTC tasting. Here we did find heritability. We went through the usual analysis, even though the distribution is not quite normal in this case, and the appropriateness of this type of statistical analysis must be questioned. In the case of PTC we do have an alternate independent reason for classifying a person on the high-low threshold continuum although the distribution is continuous. I am not sure what we would have done without any prior knowledge of possible bimodality and it is doubtful whether it would have occurred to us to look for it. It takes a fairly large body of independent information to be sure about the shape of the population distribution for a given variable. So I think what we can learn by the twin study methods is at most a clue to traits which are likely to have a large genetic component in the population under investigation, and in a particular environment. For this reason I am a little bit pessimistic about the effort involved in much more elaborate types of investigation, because the figure you wind up with is still a non-generalizable estimate of genetic variance.

On a different level, I would like to emphasize points that have already been made. The study of metabolism was tremendously advanced by the use of genetic blocks as a means of isolating steps in metabolism and, I think the same sort of thing may prove true in the psychological studies. We might compare some of these psychological variables to a blood clot which is due to an exceedingly complex process, each part of which is under genetic control. If we merely look at clots, we will never find out what is going on; it is really the *failure* to clot which yields useful information. One final remark: I mentioned to Dr. Lasker last night one correlation which I have not heard mentioned since its original publication; perhaps some of you know more about it. It seems to be an important thing to follow up. There was a report about 4 years ago in *Science* relating intelligence or performance on Army classification tests with uric acid levels in the blood. Now there's good evidence that uric acid levels are under genetic control; the condition hyperuricemia is inherited as a messy dominant and the correlation with performance on this particular test was clearly significant. As far as I could see it was a very good study, but it has not been followed up. I should think that this would be a very fertile association to explore, and perhaps it has been and I don't know anything about it.

Harris: Well, the study has been repeated, but that has not been published.

Sutton: This is what happens sometimes. I think it was DeWitt Stanton, Jr. who did it. But, I've had no contact with the people who did it, so I don't know if any further work has been done.

Vandenberg: Well, it may be like the situation with Tanner's (1950) correlation between critical flicker fusion and intelligence. There was a good correlation. It has been checked by a number of people but the negative findings are generally not published.

Harris: I have a few comments and reactions to various comments that have been made. One thing is this: you people have done quite a bit of work in which you looked at the phenotypic correlations between two traits, how these two traits or these two tests correlate over a population of individuals. You have also spent quite a bit of time looking at heritabilities. Another statistic in this same general area is the genetic correlation between the two traits. It seems to me that both Dr. Loehlin's results yesterday and some of the comments here, would suggest an interest in breaking down this phenotypic correlation into its parts, which the people in animal and plant breeding have been doing for some time. The phenotypic correlation does not tell the whole story. You have the correlations between the genetic portions of the phenotype, the correlations between the environmental portions of the phenotype and in some human data due to the family structure you may have cross correlations between genetic portion for one trait and environmental portion for the other. Following the path coefficient procedure of Wright, you can't break this phenotypic correlation down into four terms. If you consider most simply that phenotype equals genotype plus environment for two traits, these four terms will involve the square root of the heritabilities, the square root of one minus the heritabilities, genetic correlations, environmental correlations, and the cross correlations between genotype and environment. Just to get the ball rolling, I have dreamed up a procedure for estimating genetic correlations between two traits from twin data. What I have is analogous to the Holzinger method of estimating heritability and suffers from the same inadequacies. You take your twins and arbitrarily call one twin 1 and the other 2. This actually was suggested by Hazel for estimating genetic correlations for parent-offspring data. The appropriate formula for estimating the genetic correlation between traits i and j is

$$\sqrt{\frac{(\text{Cov}^M{}_{i_1 j_2} - \text{Cov}^D{}_{i_1 j_2})(\text{Cov}^M{}_{i_2 j_1} - \text{Cov}^D{}_{i_2 j_1})}{(\text{Cov}^M{}_{i_1 i_2} - \text{Cov}^D{}_{i_1 i_2})(\text{Cov}^M{}_{j_1 j_2} - \text{Cov}^D{}_{j_1 j_2})}}$$

Where $\text{Cov}^M_{i_1 j_2}$ is the covariance between the i_{th} trait in twin 1 and the j_{th} trait in twin 2 for monozygotic pairs, $\text{Cov}^D_{i_1 j_2}$ is a similar value for dizygotic pairs, etc. This formula would estimate

$$\frac{\text{Cov}_{ij}(G) - \text{Cov}_{ij}(F.S.)}{\sqrt{[\sigma^2_{Gi} - \text{Cov}_i(F.S.)] \, [\sigma^2_{Gj} - \text{Cov}_j(F.S.)]}}$$

where $\text{Cov}_{ij}(G)$ is the genotypic covariance between trait i and j, $\text{Cov}_{ij}(F.S.)$ is the covariance between traits i in an individual and trait j in its full sib, and $\text{Cov}_i(F.S.)$ is the covariance for trait i in pairs of full sibs. I think it might be informative to see how the relationship you observe between these two traits, these two tests, is determined by genetic forces and how by environmental forces.

Let me make another general comment. In biometrical genetics we consider that we have a continuous distribution and when we go into this sort of analysis, we have little hope of understanding the basic mechanisms behind the expression of this trait. In assuming a large number of loci with small effects we are saying that this trait is too complex to treat at the single gene level. Certainly if the trait is that simple, you gain by looking at it in that way. Look back in history. Galton was supposedly studying inheritance by looking at the same sort of things we calculate here, correlations, regressions, etc. Although his work developed a lot of statistical tools, he didn't get us very far in understanding the basic mechanism of inheritance. On the other hand, Mendel, by looking at some very simple traits, was able to get at the basic mechanism of inheritance, and the premise of biometrical genetics was that this knowledge of the mechanism should help us in further interpretations. When we start looking at a new trait we should see if we can classify it into classes. Do we have a bimodal or trimodal type of distribution? Do we have unusual individuals way out on the end? They might be quite informative and quite interesting, but the role of these biometrical procedures is in describing the population when we have truly a continuous variation.

Cattell: A constant difficulty I have experienced in discussions with animal geneticists is the fact that with their traits the genotype is much more obvious in the phenotype, than it is in psychology. Your last few remarks bothered me a little. You suggested that by looking at the genotype we might do a little better, but I think that a bimodal distribution in the genotype is most likely to be completely obliterated by the time the environment has had its play in behavioral traits. I wonder if you will tell us a little more about the kinds of genetic variances that would be most useful for you to examine with the methods you described earlier. If we can get separation of genetic and environmental variances in continuous traits what groups would you want? What should we be aiming to supply? Sibs from cousin marriages, sibs from non-cousin marriages, what kinds of variances?

Harris: Well, as many as you can. What I presented earlier was a parameterization for the covariances between relatives. If we have several covariances between relatives which are involved in these within and between family variances we could equate your sigma squared BH and sigma squared WH, which could be expanded into these terms that we had here yesterday.

Cattell: But by the restriction of our kind of investigation we are almost bound to have people of the same generation. They would be all of the child generation, for example. We would not have the covariance of the parents and children or mother and son. If we are restricted entirely to the child generation, what can we say about this?

Harris: Well, the types of covariances that you can get out are monozygotic twins, dizygotic twins, and full sibs. I am not familiar enough with your data to know how readily available covariances of half-sibs or cousins, etc. would be.

Cattell: As I see it, these things are immediately derivable from one another in our data today. Identical twins, fraternal twins, sibs and half-sibs all will have a standard relation to one another. The only different groups would then have to be sibs and half-sibs with a different degree of relationship with their parents. There is going to be a standard relationship of the genetic variance of half-sibs and full sibs with whatever trait we study.

Harris: If the mechanism of gene action was such that all of the genotypic variance was additive genetic variance then each one of these would be just simple multiples of the others. Much of our theoretical work, and a lot of our experimental work in many fields has been based on this assumption, but I am very skeptical of this assumption. When we look at the complexity of the biochemical and metabolic reactions that must be going on in the formation of these observed traits, it seems very likely that we have high order dominance and epistatic components of variance. If things are additive that's very good; but we can't make them that way by assuming it, and in our development we should allow for the possibility that they are not. If we do have these high order components of variance, these different covariances will not be simple multiples of each other.

Cattell: Then we have an inconsistency, because in the MAVA model and also in the standard twin method we are assuming additivity. For example, in the model I described, if you use full sibs and half-sibs as a basis for separating your environmental and hereditary effects you are assuming additivity in that separation process. We would have to rewrite these equations to conform with the genetic ones which you described.

Harris: Well, I hadn't realized that. I wondered how you arrived at your results. I knew there must be some assumptions involved in boiling down all the possible between and within variances into few enough equations so that you can get solutions for them.

Cattell: Just to clarify, we assume that the variance in the genetic component with full sibs is half of what you have with half-sibs.

Harris: So one of your assumptions is that all of the genotypic variances are of the additive type. Well, I'm skeptical of that assumption.

Cattell: It would be very difficult to rewrite the equations to handle variances in any other way, except in a trial and error fashion where you'd put in different constants for different degrees of epistacy and just see what happens with these different figures. You'd have a whole series of solutions which is perhaps not really too bad. It would be quite a job, but it ought to work. Each of these solutions for nature-nurture ratio and for the hereditary variance contribution would have to be tried again in your various genetic models and this lands you into the problem of varying two assumptions at once.

Lindzey: I think that you implied that there is enough genotype-environment interaction so that it does not matter too much what the underlying genetic mechanism may be; in the end you are going to get something like continuous variation. In any case it seems that there is a clear research strategy difference here and I think it is a very important issue which runs through all the studies we have talked about. One of the dangers which any psychologist runs in this kind of research, is being lured by the romantic appeal or general interest of the variable, rather than being influenced by the likelihood that the variable lends itself to some reasonable genetic analysis. I think immediately of race differences in behavior. This is one of the major areas in which social scientists and geneticists have worked together on a common problem. It seems clear to me that the most illuminating data on race differences and behavior, with a few very recent exceptions, were produced 60 years ago, before the Binet revolution. This is before psychologists moved to intelligence as a unitary variable, and before they moved to projective techniques and various other complex achievement variables. I wouldn't say all of the intelligence data are worthless but they are a poor return for the number of man hours that have been put into this particular kind of questioning, particularly in comparison to the very high rate of return for a very small number of studies using simple sensory-motor measures.

Cattell: Yes, I think that this is a very central problem. There is a growing difference of emphasis and one must choose either the usual taxonomic or individual differences approach or a rather special one for genetic purposes. While I certainly wouldn't close my own mind to an approach through genetics, I do want to make an appeal for the advantage of keeping the common field of discourse which one gets in dealing with standard factors derived from individual differences. In this connection I believe that three distinct sets of variables will be proposed and worked upon. First, there will be factor analytically derived concepts. Second, there are concepts which are derived from addiction to a problem, i.e. a person has chosen for some reason a particular theme, or a particular problem and wants to center his research on it. Third, one simply takes a variable from the personality sphere by some type of random sampling. Now I would assume that in genetics we would want to avoid this third approach, i.e. research on a highly specific variable, because there are an infinite number of them and no two psychologists seem to mesh under such conditions. Let me give an illustration. Six or seven years ago when Scheier and I started working on anxiety, we went through about 400 studies in the literature and found about 280 ways of measuring anxiety. I think there were only 3 or 4 repetitions of the same variable. This third approach of just taking a variable and hoping that someone else will think it is important enough to work on is pretty hopeless. This leaves one with the other two. Even if you had a concept which is not factor analytically derived, you are eventually going to have to test whether this concept is a factor analytic unit. If you are interested in a cognitive style, or a response pattern in tests as proposed by Messick, sooner or later you are going to ask yourself, "Am I talking about one thing or am I talking about two or three?" If it turns out to be three different things operating together, all the work must be redone, because you do not know how much of the association you found with various criteria was due to component A, component B, or component C. The factor analytic treatment of a concept is fundamentally axiomatic and leads to variables which are just as reliable as if they were directly tested. In that case, I think, one would either choose a problem by "addiction" and then factor analyze it or get the factors on the basis of a stratified, random, or some sort of representative sample of the variables. This would give us personality factors like dominance, submission, cyclothymia, schizothymia, and anxiety, etc. It seems very worthwhile to me to establish nature-nurture ratios for these well-known dimensions, because they are going to be reference points anyway, in a lot of clinical, educational and other work. An increasing number of criterion relations will be hung on them and the concepts are going to become enriched, so they make a natural point of cross reference for genetic investigation. But then—and now I would veer to the position of Vandenberg, Harris, and Loehlin—it is always possible that we will have just a series of nature-nurture ratios. If these dimensions are related, we are in trouble. Another approach might give us some patterns which are not purely genetic. Here we get into these various methods which I think have been reported from at least three different points of view. Dr. Guttman's method of looking for factor structures which persist and Dr. Vandenberg's factoring of twin differences of fraternal and identical twins for differences in the factor structures, both look for structure in the genetic part of the variance and the environment part of the variance separately. This is very good, but I don't think this is so fundamentally different. It is perhaps only a matter of order; whether you first factor observed phenotypic variables and then ask for the nature-nurture ratios on these factors, or whether, on the other hand, you first find out whether variables are highly genetically or highly environmentally determined and then factor them separately. It is not quite this simple a dichotomy, of course there are variations. Thompson, a few years ago, wrote a short article which appeared in the *Eugenics Quarterly* in which he proposed that a factor might be the effect of a single gene. A factor analysis might pick up pleiotropic effects and assign all these diverse bits of behavior to the one gene which is causing them.

Well, things could turn out this way, but I don't think so. On the other hand, Dr. Lindzey, to pick up your last example, we have been factoring intelligence material somewhat more sensitively than has been done before. Using strategic choice of variables and great care in rotation we've pulled out two general factors instead of one. There is a good chance that intelligence will appear as two factors: one seems to load cultural variables a good deal, while the other is the more traditional variable of innate intelligence. If this is the case, it might turn out that the 80 : 20 ratio which has typically been reported by Burt and others might become 100% for the fluid general ability; so at this juncture I am not quite as pessimistic about finding pure genetic determination of some of these phenotypic traits in the population.

Lindzey: It seems to me, there are really three different potential areas of disagreement here. The first has to do with the level of analysis, that is the complexity or simplicity of the response variables you work with. I would say that all the variables you have in mind are complex variables, what I would call complex achievement variables. The second has to do with the method of choice for identifying variables. The third area of potential disagreement relates to Dr. Guttman's work, and deals with the difference between looking for a genetic mechanism in connection with the *variables* as opposed to looking for a *structure* or *interrelationship* among the variables and being concerned with the possibility of genetic mechanism for the structure.

Cattell: In which direction are you using the word complex? Is a factor complex because it is a variable estimated and derived from many variables, or is a single variable complex because it is derived from much abstraction and really is a mixture of many things?

Lindzey: It would depend upon the variable. For example, I would consider any factor to be complex in comparison to visual acuity.

Guttman: May I say something about studying the simple variable? Try to find as simple a variable or group of variables as you can. I say this looking at the history of genetics and what we have achieved today, going as far as the analysis of the genetic code. Remember how this was built up. This beautiful integration was achieved by amassing a great amount of knowledge about what Dr. Fuller has called trivial traits, such as the number of bristles in the sex comb, things that are of use only as markers to label chromosomes. But with enough knowledge someone else can compound a theory and get down to a more molecular level. Take the history of eye color in *Drosophila* for instance. When they started to test eye colors, they didn't know they were going to run into a multiple allelic system and really discover something terribly important, they just studied different eye colors as they found them. When Müller first did his radiation experiments he was ready to take anything that came, simply for the sake of learning more about it. Of course this may not be a parsimonious way of doing it. We have no certainty, especially with psychological traits, that a simple trait which I can measure beautifully is going to be genetically determined by one or even by five or six genes, but I think it is our best chance of arriving at a solution.

Lindzey: It should not be our exclusive choice.

Guttman: Not at all.

Lindzey: If I understand Dr. Guttman correctly, she is suggesting that the simple process may be neglected simply becauce psychologists in general are less interested in these.

Cattell: There is just one point here, possibly there has been misuse of history in an argument. You are saying this is what happened, what was successful. But, of course, factor analysis and measurement of factor analytically derived variables was only possible recently, so it could not have been applied. This is like saying that few people successfully got to their destination by automobile before 1850.

Guttman: No! I didn't mean that. I meant that the approach used then is the approach which will be successful again. I really believe that.

Vandenberg: Maybe it would be better to use the contrast between observed data and abstract concepts instead of using the dichotomy simple or complex. Although we cannot work without some concepts, for certain types of work it is usually more profitable to stay close to the observable data. In psychology we have no direct data. It is always abstracted to some degree, but the degree of abstraction varies widely. A measure of visual acuity is for instance a lot less removed from the original data than a factor in a factor analysis of personality ratings.

Cattell: But, if you want a concept, Dr. Vandenberg, if you are going to get far enough away from the data for an abstraction, I suggest that a factor concept is more standard. It is something that will last even though minor changes in interpretations may occur.

Vandenberg: Well, I'm the last one to argue against the value of factor analysis, but I'm not sure that a factor is always more stable than the basic variables which went into it. I think some of them are more stable.

Lindzey: Actually, Dr. Cattell, the history of psychometrics within psychology doesn't really support you. I am not arguing against factor analysis or against factor analytically derived tests, but I am arguing for diversity. It seems to me, that it is thoroughly legitimate for people to adopt an approach to variable identification and definition that does not involve factor analysis. If we look at the history of variables and tests in psychology, the outcome from factor analytically derived tests is no more impressive than the outcome from item analysis, ordinary methods of item homogeneity, etc.

Cattell: I would agree with that, but I still would say that the best approach lies somewhere in between Minnesota and Harvard. There is the added problem of the lack of follow-up studies. It's tough going; I don't know how you are going to get people to do them.

Nichols: This brings up another criterion for the kinds of variables to work with. You are suggesting, it seems to me, that the factored test may not be superior because it does not necessarily correlate any higher with things of social importance than do some other kinds of variables. The criterion of social importance is one we should look for in human behavior genetics; the only reason I can see for working with human behavior at all, is its importance to people. If you are interested in genetic mechanisms, the problems in human behavior make it almost impossible to work with them. Trying to find simple measures because these may be closer to the genetic mechanisms doesn't seem to be a compelling argument. If this is your interest, you should not be working with behavior in the first place. Perhaps in human behavior genetics we should be interested in the genetic components of variables which are important to people.

Lindzey: By people, do you mean professional people, or do you mean people in general?

Nichols: Well, I think either of these criteria would be justifiable. For instance if Dr. Cattell is interested in surgency, even though this may not be important to everybody else, he would be justified in wondering what the nature-nurture ratio was. If someone else were interested in an "ability" for inventing patentable devices because this is of social importance, even though such a variable may be completely factorially complex, he would be justified in studying the nature-nurture ratio of this variable.

Guttman: Even there you would have to go down to the lower level. In the process of producing better plants and animals, which is certainly important to society, the plant and animal breeders have had to consider the length of the shank and other measures which, at first, you would not have considered to be of importance; but they were convenient to work with.

Lindzey: You would not *have* to go down to the lower level. I am not arguing for a molecular approach as the only approach, and we could probably agree on some complex socially important variables which should be studied. For example, someone could make an important contribution to behavioral science in a study of schizophrenia. Collecting new data is the most important thing, since enough is known now so that data could be collected which would remove most of the major ambiguities for genetic analysis. Somebody who gave a full and sophisticated statement of what is known about genetics and schizophrenia would also be playing a very important role. I'm not arguing against complexity, but I would argue strongly against taking society's definition of problems; there is no reason to think that society knows what a variable is and Dr. Cattell and I agree completely that the identification of an important variable is a very difficult and exacting task.

Cattell: Psychologists seem to take pride and joy in choosing some variable that no one else has ever worked with.

Vandenberg: It's so safe.

Fuller: This is true in animal studies too, Dr. Cattell, where we presumably are dealing with less complex behavior. One point which I would like to make may be obvious, but does get overlooked. If you take a single gene character such as phenylketonuria, it is very simple genetically: it's an enzyme deficiency. However, it is simple only in one sense, for the phenylketonuric apparently differs on almost every one of your factors on personality, or intelligence. He has convulsions and shows physiological differences. Thus genetic simplicity does not imply phenotypic simplicity and I think the reverse is true also. Phenotypic simplicity from the behaviorist's point of view does not necessarily imply simplicity. If we don't keep this in mind I'm afraid that we are going to get so badly fooled that we will be completely lost.

Cattell: I think there is a model to handle this in factor analysis. The factor order concept states that in this order you have these factors and behind them you have these higher order factors, which are operating on different levels from the first set. Moreover, I think that this is a very promising model, since in some instances we found very much the same thing you are describing. The enzyme is a single factor at the physiological level; but at the behavior level it is effecting a distinct behavior structure. Within the same framework, we can handle these things at different order levels.

Lasker: A point made by Dr. Nichols is very pertinent to what I have to say, although my point of view happens to be just the reverse. We are not interested in simplicity from a psychological point of view if we are going to understand genetic mechanisms; we are interested in genetic simplicity, otherwise we are not going to get anywhere. Why is it, for instance, that nobody here has taken the measurements of phenylketonurics for a whole variety of psychological variables? Why isn't this an immediate thing? Of course, the results may be negative. It has also been suggested that the high uric acid correlation is important. Maybe you do not get differences soon, but you have to try many times before you find anything. Once you get a few physiological variables all the different psychological variables which may be related could be studied. This method is at an analytic level and not some abstract factor analysis that's purely intuitive.

Cattell: Oh! But the very method which you are suggesting is part of this model I have just described. You are saying these are phenotypic psychological factors and at another level phenylketonuric chemistry produces the changes.

Lasker: We could use both factors and variables, but we should start with the abnormality.

Cattell: That would be a high order factor. It would be a factor among factors; it would come out as an influence upon several of your primary behavioral factors.

Lasker: It could be measured as a comparison. The parents of the patients who carry the genes could be compared to the controls in respect to both variables and factors. It would be more analytic to begin with the raw variables and see whether phenylketonuria contributed to any variables at all.

Lindzey: I certainly think there is agreement about the importance of starting with genetic entities as one approach. I don't think anyone will argue against this, but I would argue against the notion that we are only looking for genetic simplicity because I think this is unlikely to be observed for many psychological variables.

Lasker: I am sure that is correct. There are other methods of approach besides the twin method. Some of them have not been worked into a formal model. I will mention one, and introduce something completely new to the discussion. If some of the more extreme behavioral manifestations turn out to be related to homozygosity, one property of the homozygosity might be considered. The homozygous individual inherits the gene from both parents either because of a single mutational event in their common ancestry, or because of diverse mutational events. The possibility of its being a single mutational event is always increased in cases in which we are dealing with a rare condition. I had originally thought of using this property in reference to the question of non-traceable consanguinity; in my hypothesis, the distance between the birthplaces of the parents of individuals with a known homozygous condition might be less than that for those who are not homozygous for much more common conditions. I don't think this has been studied. It could be tested on some known genetic entities (such as the blood groups) and then expanded to some of the rare psychological configurations in a mass survey. When you approach very large populations with the rarer configurations, you might find out if the parents' birthplaces were closer together. There are also environmental factors involved, so that it is not completely perfect analytically; but it is another scheme for getting genetic correlations. Some subsidiary questions about mutation rates, selection pressures, and other factors that would influence these rare characteristics would have to be considered, but this is an approach which has not been mentioned previously.

Gottesman: I'd like to mention some data from the area of intelligence. Fortunately my statement does not concern any of the tests which any of the people here are particularly concerned with, because it weakens the idea that a factor may elicit more genetic variability than the mass of information found in general intelligence tests. The heritabilities calculated in twin studies for tests like the Otis IQ, or the Stanford-Binet are around the eighties; if you look at the work of both Blewett (1954) and Vandenberg (1962) on Thurstone's Primary Mental Abilities Tests you find heritabilities going from zero to the high sixties; and in Blewett's work with second order factors, the heritability dropped down to the low thirties. That would argue against some of the things which Dr. Cattell was suggesting.

Cattell: You are contrasting the relatively low heritability on the Thurstone Primary Mental Abilities with the high heritability on a general intelligence test. There is a second order factor in your general intelligence test. The general intelligence test corresponds to a second order factor among the Thurstone Primaries, wouldn't you say?

Vandenberg: I think there is a complication here that we tend to forget. Whenever you calculate a factor score as Blewett does, you are getting away from the raw data and you are inevitably going to lose something when you calculate heritabilities on estimated scores. You are losing some of the information about individual twins.

Cattell: Won't your single variables measure less accurately? Surely if you hypothesize that some generalized traits are inherited, your single variable is heavily contaminated with irrelevant specific variance.

Vandenberg: Well, the specific variance is probably highly related to the underlying genetic mechanism.

Cattell: Suppose this variable can be divided into a common factor and a specific factor, which part are you interested in?

Vandenberg: Well, we don't know whether the common part is more related to heredity or to environment. Similarly, we don't know whether the specific variance is more hereditary, environmental, or mainly error.

Cattell: The problem of estimation is just as great in the one situation as in the other. In one case you add up ten variables, like ten subtests in intelligence tests, and hope your specifics will cancel or contribute little as the common variance adds up; while in the case of the single variable you are making a poor estimate of your specific because you have so much common variance in it. My guess would be that in estimation your error is greater in the single variable if you are interested in a specific in it, than it is in adding the subtests to get a common factor.

Vandenberg: If you do a factor analysis of physical measurements, factors do not have as high heritability as original measures. Some of the highest heritabilities are for the length of the nose and this has a very low communality. If you get a factor score on this that is low, your communality will not be nearly as high. Maybe you can't compare concepts of physical anthropology with concepts in psychology, but it is useful to consider.

Cattell: The nose length is specific and the total stature is a general factor in this context. Is there not a common factor?

Vandenberg: I doubt that there is.

Cattell: Is the inheritance of the nose length better than inheritance for stature?

Vandenberg: I think it is.

Cattell: Even when corrected for stature?

Vandenberg: Well, this has not been done, but there is hardly any relation. I think there is none at all.

Lasker: This points up the reason I would like to see not only general heritability from twin studies but also attempts to get more specific information like the possibility of homozygosity from consanguinity studies. In the same comparative framework you could test both factor loadings and specific variables against consanguinity and find out which ones show some tendency to run up in offsprings of consanguineous matings or even in the much less specific model, I suggested, where distances between birthplaces were used.

Cattell: On this social importance issue which Dr. Nichols raised, I think I share Dr. Fuller's views on the importance of the variable studied. But I would point out that in an agricultural institution in South Carolina they are very much concerned with the practical matter of the amount of lean meat in bacon. Perhaps we are not near enough to breakfast to feel this is important (laughter), but they factor analyzed the pig, and found out what factors were determining leaness, so I don't think these two views on the importance of the factor are antithetic.

Harris: Many of the remarks which I have made now seem to be based on the faulty assumption that psychologists are pretty well settled on the traits they were interested in and how to measure them. Certainly this is the biggest and probably the most important problem you have, i.e. deciding on the measurement of your traits. My animal and plant breeding associates, when they are interested in the trait of weight, go and measure it. It is not quite that simple in your situation and the definition of your measures seems to be where your effort has been concentrated. Now I think this brings us around to a basic contrast of two possible approaches —do we first decide what traits we are interested in and how we measure them and then go into the genetic analysis, or should we bring these genetic considerations into the determination of the measurement of the trait. I think my remarks about the correlation between two traits are relevant here. Let us consider the trait you are interested in, e.g. intelligence, and call this a primary trait, although I guess you use the word primary a bit differently. Then, consider the test which measures this trait as an indicator trait. The relationships between this indicator trait and the primary trait may be genetic or environmental in origin. Do we want to bring these genetic considerations in and find the measurement trait that measures the genotypic portion of this primary trait? I'm a little vague about how to do that, but is that what we want to do? Or is that what we want to do in certain circumstances?

Lindzey: I don't think there can be any clear answer to this question. If it were not for genotype environment interaction, there would perhaps be a clearer answer. *A priori* it seems very likely that some of the variables which are most important to the psychologists may involve such important genetic and environmental interactions that trying to have two separate representations, two separate varibles, would leave you with most of the variance slipping through your fingers.

Harris: Very likely.

Lindzey: There has been a lot of comment about the advantages and the importance of doing family analyses of the sort which has not been done very much in this area, and Dr. Cattell points out that this is difficult for most psychologists because they have their own instrument which they want to get data on here and now. This isn't altogether true, but the exceptions involve a pathological behavioral character or, in a few cases, intelligence studies, but these all present measurement problems. The only place where enough data might be available for a limited analysis would be in some of the growth studies in which there are batteries of tests administered over a period of 40 years. Here you would have parents and sibs and in some cases grandchildren. This would be worth looking into particularly for someone in the child area. People with a major interest in this area should think seriously about encouraging facilities for routine collection of behavioral data on a long term basis. Many simple behavioral measures could be acquired without great difficulty or great cost, and with a minor expenditure of time. With census data, this kind of thing goes on all the time. I think someone in Dr. Vandenberg's position, sufficiently altruistic to consider how his data might be useful to someone else later on, might find this a very important adjunct to a research program.

Vandenberg: Since you brought this up I shall try to remember Nancy Bayley's major address at the Berkeley meeting. It followed cocktails and a heavy dinner, so you will have to forgive me if I'm inaccurate. She gave a very impressive report on the mass of data from the Berkeley growth study in which they calculated correlations between parental and child attributes, and data from other studies. She had a few remarkable generalizations—one was that the effects of parental attitudes, or differences in treatment showed no correlation with the intelligence of girls, but it did with boys. Boys who had been treated somewhat less lovingly showed poorer performance.

Lindzey: Are these child rearing attitudes?

Vandenberg: Yes, primarily measured by PARI and older questionnaires, and by actual interview assessment of the parents' personalities. Since parental attitude made a difference only for boys, she concluded by speculating that there might be a genetic vulnerability to stress or to somewhat hostile rejecting attitudes. This begins to show the kind of analyses we could do with systematic planning for such studies, because she emphasized that this was an *ex post facto* attempt to find relationships which were never planned to be studied. It was quite remarkable, I think, and really enormously consistent. The changes in the correlations between the intelligence of children and parents were quite clear cut; these boys' performances were never as high as you would predict on the basis of the increased similarity between the intelligence of children and their parents. This was true for boys in families with a rather rejecting attitude; for the other boys and all the girls this correlation showed a regular increase (with time) to a very sizeable correlation.

Lindzey: This ties in very directly with our earlier discussion about the possible implications of the XY chromosome distribution in the sexes and the relative role of environmental and genetic variance. I agreed with the prevailing opinion that the contribution of environmental variance should be somewhat less in the female than the male. In the mouse, my data and most of the other evidence indicate that the female mouse is under much less stimulus control than the male mouse; it is much more difficult to control behavior or assign variance in behavior to environmental variation or stimulus variation for the female mouse than for the male mouse. Many odd behavioral studies have similar sorts of results, for example, studies in which male data yield predictable results but female data do not fit.

Freedman: How about Witkin's study? Is this welcome here? Since that study was on humans it would be pertinent as a contradiction to what we have discussed here. That's probably the biggest study directly concerned with the question, "Are males or females more determined by the environmental cues?" The females came out to be more influenced.

Vandenberg: That is a special kind of environmental cue in a laboratory situation. Perhaps you cannot generalize from that to real life situations. I am not denying that you can frequently make such a generalization, I am raising the question of whether you can in this case.

Cattell: One other implication of the remarks by Drs. Lindzey and Lasker which I would like to see realized is the notion of a subject pool. A subject pool would be an advantage in longitudinal studies with comparable variances in a population. This has been done sporadically in the twin realm. In 1936–1938 there was a group in New York who got a set of twins together on which they all experimented. Is it possible that the participants at this conference could do something like that?

Lindzey: Isn't there a major twin registry being constructed?

Vandenberg: Well, there are several besides the twin clubs. The National Institutes of Health is organizing one for twins who are discordant in mental illness.

Lindzey: But that is specific.

Vandenberg: There is one for the Veterans Administration, but I know of none for young twins.

Nichols: Perhaps we could get a few questions about twin status and discordance for certain variables in the next United States census, especially on separated twins raised apart, or twins that are discordant for certain variables. We might obtain useful information which could not be had any other way.

Vandenberg: I believe that was considered during the last census planning, but there were so many other things that it was not possible.

Lindzey: The director is very sympathetic to behavioral research. I think that is a very important suggestion, and I think immediately of the role which Bar Harbor has played in terms of making available isogenic strains and various other special varieties of mice. There is no doubt that the reason more is known about genetics and behavior in the mouse than any other mammal, is this service by Bar Harbor. You do not have to sit down and produce your isogenic strain, or go out and look for twins; you just secure mice from Bar Harbor. I think the incidence of meaningful twin studies would immediately increase manyfold with a comparable central source.

Loehlin: Does the present census list twins?

Vandenberg: I believe multiple births are listed, but not the sex nor does it say whether it is a boy-girl pair. You can get that information by checking with birth registries, etc., but this is clearly not economically feasible.

Nichols: One available source is the National Merit Exam which is given to some 600,000 high school juniors each year. It would be a simple matter to insert an item screening for any kind of rare person—"Do you have an identical twin who is reared separately from you?" This gets checked yes or no, and later the computer could pull out the people who answered yes.

Vandenberg: Returning to the question of having a national resource, I am sure you all are aware in Russia they had an institute for twin studies which was disbanded after a very few years and never really got very far. Twin camps were organized in Germany just before Hitler took over, and they actually persisted for a few years after he came to power. But, as far as I know, nothing relative to heredity was ever published. They invited twins to a camp for a period of several weeks; and they were planning to do all kinds of observation studies. The camp routines and organized games would be observed to follow normal activities, and some standard testing procedures would be used. I think that was a wonderful idea, you wouldn't need a very large group of twins to get useful results. I think that Gedda has a camp in Italy, but I know of no studies with it. It is used to induce people to cooperate with the institute, but it has not been exploited as a situation for systematic data collections. Perhaps some summer we could try such a camp with a number of people to observe these children and give whatever measures they agreed upon. If the children enjoyed the camp enough they might return, so you could keep following these same twins. You have to *give* the subject something. We have not done very much of this, but I have become very much aware that you can't keep people indefinitely otherwise. You could give them medical or psychological advice, an occasional party, at least some contact, something to show them that you are concerned with them as people.

Guttman: In the future when we start a study, whether or not we intend to use two generations, getting information on the parent generation might be worthwhile. We should make that information as specific as possible, keep the addresses in our files for future reference, because eventually we may want to know something about the parents, simply as a general technique without even knowing what we will find.

Cattell: I believe Sweden has quite a model organization and a twin registry.

Vandenberg: It is only for the lower part of Sweden though; I believe it is called the Province of Scandia.

Cattell: Florida may be doing something like that.

Lindzey: British Columbia is also; they have actually a geneticist who is working out a suitable computer system.

Cattell: Ideally you could press a button and have the IBM cards tumble out for any age or group of twins you want to study.

Lindzey: Dr. Vandenberg, there is one serious problem in all of the continuous studies, particularly when you get beyond simple responses into the kind of research which a social psychologist or a personality psychologist would be doing. There is a tremendous interaction across tests or between tests and someone in a study for a long time will develop an attitude toward the project and the investigator. This means that with time the findings become less and less generalizable even to another population of twins, let alone to the general population.

Vandenberg: That is one of the arguments that may persuade subjects to cooperate. I tell them they will have quite an advantage in college placement tests or entering the Army with experience in test taking. I know of no way to avoid these effects.

Loehlin: I would like to stress the importance of not sticking solely to twins. It seems extremely useful to me to have similar data on adopted children. Studying adopted sibs versus natural sibs would present problems, but, the twins' social situation presents problems also and these are different problems. Something that holds up in both cases is more valid than something applying in just one case or the other. An extension to the adopted sibs situation might give a useful correction.

Vandenberg: Dr. Cattell, I wonder if I am correct in believing that you have investigated the possibility of getting adoption information from agencies? If so, could you tell us a little bit about your experience?

Cattell: I'm afraid I'm not very well informed because I delegated this job to a very extrovert research assistant who talked to a variety of agencies in Chicago, Boston, and New York. He got cooperation of about two-thirds of the agencies, and perhaps ultimately one-third of them cooperated in the sense of leading to persons coming in and being tested. We got the definite impression that we were dealing with a selective group of the population somewhat toward the lower socioeconomic level in the adoption sample, because in the middle class family it is far more common for a relative to adopt a child and frequently the adoption agency steps in only if there are no relatives who can look after the child. So, I think it is practical to do this only if you are prepared to go after about three times the cases you actually want. It would be much better if one could overcome the usual attitudes and find placement agencies which are free from the particular prejudices encountered in certain sects and subgroups.

Witherspoon: There is also the question of confidentiality which adoption agencies always guard very carefully. This is important, but it makes it difficult for scientists to get information.

Nichols: Do you think some adoption agencies might be persuaded to place twins with separate families at some distance from one another? This would be wonderful 20 years from now.

Lindzey: You would need to find an adoption agency compelled by scientific motives. Most are naturally concerned primarily with the child's welfare rather than what is best for research, and if they felt that it was psychologically better for the twins to be reared together or at least know of each others' existence, I'm sure they wouldn't change their decision.

Cattell: There are patterns in social work that change.

Loehlin: If you could demonstrate conclusively that it is bad to be brought up as an identical twin then you're in business.

Freedman: I don't think that would be difficult to demonstrate.

Lasker: There might be possibilities for this adoption work far superior to the United States, places in which the whole field is more open and where there is no social stigma to adoption. This is especially true in places in which the maternal death rate is much higher. People don't have these attitudes about the real mother, when everybody has intimate friends who were raised by an aunt or stranger as happens where maternal death rates are high. I suppose Korea might be a possibility; I know Mexico is. In some countries there is a specific pattern for adoption even when the parents are both alive and everything is going well. There is a social obligation if they have lots of children to let the childless relative have a child. In China, the childless sister often gets a son from a sister who has several.

Lindzey: I'm sorry we haven't said more about Dr. Guttman's discussion of examining structure, personality organization, organization of traits and the possibility that there may be genetic covariation, which could be estimated independently of whatever the genetic determination of the individual variables or traits may be. Is that an accurate statement of what you were saying?

Guttman: A very nice statement.

Lindzey: We are involved in an intercorrelation study now in the mouse, in which we hope to be able to show that there are differences in the organization of the traits, aside from the identifiable heritability of the various traits. I think this is a fundamental problem which is tied to the whole question of generalizability of laws of behavior. If there are important structural differences between people besides the variations in the particular value of individual variables, the problem of generalization becomes much more severe and the importance of a typological model becomes pressing.

Vandenberg: I quite agree, the importance of a typological model is important. Dr. Nichols told us a little bit about the ratio between the linguistic and the numerical or quantitative score on the graduate record exam. Now it is a well-known fact that people have different values for this ratio and that there is a fairly clear-cut sex difference. You do have both males and females with a reversed ratio, but there seems to be a rather pervasive relationship to sex. This view of abilities is unique in being derived largely from a particular technique of measuring the likelihood of success in graduate school, but this may be such a structural variable since the total score is not different, but the ratios for the two sexes are.

Nichols: This ratio is highly reliable.

Vandenberg: Is this a question of training? We might have been discouraging little girls from doing certain things and little boys from doing certain other things and pushing them in different directions. Little boys *do* behave differently from the very beginning as I was told at the Harold Jones Child Study Center; they go out to the sandbox, rain or no rain, and the little girls play house, rain or no rain. No matter what you do you can't get the little girls to go out, and you can't keep the boys in. These children were 1½ to 2 years old. Now has there been parental pressure in their past? How early could you observe this? We've been talking a great deal but not doing anything about the possibility of studying very early differences in our children and trying to increase them by giving one twin systematic training and not the other, the co-twin control method. Ideally one would like to do this in a nontrivial area; perhaps giving systematic training to one twin to increase his verbal score on some later test and another type of training to the other twin to increase his numerical or quantitative score. If this could be done it would be of extreme interest and might link studies of heredity and studies of early environmental effects. I think the co-twin method is a very powerful one, but there are some terrific practical problems in organizing such a study.

Lindzey: This particular discussion points to one of the long term astigmatisms in psychology. When people in psychology talk about sex as a parameter or variable they seem unaware that it is at least in large part a genetic variable. I'm sure that the reason this variable became so important during the thirties in American psychology is because it was not considered to be a genetic variable. This leads back to our earlier discussion on selecting characters or variables. Obviously the selection should be diverse with everybody using his own strategy, but one strategy is thinking in terms of natural selection, evolution, and fitness. There are then some variables which are more suggestive, or which offer on an intuitive ground a higher likelihood of success than others. In connection with little girls, I think immediately of maternality, which obviously had to have enormous fitness for many thousands of years. This variable has not been worked with a great deal, but there are scattered findings that suggest a certain amount of identifiable constitutional control. Levy, a clinical psychoanalyst, found one stable correlation in his very complex and largely clinical study of maternal overprotection and maternality. First of all he found a great deal of stability in his ratings of maternality, so that even over time and even using as predictors behavioral events that came early in life he could predict actual behavior with babies. The only correlation that stood up was the correlation between the diameter of the pigmented area of the nipple and maternality.

Freedman: This work was repeated at Brandeis and it didn't work out terribly well, but certainly Levy is right in suggesting tremendous adaptive value. An infant has a hard time sucking on a breast with a short nipple.

Harris: A possible conceptual approach to this sex difference is suggested by considerations in dairy cattle genetics, where, of course, only the females produce milk, but the males have a genotype which has value for milk production although it is not expressed. In traits where there is a sex influence upon the manifestation of the phenotype it might be conceptually possible to consider that there are actually two traits involved, the trait in female individuals and the trait in male individuals. Each individual has a genotype and a conceptual genotypic value for both these traits, but only one of them gets expressed as a phenotype, depending on the sex of the individual. I am not sure what kind of estimation procedures would follow from this.

Lindzey: Is the male trait actually derived from data? How do you make an estimation on the male for the female trait?

Harris: You find the genotypic value of the male for milk production with a progeny test on his daughters as another indirect assessment.

Freedman: Dr. Lindzey raised an evolutionary or Darwinian point of view. What are the adaptive traits which have been important in man's evolution? One would immediately, and perhaps not illogically, make the jump to genetic control. I think the logic is pretty strong for thinking in terms of adaptive traits from an evolutionary point of view. Maternality is an excellent example since it occurs in some little girls at age 12. For example, we have some nice concordances in identical girls on the love of dolls, as opposed to other habits. It would be extremely fruitful to think of man as an essentially biological structure who is adapted to the world in his own unique way, just as the ethologists look at their birds and their fish as having particular structures. Recently ethologists have been getting interested in genetic mechanisms; this provides a fruitful approach to problems on both the animal level and the human level.

Loehlin: To get back to these cows, are there any observable differences between the bulls that carry these high milk production genes and those who didn't?

Harris: I don't think there have been any other associated traits pointed out. There might be some correlations with growth measurements, but these are not high.

Guttman: The best bulls generally produce the best daughters; they produce cows with the best milk production, they produce heavy cows, and cows with big udders and all sorts of associated traits. You're generally not looking at the bull, you're looking at his many daughters.

Harris: Actually these relationships between milk production and other physical characteristics are not as strong as some proponents of animal judging would like to think.

Nichols: This is changing the subject a little bit, but there seem to be differences in the degree to which people will rely on the genetic model in thinking about human behavior genetics. This may be analogous to the reliance on neurological structure for behavior; we know that behavior depends on neurological action, but it has proven to be expedient to study the structure of behavior itself more or less ignoring the fact that it depends on some lower level. Some people seem willing to take this attitude toward the genetic model. Dr. Vandenberg proposed earlier that we isolate the hereditary component of correlations, factor these, and determine the structural or the unitary, hereditary influences which probably would have no direct relationships to actual genes at all. Such influences might be meaningful in the same sense that Dr. Cattell's factors are meaningful, even though they are not identifiable with any more basic structures. Other people seem more interested in attaching behavior differences to a specific genetic model.

Stafford: I would like to say something about test development. We could do more about developing tests for genetic purposes.

Lindzey: I would argue strongly against exclusive emphasis on developing behavioral variables to simply reflect genetic covariation. This may be repetitious, but the genotype-environment interaction is of such crucial significance, that in devising transenvironmental measures or measuring characters so buffered that they do not interact with the environment, you may be dealing with such a special class of phenomena that you are putting yourself out of business as a psychologist. The variables which are most important psychologically are likely to be loaded with a heavy interaction component. This makes the problem difficult, but we should not resolve the difficulty by simply avoiding the problem.

Stafford: I would agree with you, I am saying let us get a few promising traits and evaluate them. The danger you mentioned has not yet approached; we do not have a lot of tests which show genetic influences and nothing else. I'll run the risk of finding a behavioral characteristic which is almost entirely genetic with very little environmental interaction and then worry about its meaningfulness to psychology.

The second thing I want to mention is the twin study in Edwin Nolan's dissertation at Princeton around 1958. Unfortunately he tested a very small number of twins, but he tried to reach a "private" level. He obtained some TAT material, and some other personality characters, and then he worked out a hierarchy of tests. For objective tests, he found identical twins closer than fraternals as usual, but as he went down his hierarchy into more private tests, he found areas where identicals were much farther apart than the fraternals.

Lindzey: Is this corrected for attenuation? As you go down this list the psychometric problems are getting more grave and the unreliability of your measures is probably increasing.

Stafford: Well, Nolan did find this divergence at the private level and Newman, Freeman, and Holzinger (1937) also have comments on this in their book on twinning. They said there are certain unique psychological processes going on especially with *identical* twins. They try to take some opposite roles, they tend to want different things, one does dominate the other a little bit. I merely suggest this as one approach which has not been brought out in this meeting, and which should be pursued with larger numbers of twins and more elaborate instruments. I already mentioned that we might consider cross-homogamy. If we're going to look at homogamy we ought to look at different traits, such as the husband high in music and his wife high in reasoning. There may be some cross-mating attractions here, so run the intercorrelations with everything; then cross-validate. Stepparents present another approach to the heredity-environment problem. Here is a parent who had the same environment. He married and raised the child just as if he were the true biological parent, but without the true biological relationship. Correlations with stepparents would tend to show an environmental effect where we could rule out the genetic effect, although we would still have to be careful about homogamy and assortive mating.

Another point I'd like to bring up as an adjunct to a registry would be data exchange; with access to the data someone else could possibly use it later in another context.

Lasker: May I interrupt to suggest a concrete instrumentality for this, the American Documentation Institute of the Library of Congress. Is that widely used?

Vandenberg: Yes, it is beginning to be used.

Loehlin: The data are likely to be intermediate rather than raw, that is correlated matrices rather than raw data.

Vandenberg: The editor decides what is going in there, the ADI doesn't make that decision.

Lasker: The important data are those that are already described and published so you have an idea what they are about.

Stafford: I have made an agreement to have my raw data put in ADI, but I'm concerned about anyone knowing it is there.

Lasker: Have it described in an article which is published in an index where people will find it.

Stafford: Perhaps the people who are interested in behavior genetics could have a list of what raw data are available, so they are a little more accessible and use the ADI as a place for safe keeping.

I have one further point which is concerned with cross-validation. In genetics there are many ways to check a hypothesis against data collected on sets of individuals having another relationship. Such cross-validation could frequently be done on one's own data. For example, I'm interested in trying to study gene frequency. One approach which I have thought about to search for a sex-linked characteristic, is to take the score distributions of males and females in a large population. If having a high score is based on an underlying recessive gene, one should be able to go down the distributions to the place where the percentage of the males is the square root of the percentage of the females. This would be another way of tapping genetic information. In other words, if I find some evidence of gene frequency from family studies, then I should find this other relationship also.

Lindzey: Let me ask you a semi-serious, but loaded question about the gene frequency issue. What conceivable difference can it make to a psychologist how many genes are acting on a behavioral attribute? I'm not denying that there is certain elegance and a pure science aspect and all that, but as a psychologist what difference does it make?

Stafford: First of all, if my hypothesis is true that there is an underlying dichotomy in a particular trait, I should be able sooner or later to design a test that will pull these groups apart, so that we will get a bimodal distribution. It's not what you usually get when you look at test scores, you usually get a distribution of unimodal kind albeit skewed. I would like to do that with a spatial ability test where perhaps we might expect an underlying dichotomy. In order to do this we've got to take the items at a specific level of difficulty; if I estimated a gene frequency of 40%, then I'm going to start taking items at the 40% level of difficulty and I'm going to build a whole test on them to try and pull people apart at that point. One practical use for such a test would be in counseling. We have found that students who scored low on a spatial ability test just do not stay in the sciences and engineering. A tremendous number of students are leaving engineering, or the sciences; and the bulk of them score below average on such tests. If I felt more confident that there was real underlying genetic dichotomy which had a demarcation line, and knew something about the percentages passing above and below that line, I would feel a little more confident in advising counselors what to tell the students.

Lindzey: Well, that immediately makes me think of an unpublished finding of Eckhart Hess. It is tied to the proposal of looking for correlates with the color of the eye. First of all, Hess compared the distribution of blue and brown eyes on such simple behavioral characters, or choice responses, as undergraduate major, whether humanities, social sciences, mathematics, natural sciences, etc. The distribution is very significantly different in these soft, humanistic, tender areas as opposed with tough, analytic "scientific" areas, with the blue eyes going with the science major. The other finding is a very interesting one. Dr. Vandenberg and I discovered the other day that he has an interest in the old Thurstone colorform test, as does Hess. Vandenberg and Hess independently discovered the response set flaw in Thurstone's test. Hess's film is a modification of that test in which he took out the source of error, the error variance. He has very interesting findings regarding a significant difference between color-form preference between these two eye groups, the brown eye group being essentially color oriented. Another interesting item, unrelated to the eye color business, concerns Hess' question whether there is directionality to the activation of the dilation of the pupil. In other words, does a person dilate the pupil in the face of adversive stimuli as well as in the face of appetitive stimuli? If every important stimulus leads to dilation then dilation is psychologically not a very useful variable. But Hess reports that when you present noxious stimulation you get constriction of the pupil. He has done studies of homosexual and normal men with pictures of naked women; with predicted results of constriction of the iris on the part of the homosexuals and dilation on the part of the normals. I would say that this is potentially an enormously interesting variable and one which does have a simplicity in terms of the underlying mechanisms, so you would not have to be concerned with the possibility of various kinds of genetic variations.

Cattell: This intrigues me very much, because about 15 years ago I published a paper which was a by-product, an aside, in which I recorded a statistically significant connection between eye color and rigidity measures and interests in arts versus sciences.

Lindzey: Is it in the same direction?

Cattell: Yes, it is in the same direction; the scientific interest was more in the light-eyed group and the arts interest more in the dark-eyed group and rigidity was higher on the dark-eyed group. It was a syndrome of eye-color, rigidity, and interest. It doesn't make any sense theoretically.

Freedman: One thing, of course, that is a little bothersome is a culture which is quite uniform for eye color and you find artists and scientists there.

Lasker: Has intra-sibship variation been looked at? Within the same family?

Lindzey: No.

Freedman: One of the problems, as I see it here, is that in many cases you have cultural inheritance following very closely the potential genetic inheritance. Schizophrenia is a perfect example of that—if you have two schizophrenic parents your chances of being schizophrenic are much higher than if you had one. Well, you can interpret that genetically but you can just as well interpret that from the familial point of view.

Lasker: But you have a pretty good chance of separating them if you took color variation within a family.

Lindzey: You are implying that it is the kind of character which you can hope to give you a solution to this problem, whereas if you are working with something that is linked to language development, then how do you ever really try to untangle these two sources of variations. Let me make one other related comment. It should not be surprising to get variation in personality in a group of individuals with no variation in eye color. This goes back to the genotype-environment interaction which has to be underlined and underlined over and over again because behavioral scientists do think in terms of genetic characters and environmental characters, so that you can pick up hundreds of books and find descriptions of eye color and many other attributes as genetically determined characters which is nonsense. If a geneticist talked that way it would be a shorthand method of speech and he would know full well that he didn't mean it, it wouldn't really be implying what the terms mean to a psychologist. With inbred animals, psychologists who are not involved in genetics of behavior, conventionally expect that if you take so-called homozygous strains in which genetic variation has been tremendously reduced you are automatically reducing variation at the behavioral level. In other words, if you work with an isogenic strain, and you are studying learning, you would be throwing out all the error variance which used to be there, and therefore it would be much easier to establish firm empirical relationships and get significance. Of course, this is anything but the case; when you look at the variance in behavior of the inbred strains in comparison to various kinds of outbred strains very often the reverse situation obtains. In spite of the reduction in genetic contribution to the variance, the inbred strain will often be more variable than the heterozygous animal, for a complex set of reasons which are all related to the genotype-environment interaction.

Freedman: Yes, but I think that you are slightly overstating that. We know that while the C57 strain of Bar Harbor mouse is highly affected by the environment, as in Ginsberg's experiments in which he finds that the slightest variation in how one handles the C57 makes a difference, this is not the case with the CH3 strain.

Lindzey: What I would say is that the buffering or stability of the difference in the responsiveness to environmental variation across inbred strains is much less than the buffering that goes with heterzygosity in an outbred animal, and this is a problem which actually led me to look into this whole issue. I have been dealing with genotype-environment interactions in inbred strains; and there

GENERAL DISCUSSION 333

is no question that there are differences in the responsiveness to environmental variation, but the amount of difference between inbred strains is much less than the amount of difference between an inbred strain and an outbred strain.

Guttman: Which means that the inbreeding has produced a sensitivity, by the fact of its homozygosity, to stimuli; is that what you would say?

Lasker: You have another dimension, if your non-inbred strain is a wild type.

Lindzey: I'm talking about F_1's, F_2's and so on, animals that can reproduce from this genetic strain.

Lasker: I'm concerned because the wild type receive a new stability from the evolutionary perspective, from a history of adaption.

Lindzey: Your point about the difference between the C57 mice and the others is still an argument for the genotype-environment interaction. Here's the C57 genotype and here's the CH3 and you are saying that depending upon the genotype a particular set of environmental factors is going to have a big or little effect.

Stafford: I want to comment on the Hess study since I was a subject in this experiment. I was a little bit uneasy. He asked the class, "How many think that behavioral traits can be inherited? Hold your hands up," and he found that more people with blue eyes than brown eyes believe this. In our group it was quite extreme, there was only one brown-eyed person and there were 5 or 6 of us who were blue-eyed raising our hands. I became interested and went to see him and said, "I'd like to do something with your data, you've never published anything," etc., but he was very reluctant and to my knowledge has never reported this.

Lindzey: I think his reluctance in the area is the obvious reluctance in this country and perhaps in any democratic society, to report data which can be used by race fanatics; which is like Dobzansky's criticism of Coon, i.e. his theories about human evolution are going to be used by racists, and they are. There is no question that in America there are data that bear upon issues of this sort, but which are suppressed either because people are personally displeased by it, or they are professionally concerned about the implications. It's a little bit like doing ESP research on the side. There are some psychologists who are playing around with ESP, but they don't publish it. They believe that there really is something there, but they wouldn't publish data because it would get them into trouble. The same thing is true in terms of something like blue and dark eyes. The question of Jewish versus non-Jewish intelligence is another fascinating problem of great relevance to the topic we are discussing here. How much is there in the literature on this? How many good, hard reportings of what every single person who has dealt with college students knows? There's very little; and it is not reported for the same good democratic, but bad scientific values.

Freedman: Yes, but there may be another reason why this blue-eyed business has not been published; simply because it has too many holes in it. I think that that has to be considered too.

Lindzey: No, there is an array of psychological journals so you can publish anything, and in just the small scale study which Hess did at the Center, in which I was a subject, the differences were extremely large. I would say that it could have been published as a note somewhere.

Freedman: I would like to hear some geneticist say what the potential is for association of a series of behavioral traits with such a thing as eye color. This is the same problem which one faces in the logic of Sheldon's notions of a particular somatotype being connected with behavior.

Guttman: Dr. Fuller had a nice page about different types of associations between characters, and there are really so many reasons for association. Why is cancer associated with blood group O? We don't know! Association is not the right word really, there is just a percentage above the ordinary incidence. Linkage is a different story! Things are linked when they are on the same chromosome. When things are associated, possibly for some evolutionary chance reason, it is a very nice thing to study; but "Why, is *this* specifically associated with *that?*" is a question we cannot answer at this point. There may be no causal relationship.

Freedman: According to Paul Scott, who went over the animal material, there is only one case of specific association of a physical trait with a behavioral trait in animals that has been demonstrated. It is a mouse which has a neurological pattern and also has a certain coat color; that's the only definite demonstration.

Lindzey: Even if you include *Drosophila?* Did you say mammal? Or did you say animal?

Guttman: Even in mice there are many more.

Freedman: In *Drosophila* there's the so-called yellow gene with a certain mating behavior, but that's a very difficult study that has never been replicated.

Guttman: That's not so! There are twenty linkage maps of the mouse chromosome and behavioral traits are all over the place.

Lindzey: I think they are all pathological without an exception. I was talking to Dr. Fuller about it. We have what looks like clear mapping of water escape learning near the albinism gene, which would be just the kind of thing that you are talking about. Dr. Fuller says that he doesn't know about any normal character.

Guttman: This still does not imply a causal relationship.

Lindzey: No, and I really don't like to talk about causes. The question is whether there is a steady predictable relationship that stands up against all the kinds of genetic analyses. If you perform the standard crosses and these two variables stay together as predicted and go apart when predicted; whether you want to call it cause or not, there is an association. In any case, the absence of these associations, so far, can not be generalized to a statement that they don't exist. The behavior of the mouse has hardly been studied. Who's studied the behavior of the mouse? It's only in the last 10 years that anybody has been interested in it.

Harris: The two main concepts which we can use to explain these associations between a marker gene and a quantitative trait would be (1) pleiotropy, this marker gene has, through its gene product, an influence upon this quantitative trait besides the visible qualitative trait, and (2) linkage disequilibrium in the population. If we have a true random mating population, of course, we don't have these linkage disequilibriums. Even though there may be a locus which affects the quantitative trait quite closely linked, these traits would be combined at random so you wouldn't pick up an association. If you form a population by combining two populations, you would have this linkage disequilibrium between pairs of loci on the same chromosome.

Lindzey: Linkage couldn't be of great interest to a psychologist interested in behavior genetics. Is it not guaranteed that whatever the characters are, they are going to be pulled apart over time in a random breeding population?

Harris: Yes, this linkage disequilibrium is a very temporary thing, and, as you continue the random mating, it will disappear.

Lindzey: So it is of passing interest.

Guttman: It could be a tool for studying change.

Cattell: In that connection, I think my data are rather clear because I had a culturally highly homogeneous group, British university students, and there was very little minority representation. There had been group separation for a long, long time. Still, eye coloring and rigidity connected statistically. I never followed it up, I just put it aside as a curiosity which did not fit any theory.

Lasker: What was it that was associated with the eye color?

Cattell: A measure of perceptual and motor rigidity in switching over from one thing to another.

Vandenberg: At the conclusion of this conference, I want to thank all the participants, the National Institutes of Health for making this possible, and Mrs. Long and Mrs. Gliessner for their able assistance. I hope that you all feel as I do that this exchange of ideas has been very valuable and has given each of us new enthusiasm. Thank you again.

References

Blewett, D. B. An experimental study of the inheritance of intelligence. *J. Ment. Sci.*, 1954, **100**, 922–933.

Hunter, S. The inheritance of mesiodistal tooth diameter in twins. Unpublished doctoral dissertation, Univer. of Michigan, 1959.

Newman, H. H., Freeman, F. N., & Holzinger, K. J. *Twins: a study of heredity and environment.* Chicago: Univer. of Chicago Press, 1937.

Post, R. H. Population differences in red and green color vision deficiency: a review, and a query on selection relaxation. *Eugen. Quart.,* 1962, **9,** 131–146. (a)

Post, R. H. Population differences in vision acuity: a review with speculative notes on selective relaxation. *Eugen. Quart.,* 1962, **9,** 189–212. (b)

Tanner, W. P., Jr. A preliminary investigation of the relationship between visual fusion of intermittent light and intelligence. *Science,* 1950, **112,** 201–203.

Thurstone, L. L. Progress report on a Color Form test. Report no. 80 of the psychometric laboratory, University of Chicago, 1952.

Vandenberg, S. G. The hereditary abilities study: hereditary components in a psychological test battery. *Amer. J. Human Genet.,* 1962, **14,** 220–237.

Author Index

Numbers in italics refer to pages on which the complete references are listed.

A

Ahrens, R., 143, 151, *160*
Allport, G. W., 72, 73
Anastasi, A., 95, *127*
Auble, D., 145, *160*
Ax, A. F., 257, *293*

B

Baldwin, A. H., 117, *127*
Baroff, G. S., 72, 73
Barr, M. L., 7, *10*
Bartlett, M. S., 38, *40*
Bat-Miriam, M., 220, *221*
Beckett, P. G. S., 257, *293*
Beloff, J. R., 96, 101, 104, 105, 110, 120, *128*
Benedict, P. K., 259, *293*
Benjamin, J. D., 223, *228*
Bertram, E. G., 7, *10*
Beutler, E., 7, *10*
Birdsell, J. B., 21, 25
Blewett, D. B., 96, 101, 104, 105, 110, 120, *128*, 240, *242*, 314, *336*
Bohidar, N. R., 88, *93*
Bonné, B., 25, 27
Bonner, J., 8, *11*
Brace, C. L., 248, *252*
Breul, H., 124, *128*
Broadhurst, P. L., 249, *252*
Bruell, J. H., 248, *252*
Brues, A., 19, 25
Burks, B. S., 96, *127*
Burlingham, D. T., 190, *196*
Burnet, F. M., 10, *10*
Burt, C., 90, *93*, 95, 100, 101, *127*, *128*, 190, *196*
Butcher, H. J., 120, *128*
Byrd, E., 191, *196*

C

Cattell, R. B., 30, *40*, 71, 73, 95, 96, 97, 98, 99, 101, 102, 104, 105, 108, 110, 112, 113, 115, 117, 120, 121, 122, 123, 124, 126, *127*, *128*
Childs, B., 7, *10*
Clark, P. J., 25, 38, *40*
Clausen, J. A., 220, *222*
Cockerham, C. C., 84, *93*
Cohen, B. D., 257, *293*
Crocetti, G. M., 255, 265, *294*
Crow, J. F., 25, 114, *128*

D

Dahlstrom, W. G., 64, 68, 73
Darlington, C. D., 114, *128*
Davidson, R. G., 7, *10*
Davis, A., 101, *128*
Davis, E. A., 189, *196*
Day, E. J., 189, *196*
DeGeorge, F. V., 144, *161*
Dencker, S. J., 34, *40*
Dennis, W., 190, *196*, *197*
Diamond, S., 224, *228*
Dobzhansky, T., 19, 25, 69, 70, 71, 73, 245, *252*
Drew, A. L., 24, 25
Duffy, E., 224, *228*
Dunn, L. C., 73, 73
Dyk, R. B., 227, *229*

E

Edwards, A. L., 73, 75
Eells, K., 101, 128
Elley, W. B., 101, 129
Elliot, O., 159, 161
Erlenmeyer-Kimling, L., 240, 242, 267, 294
Eysenck, H. J., 96, 120, 128, 190, 197, 224, 228

F

Fairbanks, V. F., 7, 10
Farina, A., 261, 293
Faterson, H. F., 227, 229
Feuer, G., 249, 252
Fisher, R. A., 84, 90, 93, 101, 112, 114, 129
Freedman, D. G., 129, 141, 143, 144, 156, 159, 160, 161
Freeman, F. N., 46, 59, 96, 129, 188, 190, 191, 197, 240, 242, 328, 337
Fremming, K. H., 265, 293
Freud, S., 223, 228
Frohman, C. E., 257, 293
Fuller, J. L., 73, 79, 95, 129, 141, 161, 247, 248, 249, 251, 252, 253, 256, 293

G

Gabriel, R. K., 217, 221
Gardner, R. W., 226, 228, 229
George, W. C., 72, 73
Gieseking, C. J., 288, 294
Glass, B., 25, 27
Goldfarb, C., 267, 294
Goldstein, K., 153, 159, 161
Goodenough, D. R., 227, 229
Goodman, M., 25, 26
Gorsuch, R. L., 122, 124, 129
Gottesman, I. I., 63, 65, 67, 70, 72, 73, 74, 96, 129
Gottlieb, J. S., 257, 293
Griffing, B., 88, 93
Grumbach, M. M., 7, 11
Guttman, L., 200, 220, 221, 222
Guttman, R., 200, 204, 221

H

Haldane, J. B. S., 114, 129
Ham, G. C., 223, 229
Halsey, A. H., 76, 76
Hammond, S. L., 195, 197
Harris, D. L., 96, 129
Hartman, H. P., 124, 128
Hartmann, H., 223, 225, 229
Hauge, M., 47, 59
Havighurst, R. J., 101, 128
Hayman, B. I., 129, 137
Herrick, V. E., 101, 128
Hess, C., 25, 27
Hess, E. H., 123, 129
Higgins, J., 256, 294
Higgins, J. V., 106, 129
Hirsch, J., 71, 72, 73
Hogben, L., 183, 186
Hollingshead, A. B., 70, 73
Holzinger, K. J., 46, 59, 90, 93, 96, 129, 188, 190, 191, 197, 233, 240, 242, 328, 337
Holzman, P. S., 226, 228
Horn, J., 120, 128
Howard, M., 90, 93, 95, 100, 101, 127, 128
Huang, R. C., 8, 11
Hunt, J. McV., 123, 129, 134
Hunter, S., 297, 336
Husen, T., 240, 242

I

Ingram, V. M., 1, 11
Inhelder, B., 123, 129
Isaacs, J. T., 95, 101, 129
Ishikuni, N., 24, 25

J

Jackson, D. N., 226, 229
Jacob, F., 4, 11
Jahn, E. F., 25, 27
Jarvik, L. F., 240, 242
Jones, H. E., 46, 59, 191, 197
Juel-Nielsen, N., 47, 59

K

Kaila, E., 143, 151, 161
Kallmann, F. J., 67, 70, 72, 73, 263, 264, 294

AUTHOR INDEX

Kallman, H. J., 223, *229*
Kallman, J. W., 96, *129*
Karp, G. W., 8, 9, *11*
Karp, S. A., 227, *229*
Karpman, B., 190, *197*
Keller, B., *160*
Kempthorne, O., 84, 87, 91, *93*, *94*
Kent, E., 190, *197*
King, J. A., 159, *161*
Klein, G. S., 224, 226, *228*, *229*
Knapp, R. B., 122, *129*
Kristy, N., 101, 120, *128*
Kucznski, R. R., 25

L

Lacey, B. C., 274, 279, *294*
Lacey, J. I., 274, 279, *294*
Langner, T. S., 70, *74*
Lazarsfeld, P. F., 220, *222*
Lemkau, P. V., 255, 265, *294*
Lerner, M. I., 228, *229*
Lewis, A. J., 268, *294*
Lilienfeld, A., 45, 52, *59*
Linton, H. B., 226, *228*
Loehlin, J. C., 95, *129*
Lubin, A., 288, *294*
Luria, A. R., 189, 196, *197*
Lush, J. L., 89, *94*
Lyon, M. F., 7, *11*

M

McArthur, N., 19, *25*
MacArthur, R. S., 101, *129*
McCarthy, D., 189, *197*
McClearn, G. E., 247, *252*
Malécot, G., 86, *94*, 114, *129*
Marlowe, D., *74*, 75
Mather, K., 81, 84, *94*, 114, *129*
Matsumoto, Y. S., 24, *25*
Mednick, M. T., 271, *294*
Mednick, S. A., 256, 258, 264, 271, *294*
Messick, S. J., 226, *229*
Michael, S. T., 70, *74*
Monod, J., 4, *11*
Mooty, M. E., 193, 195, *197*
Morishima, A., 7, *11*

Morton, N. B., *25*
Mowrer, E. R., 190, *197*
Murray, H. A., 68, *74*

N

Naeslund, J., 35, *40*
Neel, J. V., 24, *25*
Nemoto, H., 24, *25*
Newman, H. H., 46, *59*, 96, *129*, 188, 190, 191, *197*, 240, *242*, 328, *337*
Nielsen, A., 47, *59*
Nitowsky, H. M., 7, *10*

O

Opler, M. K., 70, *74*
Osborne, R. H., 91, *93*, 144, *161*

P

Pasamanick, B., 52, *59*
Pawlik, K., 122, *129*
Penrose, L. S., 19, *25*, 65, *74*, 101, *129*, 235, *242*
Piaget, J., 123, *129*
Pickford, R. W., 33, *40*
Porteus, S. D., *25*
Post, R. H., 298, *337*
Prechtl, H. F. R., 159, *161*
Prell, D. B., 96, 120, *128*, 190, *197*, 224, *228*
Price, B., 151, 161, *161*

R

Race, R. R., 66, *74*
Radcliffe, J. A., 120, *128*
Redlich, F. C., 70, *73*
Reed, E. W., 106, *129*
Reed, S. C., 106, *129*
Reed, T. E., 250, *252*
Rennie, T. A., 70, *74*
Rodgers, D. A., 247, *252*
Roe, A., 96, *127*
Rosenthal, D., 46, *59*, 268, *294*
Russell, L. B., 7, *11*

S

Sacks, M. S., 25, 27
Sanger, R., 66, *74*

Sanua, V. D., 267, *294*
Schnell, F. W., 88, *94*
Scheier, I. H., 115, 122, 124, *128*
Schendel, L. I., 192, 195, *197*
Schlesinger, K., 247, *253*
Schull, W. J., 38, *40*
Schulz, B., 268, *294*
Schwartz, J. B., 193, 195, *197*
Scott, J. P., 248, 251, *253*
Sealy, A. P., 117, *130*
Shapiro, H. L., *25*
Sheerar, E., 51, *59*
Shields, J., 46, *59*, 190, *197*
Siegel, S., 177, *186*
Silber, D., 257, *294*
Simmons, R. T., 21, *25*
Simonds, B., 46, *59*
Simons, G., 192, 195, *197*
Simpson, G. G., 63, 64, *74*
Skeels, H. M., 191, *197*
Skipper, D. S., 195, *197*
Skodak, M., 191, *197*
Slater, E., 190, *197*
Smith, S. M., 65, *74*, 235, *242*
Sobel, D. E., *294*
Spence, D. P., 226, *228*
Spitz, R. A., 143, 151, *161*
Spuhler, J. N., *25*
Srole, L., 70, *74*
Stafford, R. E., 183, *186*
Star, S. A., 220, *222*
Stern, C., 69, 70, *74*
Stice, G. F., 101, 120, *128*
Stouffer, S. A., 220, *222*
Strömgren, E., *294*
Suchman, E. A., 220, *222*
Sutton, H. E., 3, 5, 8, 9, *11*, 38, *40*
Svalastoga, K., 266, 268, *295*
Sweeney, A. B., 120, *128*
Symonds, P. M., 117, *130*

T

Tanner, W. P., Jr., 301, *337*
Taylor, J. H., 7, *11*
Templin, M. C., 189, *197*
Thompson, W. R., 63, 73, *74*, 79, 95, 120, *129*, *130*, 141, *161*, 249, 252, 256, *294*
Thurstone, L. L., 38, *40*, *337*
Tindale, N. B., 21, 22, *25*
Tourney, G., 257, *293*
Tryon, R. C., 63, 69, 71, *74*
Tyler, R. W., 101, *128*

V

van Aarde, I. M. R., 88, *94*
Vandenberg, S. G., 38, 39, *40*, 63, *74*, 90, *94*, 95, 121, *130*, 164, *168*, 200, *222*, 226, 229, 314, *337*

W

Welsh, G. S., 64, 68, *73*
Whitehorn, J. C., *295*
Wild, C., 258, 271, *294*
Williams, H. L., 288, 289, *294*
Williams, M., 257, *295*
Witherspoon, R. L., 191, 195, *196*, *197*
Witkin, H. A., 227, *229*
Wolff, P., 151, *161*
Wright, S., *25*, 90, *94*, 101, 114, *130*

Y

Yanase, T., 24, *25*
Yeh, M., 7, *10*
Young, W. C., 249, *253*
Yerushalmy, J., 51, *59*
Yudovich, F., 189, *196*, *197*

Z

Zipf, G. K., *295*

Subject Index

A

Absolute mean difference between twins
  on household chores, 55
  on food preferences, 56
  on beverage consumption, 57
Achievement
  no general factor in, 71
  mean scores of ethnic groups in Israel, 200–204, 216–217
Activation level, 224, 225
  temperament and, 224
Adaptive value of personality trait, 68, 77
Adjective Check List, 269, 272
Adolescent preschizophrenics
  school report form for, 288–289
  general description of, 292–293
Adoption, 321–323
  attitudes of different cultures, 323
  class differences toward, 322
Adoption agencies
  attitudes toward research, 322
Alcohol preference differences, 246–248
American Documentation Institute of the Library of Congress, 328, 329
Anemia
  drug induced, 18
  sickle cell, 19
Animal studies, 245–251
  additivity hypothesis in, 248
  and alcohol preference differences, 246–248

Animal Studies—*Continued*
  and inherited physiological disorders, 246
  as models for human genetics, 245
  of physique and behavior, 248
  of sexual behavior, 249
Anthropology
  and human genetics, 17–28
Anthropometrics, 216–218
Antibodies
  synthesis of, 9
Anxiety, 256, 260, 306
Arousal response
  in schizophrenia, 256–258
Artistic expression, 188
  in twins, 193
Assortative mating, 66, 71, 78, 90, 105, 233, *see also* Random mating
  and behavioral characteristics, 43
  in different ethnic groups, 219
Avoidance learning, 259, 260–262
Autonomic imbalance, 256–257

B

β-amino-isobuteric acid, 298–299
β-galactosidase, 4
Balance theory for polygenic traits, 70
Bayley Infant Behavior profile, 142–143, 145
Bayley Mental & Motor Scales, 145
Behavioral genetics
  and democratic ideals, 333
Berkeley growth studies, 317
Between-family environmental variance, 97, 116, 117, 118, 123–124

Beverage consumption
  twin data on, 55, 56, 57
Biometrical genetics (methods of), 30
  heritability estimates, 40
Birth order
  and frequency of twinning, 50, 51
Bivariate distribution, 172, *see also* Parent-child scores
Blind infants, 144
  and social smiling, 151, 152
Blood groups, 19, 24, 125
Brain-damaged twin
  comparison of with normal twin, 35

C

California Psychological Inventory, 241
Census, United States
  and listings of multiple births, 319
Children's Apperception Test (CAT), 187, 191
China
  distribution of shovel-shaped incisors, 23
  distribution of stature, 23
Chromosomal abnormalities, 29, 41, 42
  and trisomic conditions, 42–43
Chromosome maps, 29
Chromosomes
  assigning letters to linkage groups, 41
Coat color in mice, 7
"Coefficient de parenté," 86
Cognitive style, 226
Color Form movie, 330
Color-vision anomalies, 33, 34, 95
  and selection pressure, 298
Comparative culture studies, 119
Comparative genetic studies, 119
Comparative intra-pair factoring, 121
Comparative studies
  and MAVA method, 126
Concordance
  on environmental factors for twins, 50, 58

Concordance—*Continued*
  of identical and fraternal twins, 32, 53, 54, 58
Conditioned reflex, 116
Configurational analysis, 204
Co-repressor gene, 6
Correlational analysis
  of sex differentiated traits, 172
Correlations, 82, 201
  parent-child, 215, 218–220
  sib, 215, 218–220
Covariance
  between sibs, 82
  genotypic, 88
  between traits, 218
  cross-cultural invariance, 200
  inheritance of, 215–221
  sources of, 218
  within families, 218–219
Covariance matrix
  of identical and fraternal twins, 36, 37
Critical flicker fusion, 301
Cross-cultural studies, 199–221
Co-twin control studies, 34
Culture
  inheritance of, 18
  selection for, 18

D

Deaf infants
  social smiling in, 153
Defense mechanisms, 223–226, *see also* Personality traits, 64
  in schizophrenia, 256
Deoxyribonucleic acid, 1–3, 245
Descriptive ratios
  of population genetics, 96
Determinism, 72
Dichotomic analysis, 176–183
  parent-child relation, 176
  with twin data, 180
"Dilute" gene, 246
DNA, *see also* deoxyribonucleic acid
Dominance
  deviations, 85
  variance, 85

Drosophila, 246
  studies of, 335
Drugs
  inducing anemia, 18

E

Ego organization, 226
EMG (electromyographs), 256, 257, 270
Emotionality factor, 164–167
Environment
  and twin concordance, 46, 59
  effects on males and females, 31
  effects on twin differences, 30, 241
  effects in Israeli groups, 199, 216
  twins' family, 47
Environmental mold traits, 168
Environmental response, 110
Epistasis, 90
  definition of, 173
Ergs, 168
Estimation of heritability, 232
Ethnic groups
  differences between, 200–201, 204
  in Israel, 200–220
  and test performance, 216
Ethology, 141, 160
Eugenics, 73
Evolution, 63, 141, 159
  and human behavior genetics, 141
  of function, 160
Extroversion-introversion factor, 164–167
Eye color
  and behavioral traits, 125, 330–332, 334, 336
Extrasensory perception (ESP), 334

F

F ratio, 90
Factor analysis, 163–169, 305–307, 309–312
  of physical measurements, 315
Factor identification, 169
Families
  covariation of traits within, 218–219, 220–221

Family constellation(s), 98
Family correlation coefficients
  of Mental Arithmetic Test, 184
  Identical Blocks Test, 184
  correction for attenuation, 184
Family correlations of test scores, 183, 184
Family environment
  effects on children of deviating from norm, 104–107
Family size
  and frequency of twinning, 52
Family studies
  of gene frequency, 31, 40, 174–176, 329, 330
Fear of strangers, 143, 146
  and age of child, 153–158
Female variability
  in twin pairs, 14, 15
  and X chromosomes, 13–15
Fertility
  differential, 91
Fitness, 67
Flight response, 143, 159
Folke-register in Denmark, 263, 266
Food preferences, twin data on, 55, 57, 60–61
Foster children, as biologically abnormal sample, 100
"Founding" effect in genetics of isolates, 20
Freud, 223, 225

G

G6PD, see Glucose-6-phosphate-dehydrogenase
Galvanic skin reflex, see GSR
Gene frequency
  from family studies, 174–176, 329
  ABO in man, 19
  blood groups in primates, 20
  shovel-shaped incisors, 23
  stature in China, 23
General ability factor, 120
Generalization, 256–258, 281–282
  reactiveness, 259
  testing, 271

Generations
  differences in test scores between, 204, 215
Genetic correlation, between traits, 301–302, 327
Genetic drift, 24
Genetic markers, 218, 221
Genetic mechanisms,
  as source of variance, 116
  modified by culture, 18
Genetics and learning theory, 115–116
Genotype-environment interaction, 327, 332, 333
Genotypic values, decomposition of, 85
Gesell Developmental Schedules, 191, 193
Glucose-6-phosphate-dehydrogenase, 7
  deficiency, 18
GSR, 256–257, 270
  analysis of measures, 276–284
  variables measured, 273–274

H

Haptoglobins, 8, 24
Heart rate, 270
Hemoglobins, mutant, 1
Heredity and environment
  correlations and interactions of, 100, 101
  interaction effects and formulas for, 102–103
Heredity-environment analysis of personality dimensions, 163–170
Heredity-environment association
  environmental causation of, 111–112
  genetic causation of, 109–111
Heredity
  and fear of strangers, 159
  and smiling, 159
Heritability estimates
  definition, 43, 89, 232–234
  difficulties in, 190–200
  discrepancies between studies of, 32

Heritability Estimates—*Continued*
  and selection pressures, 298
  increasing the value of, 40
  of deficiencies, 297
  of general ability, 238
  of personality measures, 63–68, 242
  of residual scores, 238
  of specific ability, 38–39, 238–239
  of tooth eruption, 297
Heterozygote, 19
Holland Vocational Preference Inventory, 241
Homogamous mating, 226
Homogamy, 185
  cross homogamy, 328
Homozygosity
  and mutations, 313
Hormones, differences in response to, 249
Household chores, twin concordance in, 55
Human Genetics Institute in Copenhagen, 263, 266
Huntington's chorea, 95

I

Inbreeding, 83
  and behavioral variation, 332–333
Infants, 143–144
  blind, *see* Blind infants
  deaf, *see* Deaf infants
Intellectual capacity, and personality organization, 224
Intelligence
  hereditary factors in, 314
  theories of, 237
Intelligence quotient, 70
  differences in twins, 195
  relation to social status, 70
Interaction (*see also* Variance)
  of environmental-hereditary variance, 298, 305
Intercorrelations (*see also* Covariance)
  among scores, 200, 204–214, 216
  constancy of, 216

Intercorrelations—*Continued*
  patterns of, 183, 204
  within families, across ethnic groups, 216
Intraclass correlation, 233
IQ *see* Intelligence Quotient
Invariance of pattern
  cross-cultural, cross-ethnic, 200, 204, 216, 218–219
Isochromosome, 43
Isolates
  genetic, 21–25
  in Australia, 23
  of islands in Japan, 24
  of Pitcairn Island, 23
  of Tustan da Cunha, 23
  of upper Xingu River, 24
Isolation
  and behavior of dogs, 252
Israel Defense Army, 216, 217
Item analysis
  of personality inventory, 164

J

Junior Personality Quiz, 163

K

Kohs Block Test, 180

L

Language development and development of mental processes, 189, 196
  retardation of, in twins, 189, 192
"Law of coercion to the biosocial mean," 102, 113
Learning theory
  defects of, 115
  and MAVA, 123
Leisure activities
  twin concordance in, 54
Letter Concepts test, 180
Linkage, 29, 41, 88, 220–221, 334
  of psychological variables with blood types, 32
Linkage disequilibrium, 335, 336

Longitudinal studies, 188
  of schizophrenia, 255–293
  of twins, 188
Lyon hypothesis, 7, 13–15

M

Malaria and sickle cell anemia, 19
Man
  as polytypic species, 19
Manic depressive disorder, 64, 96
Marker gene, 335
Maternal age and frequency of twinning, 50–51
Mating structure, 20, 83, 173
Maturation and MAVA, 123
MAVA method, 30, 95–139
  discussion of, 130–139
  and age differences, 124
  and cultural differences, 124
  contributions to learning theory and maturation theory, 123–125
  criticisms of design, 98, 99, 100
  relation to factor analytic concepts, 119–122
  sample size needed, 130
Measurement
  of behavioral traits, 92
  of indicator traits, 92
Measurement trait, 316
Memory apparatus, 225
Mendelian populations, 63
Mental illness, 69
Metabolism, 300
Methodology
  covariance of twin differences, 35–40
  heritability of single items, 164
  MAVA, 130–139
  twin studies, 31–35
  two-generation study of inheritance of covariation, 215–221
Mice
  Bar Harbor, 319, 332
  coat color in, 7
  genetics and behavior in, 319
Michigan twin study, 35, 38

MMPI, 64, 65, 74, 75, 269, 272, 292
Mode-of-inheritance, 247–248
Mosaicism, red cell, 13
Motor development
  of twins, 145, 194
Musical aptitude
  Pitch Discrimination, Test of, 179, 180
Musical stimulus
  response to, 192
Multiple abstract variance analysis
  see MAVA
Multivariate analysis
  of twin data, 35–40
  of Israeli students' scores, 200

N

National Institutes of Health Registry for twins, 35
National Merit Scholarship Corporation
  qualifying test, 236, 320
  twin study, 231–242
Natural selection
  for polymorphisms, 18–19
  against psychopathology, 67–69
Nature-nurture ratios, 96
Nonmetric technique of multivariate analysis, 204
Nose length, inheritance of, 315
Nursery school
  study of twins in, 191–192

O

Operator gene, 5
"Operon," 5

P

Panmixis, 83, 84, 86
Parent-child scores
  bivariate distribution of, 172
  correlational analysis of, 183
  dichotomic analysis of, 176
  trivariate distribution of, 172, 182
Parent Attitude Research Instrument, 317

Parents & children
  probabilities of resemblance, 175
Parental attitudes
  and intelligence of children, 318
PARI see Parent Attitude Research Instrument
Pattern
  stability of correlational, 200, 218, 221
Peabody Picture Vocabulary Test, 191, 192
Pedigree studies, 30, 31, 32
Penetrance
  of dominant gene, 173
Perceptual speed
  symbol comparison test of, 179
Personality
  hereditary factors in, 64, 65, 74, 163, 241
  tests, 64, 65, 74
Personality inventory
  item analysis of, 163
Personality theory
  genetics and, 223
Personality dimensions
  heredity-environment analysis of, 163–170
Personality traits
  as measured by questionnaire, 64, 163, 241
  as dynamic concepts, 223–225
Phenocopies, 29
Phenotypic correlation, 301
Phenotypic variables
  selection of, 250
Physical activity interest factor, 164–167
Physiological genetics, 245–246
Physiological genetics
  and animal studies, 246
Pitch perception, 33, 180
Placement correlations, 107
Planaria, 9
Pleiotrophy, 121, 335
Polygenic model of inheritance, 67, 70, 96, 171

SUBJECT INDEX

Polymorphism
  in man, 18, 71, 77
Population genetics, 125, 245, 251
  applied to families, 172
Populations
  mating structures in, 20–21
  of small islands 21
  size of, 24
Position effects, 88
Process schizophrenics, 257, 261–262, 264, 268
  birthrate among, 267
Primary Mental Ability tests (PMA), 38, 39, 180, 314
Primary trait, 316
Primates, 20
Profiles
  of traits in human populations, 199, 221
  see also Traits
Proteins
  structure of, 1
Psychoanalytic theory
  and human genetics, 223–224
Psychoactive drugs, 250
PTC (phenylthiocarbamide), 33
PTC tasting
  inheritance of, 299
Psychiatric genetics literature, 256
Psychophysiological studies
  of schizophrenia, 256, 273–274
Psychotherapy, 225

Q

Questionnaires
  reliability of mailed, 242–243

R

Race differences
  in behavior, 305, 334
Radex method of multivariate analysis, 204
Random mating, 20, 67, 83, 173, 336
Reactive schizophrenics, 257, 259–261, 264
Reciprocal augmentation, 258–259

Regression, statistical problems of, 220–221
Regulator gene, 5
Respiration, 270
Response set, 74
Ribonucleic acid
  RNA, role of in memory, 9, 11–13
  transfer, 3
  messenger, 3
Ribosomes, 3
RNA, see also ribonucleic acid

S

Schizophrenia
  and avoidant responses, 260, 292–293
  and behavioral theory, 256–259
  and family situations, 261, 264, 268, 290
  chronicity in, 259–261
  in society, 257–258
  inheritance of, 96, 331
  learned disorder of thought, 256–257, 261
  longitudinal study of, 255–293
  natural selection and, 67
  reactive and process, 260–262
  single disease or not, 77
  stimulus generalization, 256–257, 258
  teachers' judgment of students, 288–289, 290
Selective advantage
  of personality traits, 68
Self-evaluation
  co-twin comparison, 57–58
Sensory-motor measures, 305
Serum cholesterol levels, 45
SES, see also Socioeconomic status
Sex-chromatin body, 7
Sex differences
  in early childhood, 324–326
  on Graduate Record Exam, 323–324
  X-linkage hypothesis, 184

Sex-linkage, 29, 88, 184–185
Shovel-shaped incisors, 23
Simplex pattern of intercorrelations, 201, 204
Smiling
  social, 151
  development of, 143, 147
Social class, 69, 70, 76
Social development of twins, 190
Social psychiatry, 70
Socioeconomic status, 52, 59
  and frequency of twinning, 52–53
  measurement of, 51–52, 53
Socioenvironmental factors, comparison for MZ and DZ twins, 45–61
Source trait, 116, 120
  "constitutional", 125
  "environmental mold", 125
Spatial ability, 39, 184, 329, 330
Spelling ability
  evidence of bimodality, 181
  test of sex differences in, 185
Stress, reaction to, 256
Structural genes, 4
Subcultures (*see also* Ethnic groups)
  and test performance of Israeli children, 204
Surface traits, 120, 122

T

Teaching methods, effectiveness of, 35
Temperament, and personality organization, 224
Templin-Darley Articulation Test, 191, 192
Templin Sound Discrimination Test, 191, 192
Thurstone Temperament Survey, 163
Transmission of traits, determined by gene on X chromosome, 183
Trivariate distribution, *see* Parent-child scores
Tritium labeling, 7
Twin registry, 319, 321, 328
"Twin situation," 187, 189, 193, 196

Twin studies
  co-twin control method, 34, 324
  criticisms of, 98, 102
  differences in the results of, 32
  in Germany, 320
  intercultural, 227
  Michigan, 164
  need for replication of, 32
  number of females in, 47
  obtaining more information from, 33
  and psychoanalytic theory, 227
  review of literature, 188–191
  sample size in replication, 40
  socioenvironmental factors in, 45–61
Twins
  divergence in, 328
  frequency of, 50–51
  intra-pair similarities for MZ and DZ, 53, 54, 58
  motor development, 194
  personal and social characteristics of, 46, 58–59
  raised apart, 319, 320
  response to music stimuli, 185, 193
  social interaction of, 190, 194
  study of infant, 143–160
Typologies, 220

U

Uric acid levels in blood, 300

V

Variables
  selection of, for study, 305–311, 324–325
  social importance of, 310–311
Variance
  between family, 97, 116, 117, 118, 123–124
  environmental, 307, 318
  genetic, 307, 318
  interaction of hereditary-environmental, 298, 305
  measures in twin data, 233, 234

SUBJECT INDEX 351

Variance—*Continued*
  sources of in twin data, 232–233
  within & between families, 82
  within family, 97, 116, 117, 118, 123–124

W

WISC, 36, 269, 271, 273
  results with schizophrenics, 286–288
Within-family environmental variance, 97, 116, 117, 118, 123–124
Word Association Test (WAT), 269, 272–273
  results with schizophrenics, 284–286

X

X chromosome
  inactivation of, 6
X-linked traits, *see* Sex-linkage

Z

Zygosity
  determination of, 32, 38, 47, 48, 65, 145, 161, 188
  diagnosis by questionnaire (for large samples), 234–237
  diagnosis by physical similarity index, 235–237
  parental misclassification of, 47, 49, 50

Table A5.1 SMSAs arrayed by size, 1977 (all populations in thousands).

SMSA	Population
New York, New York/New Jersey	9,387
Los Angeles–Long Beach, California	7,031
Chicago, Illinois	7,017
Philadelphia, Pennsylvania/New Jersey	4,794
Detroit, Michigan	4,370
Boston–Lowell–Brockton–Lawrence–Haverhill, Massachusetts	3,898
San Francisco–Oakland, California	3,182
Washington, DC/Maryland/Virginia	3,033
Nassau–Suffolk, New York	2,688
Dallas–Fort Worth, Texas	2,673
Houston, Texas	2,512
St. Louis, Missouri/Illinois	2,380
Pittsburgh, Pennsylvania	2,294
Baltimore, Maryland	2,147
Minneapolis–St. Paul, Minnesota/Wisconsin	2,037
Newark, New Jersey	1,969
Cleveland, Ohio	1,950
Atlanta, Georgia	1,832
Anaheim–Santa Ana–Garden Grove, California	1,801
San Diego, California	1,683
Denver–Boulder, Colorado	1,464
Miami, Florida	1,441
Milwaukee, Wisconsin	1,427
Seattle–Everett, Washington	1,427
Tampa–St. Petersburg, Florida	1,380
Cincinnati, Ohio/Kentucky/Indiana	1,375
Buffalo, New York	1,313
Riverside–San Bernadino–Ontario, California	1,306
Kansas City, Missouri/Kansas	1,293
Phoenix, Arizona	1,254
San Jose, California	1,217
Indianapolis, Indiana	1,144
New Orleans, Louisiana	1,133
Portland, Oregon/Washington	1,121
Columbus, Ohio	1,087
Hartford–New Britain–Bristol, Connecticut	1,052
San Antonio, Texas	1,025
Rochester, New York	970
Sacramento, California	929
Memphis, Tennessee/Arkansas/Mississippi	886
Louisville, Kentucky/Indiana	883
Fort Lauderdale–Hollywood, Florida	864
Providence–Warwick–Pawtucket, Rhode Island	852
Dayton, Ohio	833
Salt Lake City–Ogden, Utah	822
Bridgeport–Stamford–Norwalk–Danbury, Connecticut	807

Table A5.1 SMSAs arrayed by size, 1977 (all populations in thousands) – *continued.*

SMSA	Population
Birmingham, Alabama	805
Norfolk–Virginia Beach–Portsmouth, Virginia/ North Carolina	803
Albany–Schenectady–Troy, New York	794
Toledo, Ohio/Michigan	777
Greensboro–Winston–Salem–High Point, North Carolina	774
Nashville–Davidson, Tennessee	773
Oklahoma City, Oklahoma	769
New Haven–Waterbury–Meriden, Connecticut	759
Honolulu, Hawaii	723
Jacksonville, Florida	694
Akron, Ohio	661
Syracuse, New York	647
Gary–Hammond–East Chicago, Indiana	644
Worcester–Fitchburg–Leominster, Massachusetts	643
Northeast Pennsylvania	629
Allentown–Bethlehem–Easton, Pennsylvania/New Jersey	624
Tulsa, Oklahoma	610
Richmond, Virginia	603
Charlotte–Gastonia, North Carolina	597
New Brunswick–Perth Amboy–Sayreville, New Jersey	594
Orlando, Florida	593
Springfield–Chicopee–Holyoke, Massachusetts/Connecticut	589
Omaha, Nebraska/Iowa	581
Grand Rapids, Michigan	576
Jersey City, New Jersey	564
Youngstown–Warren, Ohio	541
Greensville–Spartanburg, South Carolina	526
Wilmington, Delaware/New Jersey/Maryland	516
Flint, Michigan	514
Long Branch–Asbury Park, New Jersey	492
Raleigh–Durham, North Carolina	487
West Palm Beach–Boca Raton, Florida	477
Austin, Texas	474
Fresno, California	472
Fall River–New Bedford, Massachusetts	468
Oxnard–Simi Valley–Ventura, California	468
Paterson–Clifton–Passaic, New Jersey	462
Lansing–East Lansing, Michigan	455
Tucson, Arizona	455
Knoxville, Tennessee	449
Baton Rouge, Louisiana	435
El Paso, Texas	435
Harrisburg, Pennsylvania	429
Mobile, Alabama	425
Tacoma, Washington	423
Johnson City–Kingsport–Bristol, Tennessee/Virginia	408

Table A5.1 SMSAs arrayed by size, 1977 (all populations in thousands)
– *continued.*

SMSA	Population
Chattanooga, Tennessee/Georgia	403
Albuquerque, New Mexico	402
Canton, Ohio	401
Wichita, Kansas	395
Charleston–North Charleston, South Carolina	385
Davenport–Rock Island–Moline, Iowa/Illinois	375
Columbia, South Carolina	374
Fort Wayne, Indiana	370
Little Rock–North Little Rock, Arkansas	369
Beaumont–Port Arthur–Orange, Texas	364
Newport News–Hampton, Virginia	364
Bakersfield, California	363
Peoria, Illinois	362
Las Vegas, Nevada	360
Shreveport, Louisiana	356
York, Pennsylvania	352
Lancaster, Pennsylvania	333
Des Moines, Iowa	327
Utica–Rome, New York	317
Trenton, New Jersey	317
Madison, Wisconsin	313
Stockton, California	311
Spokane, Washington	310
Binghamton, New York/Pennsylvania	306
Corpus Christi, Texas	303
Reading, Pennsylvania	302
Huntington–Ashland, West Virginia/Kentucky/Ohio	297
Jackson, Mississippi	296
Lexington–Fayette, Kentucky	294
Vallejo–Fairfield–Napa, California	294
Evansville, Indiana/Kentucky	291
Huntsville, Alabama	291
Appleton–Oshkosh, Wisconsin	289
Colorado Springs, Colorado	288
Augusta, Georgia/South Carolina	287
Santa Barbara–Santa Maria–Lompoc, California	287
Lakeland–Winter Haven, Florida	277
South Bend, Indiana	277
Salinas–Seaside–Monterey, California	275
Pensacola, Florida	274
Erie, Pennsylvania	272
Johnstown, Pennsylvania	268
Kalamazoo–Portage, Michigan	268
Rockford, Illinois	268
Duluth–Superior, Minnesota/Wisconsin	267
Lorain–Elmira, Ohio	264
Santa Rosa, California	263
Charleston, West Virginia	261

Table A5.1 SMSAs arrayed by size, 1977 (all populations in thousands) – *continued*.

SMSA	Population
Manchester–Nashua, New Hampshire	255
Montgomery, Alabama	251
Hamilton–Middletown, Ohio	251
Ann Arbor, Michigan	250
Eugene–Springfield, Oregon	249
New London–Norwich, Connecticut	245
Macon, Georgia	242
Modesto, California	236
Melbourne–Titusville–Cocoa, Florida	233
McAllen–Pharr–Edinburg, Texas	232
Portland, Maine	232
Poughkeepsie, New York	232
Fayetteville, North Carolina	231
Columbus, Georgia/Alabama	229
Saginaw, Michigan	227
Salem, Oregon	217
Roanoke, Virginia	216
Savannah, Georgia	215
Daytona Beach, Florida	214
Lima, Ohio	210
Killeen–Temple, Texas	209
Lubbock, Texas	200
Galveston–Texas City, Texas	195
Springfield, Missouri	193
Atlantic City, New Jersey	190
Fort Smith, Arkansas/Oklahoma	187
Topeka, Kansas	187
Springfield, Illinois	186
Lincoln, Nebraska	184
Battle Creek, Michigan	182
Springfield, Ohio	182
Anchorage, Alaska	181
Wheeling, West Virginia/Ohio	181
Muskegon–Northon Shores–Muskegon Heights, Michigan	179
Racine, Wisconsin	177
Brownsville–Harlingen–San Benito, Texas	176
Provo–Orem, Utah	176
Biloxi–Gulfort, Mississippi	175
Green Bay, Wisconsin	175
Terre Haute, Indiana	173
Santa Cruz, California	169
Asheville, North Carolina	168
Champaign–Urbana–Rantoul, Illinois	168
Cedar Rapids, Iowa	167
Sarasota, Florida	167
Fort Myers–Cape Coral, Florida	165
Steubenville–Weirton, Ohio/West Virginia	165
Waco, Texas	162

Table A5.1 SMSAs arrayed by size, 1977 (all populations in thousands) – *continued.*

SMSA	Population
Amarillo, Texas	158
St. Cloud, Minnesota	158
Yakima, Washington	157
Lake Charles, Louisiana	156
Reno, Nevada	156
Fayette–Springdale, Arkansas	155
Parkersburg–Marietta, West Virginia/Ohio	154
Jackson, Michigan	150
Boise City, Idaho	146
Lynchburg, Virginia	146
Clarksville–Hopkinsville, Tennessee/Kentucky	145
Pittsfield, Pennsylvania	143
Alexandria, Louisiana	140
Tallahassee, Florida	138
Anderson, Indiana	137
Waterloo–Cedar Falls, Iowa	137
Altoona, Pennsylvania	134
Lafayette, Louisiana	132
Wichita Falls, Texas	132
Abilene, Texas	131
Fargo–Moorhead, North Dakota/Minnesota	131
Longview, Texas	131
Vineland–Millville–Bridgeton, New Jersey	131
Mansfield, Ohio	130
Wilmington, North Carolina	130
Monroe, Alabama	129
Petersburg–Colonial Heights–Hopewell, Virginia	129
Decatur, Illinois	128
Muncie, Indiana	128
Gainesville, Florida	127
Florence, Alabama	126
Bradenton, Florida	125
Fort Collins, Colorado	125
Kenosha, Wisconsin	125
Tuscaloosa, Alabama	124
Eau Claire, Wisconsin	123
Pueblo, Colorado	123
Bay City, Michigan	120
Sioux City, Iowa/Nebraska	120
Bloomington–Normal, Illinois	119
Texarkana, Texas/Arkansas	119
Lawton, Oklahoma	118
Richland–Kennewick, Washington	117
Lafayette–West Lafayette, Idaho	115
Pascagoula–Moss Point, Mississippi	113
Williamsport, Pennsylvania	113
Anniston, Alabama	112
Tyler, Texas	111

Table A5.1 SMSAs arrayed by size, 1977 (all populations in thousands) – *continued.*

SMSA	Population
Greeley, Colorado	110
Albany, Georgia	106
Kokomo, Indiana	104
Odessa, Texas	103
Sioux Falls, South Dakota	103
Billings, Montana	101
St. Joseph, Missouri	100
Burlington, North Carolina	99
Elmira, New York	99
Grand Forks, North Dakota/Minnesota	99
Dubuque, Iowa	97
Gadsden, Alabama	97
Kankakee, Illinois	96
Lewiston–Auburn, Maine	95
Bloomington, Indiana	92
Panama City, Florida	92
Rochester, Minnesota	90
Columbia, Missouri	89
La Crosse, Wisconsin	86
Laredo, Texas	85
Great Falls, Montana	84
Pine Bluff, Arkansas	84
Sherman–Denison, Texas	84
Owensboro, Kentucky	80
Bryan–College Station, Texas	77
San Angelo, Texas	77
Midland, Texas	73
Lawrence, Kansas	65

and low levels of unionization, a good climate, low housing prices, an educational options were not enough to capture a share of high tec activity.

On the other hand, in the mid-1970s, the forces accounting for change in high tech plant and job distribution appear to have changed in ways n well captured by our model. The business services variable was the on one which continued to show a significant relationship with both job an plant change. Unionization rates and wage rates both assumed the hyp thesized negative relationship, a switch from their signs in the distr butional regressions for this group, but were not significant except in th case of unionization rates' association with plant growth. Climate als could be detected as a significant positive force in plant change, thoug not for job change. In the only exception to a persistent pattern of raci discrimination, the percent black variable actually produces a positive (though not statistically significant) parameter in the job change regression.

Overall, the results for smaller cities are more rewarding in confirming notions about longer-term evolution of high tech sectors than they are in explaining the forces behind more recent changes. In general, the level of explained variation was lowest for this set of regressions, ranging from $R^2 = 0.36$ in the case of plant distribution in 1977 to $R^2 = 0.12$ in the job change regression. This suggests that the greatest caution should be taken in inferring development strategies for smaller metropolitan areas from the results of our locational analysis. Rather, this set tells us more about the threshold conditions for entering the high tech race than for success in the competition.

Notes

1. The unsatisfactory t-statistics for this set can be explained by a somewhat higher degree of multicollinearity within this set; nevertheless, the R^2 for plant and job distribution regressions were 0.67 and 0.58 respectively, indicating that, as a whole, the model performed remarkably well.
2. The reader should note that only 28 of the 189 small SMSAs had any Fortune 500 headquarters, so that this variable acted more like a dummy variable than it did on the other regressions.

References

Adams, G. 1981. *The iron triangle: the politics of defense contracting.* New York: Council on Economic Priorities.
Allen, G. C. 1929. *The industrial development of Birmingham and the Black Country 1860–1927.* London: Allen & Unwin.
Arrington, L. and G. Jensen 1965. *The defense industry of Utah.* Salt Lake City: Utah State University, Department of Economics, for the Utah State Planning Program, Economic and Population Studies.

Baldwin, W. 1967. *The structure of the defense market, 1955–1964.* Durham, NC: Duke University Press.
Berry, B. J. L. 1970. The geography of the United States in the year 2000. *Institute of British Geographers, Transactions* **51**, 21–53.
Birch, D. L. 1979. *The job generation process.* Cambridge, MA: MIT Program on Neighborhood and Regional Change, Massachusetts Institute of Technology.
Bluestone, B. and B. Harrison 1982. *The deindustrialization of America: plant closings, community abandonment, and the dismantling of basic industry.* New York: Basic Books.
Bolton, R. E. 1966. *Defense purchases and regional growth.* Washington DC: Brookings Institution.
Boyer, R. and D. Savageau 1981. *Places rated almanac: your guide to finding the best places to live in America.* Chicago: Rand McNally.
Bowles, S., D. M. Gordon and T. E. Weisskopf 1983. *Beyond the waste land: democratic alternative to economic decline.* Garden City, NY: Anchor/Doubleday.
Breheny, M., P. Cheshire and R. Langland 1985. The anatomy of job creation: Industrial change in Britain's M4 Corridor. In *Silicon landscapes*, P. Hall and A. Markusen (eds.), 118–33. Boston: Allen & Unwin.
Buswell, R. J. and E. W. Lewis 1970. The geographical distribution of industrial research activity in the United Kingdom. *Regional Studies* **4**, 297–306.

Carus-Wilson, E. M. 1941. An industrial revolution of the thirteenth century. *Economic History Review* **11**, 39–60.
Census of Manufacturers. Location of manufacturing plants. 1972 and 1977. Computer Tape Documentation. US Department of Commerce, Bureau of the Census. Washington, DC: Industry Division of the Census.
Census of Manufacturers. Industry statistics by employment size of establishment. 1972 and 1977. US Department of Commerce, Bureau of the Census. Washington DC: Government Printing Office.
Checkland, S. 1975. *The Upas tree.* Glasgow: Glasgow University Press.
Chinitz, B. 1961. Contrasts in agglomeration: New York and Pittsburgh. *American Economic Review* **50**, 279–89.
Chisholm, M. and J. Oeppen 1973. *The changing pattern of employment: regional specialisation and industrial localisation in Britain.* London: Croom Helm.

References

Clapham, J. H. 1910. The transference of the worsted industry from Norfolk to the West Riding. *Economic Journal* **20**, 195–210.
Clark, J., C. Freeman and L. Soete 1981. Long waves, inventions and innovations. *Futures* **13**, 308–22.
Clayton, J. 1962. Defense spending: key to California's growth. *Western Political Quarterly* **15**, 280–93.
Conroy, M. 1975. *Regional economic diversification*. New York: Praeger.
County Business Patterns. Establishment, employment and payroll by employment size class. 1972, 1974 and 1977. US Department of Commerce, Bureau of the Census. Washington, DC: Government Printing Office.

DeGrasse, R., Jr. 1983. *Military expansion, economic decline: the impact of military spending on US economic performance*. New York: M. E. Sharpe.
Deitrick, S. 1984. Unpublished interview with Lockheed Corporation. November.
Dempsey, R. and D. Schmude 1971. Occupational impact of defense expenditures. *Monthly Labor Review* **94**(12), 12–15.
Dicken, P. and P. E. Lloyd 1978. Inner metropolitan industrial change. Enterprise structures and policy issues: case studies of Manchester and Merseyside. *Regional Studies* **12**, 181–97.
Dorfman, N. S. 1983. Route 128: the development of a regional high technology economy. *Research Policy* **12**, 299–316.
Dumas, L. 1982. Military spending and economic decay. In *The political economy of arms reduction*, L. Dumas (ed.), 1–26. AAAS Selected Symposium. Boulder: Westview.

Economic Development Commission of the City of Chicago 1980. Chicago: a place for profit in the machine tool industry. Mimeo. Chicago: Economic Development Commission.

Feldman, M. 1985. Biotechnology and local economic growth: the American pattern. In *Silicon landscapes*, P. Hall and A. Markusen (eds.), 65–79. Boston: George Allen & Unwin.
Feller, I. 1975. Invention, diffusion and industrial location. In *Locational dynamics of manufacturing activity*, L. Collins and D. Walker (eds.), 83–107. London: Wiley.
Freeman, C. 1982. *The economics of industrial innovation*, 2nd edn. Cambridge, MA: MIT Press.
Freeman, C., J. Clark and L. Soete 1982. *Unemployment and technical innovation: a study of long waves and economic development*. Westport, CN: Greenwood Press.
Friedmann, J. 1972. A general theory of polarized development. In *Growth centers in regional economic development*, N. M. Hansen (ed.), 82–107. New York: Free Press.

Gansler, J. 1980. *The defense industry*. Cambridge, MA: MIT Press.
Glasmeier, A. 1985. Innovative manufacturing industries: spatial incidence in the United States. In *High Technology, space, and society*, M. Castells (ed.), 55–80. Beverly Hills: Sage.
Glasmeier, A. 1986b. *The structure, location and role of high technology industries in US regional development*. Unpublished PhD Dissertation, University of California, Berkeley.

Glasmeier, A., P. Hall and A. Markusen 1983. Recent evidence on high-technology industries' spatial tendencies: a preliminary investigation. In *Technology, innovation and regional economic development*, Appendix C, 145–167, Office of Technology Assessment. Washington, DC: OTA.

Glasmeier, A., P. Hall and A. Markusen. 1984. Metropolitan high-technology industry growth in the mid 1970s: can everyone have a slice of the high-tech pie? *Berkeley Planning Journal* 1, 131–42.

Gordon, R. and L. Kimball 1986 (forthcoming). Industrial structure and the changing global dynamics of location in high technology industry. In *The spatial impact of technological change*, J. Brotchie, P. Hall, P. W. Newton (eds.), forthcoming. London: Croom Helm.

Great Lakes Commission 1984. *The Great Lakes Governors' Commission on the Machine Tool Industry. Final report and recommendations*. Ann Arbor: Great Lakes Commission.

Hall, P. 1962. *The industries of London since 1861*. London: Hutchinson.

Hall, P. and A. Markusen (eds.) 1985. *Silicon landscapes*. Boston: George Allen & Unwin.

Hall, P., A. R. Markusen, R. Osborn and B. Wachsman 1985. The American computer software industry: economic development prospects. In *Silicon landscapes*, P. Hall and A. Markusen (eds.), 49–64. Boston: Allen & Unwin.

Harrington, P. and R. Vinson 1983. High technology industry: identifying and tracking an emerging source of employment strength. Mimeo, Center for Labor Market Studies, Northeastern University.

Harrison, B. 1982. *Rationalization, restructuring, and industrial reorganization in older regions: the economic transformation of New England since World War II*. Cambridge, MA: Joint Center for Urban Studies of the Massachusetts Institute of Technology and Harvard University.

Heckman, J. 1978. An analysis of the changing location of iron and steel production in the twentieth century. *American Economic Review* **68**, 112–33.

Henry, D. K. 1983. Defense spending: a growth market for industry. *US Industrial Outlook* 1983, XXXIX–XLVII.

Hirsch, S. 1967. *Location of industry and international competitiveness*. Oxford: Oxford University Press.

Howells, J. R. L. 1984. The location of research and development: some observations and evidence from Britain. *Regional Studies* **18**, 13–29.

Hoover, E. M. 1948. *The location of economic activity*. New York: McGraw-Hill.

Hoover, E. M. and R. Vernon 1959. *Anatomy of a metropolis: the changing distribution of people and jobs in the New York Metropolitan Region*. Cambridge, MA: Harvard University Press.

Hymer, S. H. 1972. The multinational corporation and the law of uneven development. In *Economics and the world order*, J. N. Bhagwati (ed.), 113–40. New York: Macmillan.

Hymer, S. H. 1979. The multinational corporation and the international division of labor. In *The multinational corporation: a radical approach*, S. H. Hymer, R. Cohen et al. (eds.), 140–64. Cambridge: Cambridge University Press.

Joint Economic Committee 1982. *Location of high technology firms and regional economic development*. Washington, DC: Government Printing Office.

Karaska, Gerald 1967. The spatial impacts of defense–space procurement: an analysis of subcontracting patterns in the United States. *Peace Research Society Papers* **8**, 108–22.

References

Karlson, S. 1983. Modeling location and production: an application to US fully integrated steel plants. *Review of Economics and Statistics* **45**, 41–50.

Keeble, D. E. 1976. *Industrial location and planning in the UK*. London: Methuen.

Kondratieff, N. D. 1935. The long waves in economic life. *Review of Economic Statistics* **17**, 105–15.

Kuttner, R. 1984. *The economic illusion: false choices between prosperity and social justice*. Boston: Houghton Mifflin.

Kuznets, S. S. 1940. Schumpeter's business cycles. *American Economic Review* **30**, 250–71.

Kuznets, S. S. 1946. *National product since 1869*. Publication No. 46. New York: National Bureau of Economic Research.

Kuznets, S. S. 1966. *Modern economic growth: rate, structure and spread*. New Haven and London: Yale University Press.

Lawrence, R. Z. 1984. *Can America compete?* Washington, DC: Brookings Institution.

Lloyd, P. E. and C. M. Mason 1978. Manufacturing industry in the inner city: a case study of Manchester. *Institute of British Geographers, Transactions*, N. S. **3**, 60–90.

Lösch, A. 1954. *The economics of location*. New Haven: Yale University Press.

McDonald, J. F. 1984. *Employment location and industrial land use in metropolitan Chicago*. Champaign, IL: Stipes.

Malecki, E. J. 1980a. Corporate organization of R and D and the location of technological activities. *Regional Studies* **14**, 219–34.

Malecki, E. J. 1980b. Science and technology in the American urban system. In *The American metropolitan system: past and future*, S. D. Brunn and J. O. Wheeler (eds.), 127–44. New York: Wiley.

Malecki, E. J. 1980c. Technological change: British and American research themes. *Area* **12**, 253–60.

Malecki, E. J. 1981a. Government-funded R & D: some regional economic implications. *Professional Geographer* **33**, 72–82.

Malecki, E. J. 1981b. Public and private sector interrelationships, technological change, and regional development. *Papers of the Regional Science Association* **47**, 121–38.

Malecki, E. J. 1983. Federal and industrial R & D: locational structures, economic effects and interrelationships. Final Report, National Science Foundation, Division of Policy Research and Analysis.

Malecki, E. J. 1986 (forthcoming). Research and development and the geography of high-technology complexes. In *Technology, regions and policy*, J. Rees (ed.), 51–74. Totowa: Rowman and Littlefield

Mandel, E. 1975. *Late capitalism*. London: Verso.

Mandel, E. 1980. *Long waves of capitalist development*. Cambridge: Cambridge University Press.

Markusen, A. 1985a. *Profit cycles, oligopoly and regional development*. Cambridge, MA: MIT Press.

Markusen, A. 1985b. Defense spending: a successful industrial policy? In *The politics of industrial policy*, S. Zukin (ed.), 70–84. New York: Praeger.

Markusen, A. 1986. Defense spending and the geography of high tech industries. In *Technology, regions and policy*, J. Rees (ed.), 94–119. Totowa: Rowman and Littlefield.

Marshall, A. 1890. *Principles of economics*. London: Macmillan, 1920.

Martin, J. E. 1966. *Greater London: an industrial geography*. London: Bell.

Martin, R. L. 1972. Information theory and employment location: trends in East Anglia. Mimeo. Cambridge: Cambridge University, Department of Geography.
Marx, K. 1867. *Capital: a critique of political economy*. Harmondsworth: Penguin, 1976.
Massey, D. and R. Meegan 1982. *The anatomy of job loss*. London: Methuen.
Melman, S. 1974. *The permanent war economy: American capitalism in decline*. New York: Simon & Schuster.
Mensch, G. 1979. *Stalemate in technology: innovations overcome the depression*. Cambridge, MA: Ballinger.
Mollenkopf, J. 1983. *The contested city*. Princeton: Princeton University Press.
Mueller, W. F. 1970. *Monopoly and capitalism*. New York: Random House.

Norton, R. D. and J. Rees 1979. The product cycle and the decentralization of American manufacturing. *Regional Studies* **14**, 141–51.

Oakey, R. P., A. T. Thwaites and P. A. Nash 1980. The regional distribution of innovative manufacturing establishments in Great Britain. *Regional Studies* **14**, 235–53.
OTA 1983. *Technology, innovation and regional economic development*. Washington, DC: Office of Technology Assessment.

Perry, D. C. and A. J. Watkins 1978. The rise of the Sunbelt cities. *Urban Affairs Annual Reviews* **14**. Beverly Hills and London: Sage Publications.
Pred, A. 1975. Diffusion, organization, spatial structure, and city-system development. *Economic Geography* **51**, 252–68.

Rees, J. 1982. Defense spending and regional industrial change. *Texas Business Review* **56**, 40–4.
Rees, J. and H. Stafford, 1984. High technology location and regional development: the theoretical base. In *Technology, innovation and regional economic development*, Office of Technology Assessment, 97–107. Washington, DC: OTA.
Richardson, H. 1973. *Regional growth theory*. London: Macmillan.
Riche, R. W., D. E. Hecker and J. U. Burgan 1983. High technology today and tomorrow: a small slice of the employment pie. *Monthly Labor Review* **106**:11, 50–8.
Rogers, E. M. and J. K. Larsen 1984. *Silicon Valley fever: growth of high-technology culture*. New York: Basic Books (London: George Allen & Unwin).
Rothwell, R. 1982. The role of technology in industrial change: implications for regional policy. *Regional Studies* **16**, 361–9.
Rutzick, M. A. 1970. Skills and location of defense-related workers. *Monthly Labor Review* **93**(2), 11–16.

Saxenian, A. 1981. *Silicon chips and spatial structure: the industrial basis of urbanization in Santa Clara County, California*. Berkeley: University of California, Institute of Urban and Regional Development, Working Paper No. 345.
Saxenian, A. 1985. The genesis of Silicon Valley. In *Silicon landscapes*, P. Hall and A. Markusen (eds.), 20–34. Boston: Allen & Unwin.
Scherer, F. M. 1980. *Industrial market structure and economic performance*. Chicago: Rand McNally.
Schumpeter, J. A. 1911. *The theory of economic development*. Cambridge, MA: Harvard University Press, 1961.

Schumpeter, J. A. 1939. *Business cycles: a theoretical, historical and statistical account of the capitalist process*, 2 vols. New York and London: McGraw Hill. Reprinted Philadelphia: Porcupine Press, 1982.
Schumpeter, J. A. 1942. *Capitalism, socialism and democracy*. New York: Harper & Row.
Shutt, R. 1984. *The military in your backyard: a workbook for determining the impact of military spending in your community*. Mountain View, CA: Mid-Peninsula Conversion Project.
State of California 1981. *An industrial strategy for California in the eighties*. Department of Economic and Business Development, Office of Economic Policy, Planning, and Research, Executive Summary, October 23, 1981.
Storper, M. and R. Walker 1984. The spatial division of labor: labor and the location of industries. In *Marxism and the metropolis*, W. K. Tabb and L. Sawyer (eds.), 19–47. New York: Oxford University Press.

Taylor, M. J. 1975. Organizational growth, spatial interaction and locational decision-making. *Regional Studies* **9**, 313–23.
Technical Marketing Associates 1977. *High technology enterprise in Massachusetts*. Cambridge, MA: Technical Marketing Associates.
Technical Marketing Associates 1979. *High technology enterprise in Massachusetts – its role and its concerns*. Cambridge, MA: Technical Marketing Associates.
Theil, H. 1967. *Economics and information theory*. Amsterdam: North Holland.
Thomas, M. 1975. Growth pole theory, technological change and regional economic growth. *Regional Science Association, Papers and Proceedings* **34**, 3–25.
Thomas, M. 1981. Growth and change in innovative manufacturing industries and firms. Working Paper, February. Laxenburg, Austria: International Institute for Applied Systems Analysis.
Thompson, W. 1962. Locational differences in inventive effort and their determinants: Economic and social factors. In *The rate and direction of inventive activity*, National Bureau of Economic Research, 253–71. Princeton, NJ: Princeton University Press.
Thompson, W. 1975. Internal and external factors in urban economies. In *Regional policy: readings in theory and applications*, J. Friedmann and W. Alonso (eds.), 201–20. Cambridge, MA: MIT Press.
Thurow, L. 1980. *The zero sum society*. New York: Basic Books.

US Department of Commerce, Bureau of the Census. Annual. *Annual survey of manufacturers*. Washington DC: Government Printing Office.

van Duijn, J. J. 1983. *The long wave in economic life*. London: Allen & Unwin.
Vernon, R. 1966. International investment and international trade in the product cycle. *Quarterly Journal of Economics* **80**, 190–207.
Vinson, R. and P. Harrington 1979. *Defining "high technology" industries in Massachusetts*. Boston, MA: Commonwealth of Massachusetts, Department of Manpower Development.
von Thünen, J. H. 1966 (1826). *Von Thünen's isolated state*. Edited by P. Hall, translated by C. M. Wartenberg. Oxford: Pergamon, 1966.

Weber, A. 1929. *Theory of the location of industry*. Chicago: University of Chicago Press.
Weinstein, B. L., H. T. Gross and J. Rees 1985. *Regional growth and decline in the United States*, 2nd edn. New York: Praeger.

Weiss, M. 1985. High technology industries and the future of employment. In *Silicon landscapes*, P. Hall and A. Markusen (eds.), 80–93. Boston: Allen & Unwin.

Wise, M. J. 1949. On the evolution of the jewellery and gun quarters in Birmingham. *Institute of British Geographers, Transactions and Papers* **15**, 57–72.

Wells, L. T. Jr. 1972. *The product life cycle and international trade.* Cambridge, MA: Harvard University Press.

Index

Numbers in brackets after individual industries denote Standard Industrial Classification (SIC).

access factors 5, 8, 76, 98, 132–3, 136, 143, 145, 147, 149–53, *Table* 9.1, 156–9, 161, 163, *Table* 9.2, 165–8, 169n, 175–6, 179, 193–4, 196–7, *Table* A4.1, 204
adhesives, sealants (2891) *Tables* 2.6, 3.1, 3.4, 5.1, 5.4, 6.4, 6.5
aerospace industry 13, *Table* 2.3, 86, 103, 105, 113, 125, 140, 143, 180; *see also* entries for aircraft; space vehicles
agglomeration 4–5, 7–8, 42, 72–5, 77, 79, 94–5, 97–131, *Tables* 7.2–7.6, *Figs* 7.3–7.10, 132–3, 135, 137, 139, 145, 147–51, 153–4, *Table* 9.1, 156–7, 159, 161–3, *Table* 9.2, 165–7, 169n, 175, 186, 202–4; *see also* concentration, geographical
agricultural chemicals (287) 17, *Tables* 2.5, 4.1, 55, *Fig* 4.4, 165, 184
air, gas compressors (3563) *Tables* 2.6, 3.1, 3.4, 3.5, 5.1, 5.3, 6.4, 6.5
aircraft and parts (372) *Tables* 2.1, 2.4, 2.5, 31, 46, *Table* 4.1, *Fig* 4.3, 54, 73, 128, 166, 171
aircraft (3721) *Tables* 2.6, 3.1, 29–30, 32, *Tables* 3.2–3.5, 5.1, 5.2, 68, *Tables* 5.5, 6.4, 6.5, 86, 88, 103–4
aircraft engines, parts (3724) 32, *Tables* 2.6, 3.1–3.5, 5.1, 5.2, 5.5, 80, *Tables* 6.4, 6.5, 88, 104
aircraft industry 41, 46, 67, 179
aircraft parts, auxiliary equipment, NEC (3728) 32, *Tables* 2.6, 3.1–3.5, 5.1, 5.3, 5.5, 6.4, 6.5, 88, 104
airport access 133, 145, 147, 151–3, *Table* 9.1, 156–7, 159, 161, *Table* 9.2 165–7, 169n, 175–6, 193, 197, *Table* A4.1, 204
Alabama *Table* 7.4, 112
Alaska *Fig.* 7.2, *Table* 7.4, 111
alkalies and chlorine (2812) 32, *Tables* 2.6, 3.1, 3.4, 5.1, 5.2, 6.3–6.5, 87, 171
Allen, G. C. 139

amenity factors 8–9, 76, 132, 134–5, 142, 144, 146–7, 149–53, *Table* 9.1, 156–8, 160–1, 163–8, *Table* 9.2, 169n, 175–6, 194–5, 197, *Table* A4.1, 202–4, 210
ammunition, except small arms, NEC (3483) 32, *Tables* 2.6, 3.1, 3.4, 5.1, 5.2, 5.5, 6.4–6.6, 171
Anaheim-Santa Ana-Garden Grove, California *Table* 7.6, 116–19, *Fig* 7.3, 172–3, *Table* A5.1
Anchorage, Alaska 157, *Table* A5.1
apparel (23) *Tables* 2.2, 2.4
Arizona 99, *Tables* 7.1, 7.2, *Figs* 7.1, 7.2, 101, 103, *Tables* 7.4, 7.5, 111, 128, 130, 131n, 172
Arkansas *Table* 7.4, 111, 131n
Austin, Texas 72, 77, 128–9, 140, 153, 174, 201, *Table* A5.1
auto, controls regulators, resistors, communications, environmental applications (3822) *Tables* 2.6, 3.1, 3.4, 5.1, 5.2, 5.5, 70, *Tables* 6.4, 6.5
automobile industry 11, 41, 44–5, 76, 103, 112

Bakersfield *Fig* 7.3, 119, *Table* A5.1
ball, roller bearings (3562) *Tables* 2.6, 3.1, 3.4, 5.1, 5.2, 6.4, 6.5, 104
Bay City, Michigan 166, *Table* A5.1
Beaumont-Port Arthur 128, *Table* A5.1
Billings, Montana *Tables* 7.6, A5.1
biological products (2831) 32, *Tables* 2.6, 3.1, 3.3–3.5, 5.1, 5.4, 6.4–6.6, 89, 104
biotechnology 2, 12–13, *Table* 2.3, 17, 141, 178, 180, 184
Birch, D. L. 137
Birmingham, England 136, 139
black, percent 8, 144, 146, 151–3, *Table* 9.1, 156–7, 159–60, 162, *Table* 9.2, 165–8, 175–6, 193, *Table* A4.1, 202–4, 210–11
blowers, exhaust, ventilation fans (3564) *Tables* 2.6, 3.1, 3.4, 5.1, 5.4, 6.4, 6.5

Boise, Idaho Table 7.6, 116, Table A5.1
Boston-Lowell-Brockton-Lawrence-
    Haverhill Massachusetts Table 7.4,
    116, 122, Fig 7.6, 130, 140–1, 172–3,
    180, Table A5.1
branch plants 77–8, 112, 116–17, 129,
    132, 173–4, 180, 187, 204
Bridgeport-Stamford-Norwalk 119, Fig
    7.4, Table A5.1
Bureau of Industrial Economics (BIE) 54,
    58n
business climate 5, 8, 43, 132, 139, 145,
    153, 168
business headquarters see Fortune 500
business machines (357) 31
business management and consulting
    services (7392) Table 2.1
business services 5, 8, 42, 74–5, 87, 98,
    116–17, 136, 145, 147, 151, 153,
    Table 9.1, 156, 159, 161–2, Table
    9.2, 165–8, 173–6, 185, 196–7, Table
    A4.1, 202–4, 210
Buswell, R. J. and E. W. Lewis 141

calculating accounting machines, except
    electrical computer equipment
    (3574) Tables 2.6, 3.1, 3.4, 5.1, 5.2,
    6.3–6.6, 89
California 14, Table 2.3, 76, 95, 99, 103,
    Tables 7.1–7.5, Figs 7.1, 7.2, 106,
    111–12, 130, 131n, 172, 179, 197
Californian Commission on Industrial
    Innovation 2, 14, Table 2.3
capitalist system 136–8
carbon, graphite products (3624) Tables
    2.6, 3.1, 3.4, 5.1, 5.2, 6.3–6.5, 87
carbon black (2895) Tables 2.6, 3.1, 3.4,
    5.1, 5.2, 6.3–6.5, 87
cathode ray tubes, NEC (3671) Tables 2.6,
    3.1, 3.3–3.5, 5.1, 5.2, 5.5, 6.4–6.6,
    70, 89, 192n
Cedar Rapids, Iowa Table 7.6, 174, Table
    A5.1
cellulosic man-made fibers (2823) Tables
    2.6, 3.1, 3.4, 5.1, 5.2, 6.3–6.5, 87
Census of Manufactures, United States
    57n, 71, 148; Industry statistics
    188–9, Table A2.1, 195; Location of
    manufacturing plants tapes 152,
    186–7, 191n–2n; Standard
    Industrial Classification (SIC) 6,
    22n–3n, 41, 57n, Appendix 1,
    Appendix 2
Champagne-Urbana, Illinois Table 7.6,
    116, Table A5.1
Checkland, S. 138–9
chemicals (28) 12, Tables 2.2, 2.4, 15, 78,
    163, 165

chemicals, chemical preparations, NEC
    (2899) 29, Tables 2.6, 3.1, 3.2, 3.4,
    5.1, 5.4, 6.2, 6.4, 6.5, 87
Chesapeake/Delaware River region
    Tables 7.2, 7.5, 104, 111–12, 172–3
Chesapeake to Hudson 119, Fig 7.4
Chicago, Illinois 7, 44, 94, 101, 110, Table
    7.6, 116, 122–5, Fig 7.7, 173, 178,
    Table A5.1
Chinitz, B. 139
Cleveland, Ohio Table 7.6, 141, 173, 178,
    Table A5.1
climate 134, 142, 144, 147, 150, 152,
    Table 9.1, 156, 158, 161, Table 9.2,
    165–7, 169n, 175, 194, Table A4.1,
    202, 204, 210
Colorado 99–101, Tables 7.1–7.4, Figs
    7.1, 7.2, 106, 112–13, 166
Colorado front range 125–6, Fig 7.9, 130
Colorado Springs 77, 125–6, Fig 7.9, 179,
    Table A5.1
Columbia, Missouri 157, Table A5.1
commercial R & D laboratories (7391,
    7397) Table 2.1, 17, 73, 185
Commission on Industrial Innovation 2,
    14, Table 2.3
communications equipment (366) Tables
    2.1, 2.3–2.5, 31–2, 46, 49, Fig 4.1,
    Table 4.1, 54, 163, 165
competition 4–6, 42–4, 48, 59, 65, Table
    5.4, 67–8, 70, 73, 87, 137, 171
computer industry 6, Table 2.3, 67, 75,
    87, 95, 171, 180
computer programming services (737)
    Table 2.1
computer software and data processing
    services 2, Table 2.3, 17, 75, 185
concentration, geographical 4–5, 7, 42,
    44, 45, 47, 71, 78–88, Tables 6.3,
    6.4, 93, 95, 132, 137;
    reconcentration 44–5, 89, 92, 112,
    131n; see also agglomeration
concentration ratio 59–70, Tables
    5.1–5.5
Connecticut 99, Tables 7.1, 7.2, Figs 7.1,
    7.2, 101, 104, Tables 7.4, 7.5, 111,
    172
connectors, electronic applications (3678)
    32, Tables 2.6, 3.1, 3.3–3.5, 5.1, 5.3,
    5.5, 6.4, 6.5
Conroy, Michael 187–8, 192n
construction equipment (353) Tables 2.5,
    4.1, 52, Fig 4.2, 166
construction machine equipment (3531)
    29, Tables 2.6, 3.1–3.4, 5.1, 5.3, 6.2,
    6.4, 6.5, 87, 104–5
conveyors, conveying equipment (3535)
    Tables 2.6, 3.1, 3.4, 5.1, 5.4, 6.4, 6.5

Index

cores, high tech 7–8, 42–3, 72–7, 97–131, *Tables* 7.1–7.6, *Figs* 7.1–7.10, 172–4, 177–80, 197
costs 4, 42–3, 46–8, 76, 95, 135, 137, 148, 163, 168
County Business Patterns (CBP) 39n, 188–9, *Table* A2.1
cultural factors 134–5, 175
cyclic crudes, intermediates, dyes (2865) *Tables* 2.6, 3.1, 3.4, 5.1, 5.3, 6.4, 6.5, 104

Dallas-Fort Worth, Texas *Table* 7.6, 116, 128, 140, 153, *Table* A5.1
Daytona Beach *Fig* 7.8, *Table* A5.1
decentralization *see* dispersal
defense: –related industries 6, 13, 29, 32, 41, 45–7, *Table* 4.1, 52–5, *Fig* 4.3, 58n, 67–9, *Table* 5.5, 73, 81, 86, 88, 117, 119, 125, 128, 163–6, *Table* 9.2, 171, 174, 176–80; spending 2, 8–9, 40, 73, 132, 140–2, 145, 148–51, 153, *Table* 9.1, 156–7, 159, 162–3, *Table* 9.2, 165–8, 169n, 171, 175–81, 194, 196, *Table* A4.1, 202–4
definition, high tech industry 3, 6, 10–23
dental equipment, supplies (3843) *Tables* 2.6, 3.1, 3.4, 3.5, 5.1, 5.4, 6.4, 6.5
Denver-Boulder *Fig* 7.9, *Table* A5.1
Detroit, Michigan 45, 94, *Table* 7.6, 116, 139, 173, *Table* A5.1
dispersal 4, 7–8, 42–3, 47, 71–2, 75–6, 78–96, *Tables* 6.2, 6.4, 6.5, 105–6, 110–13, 117, 119, 122, 128–32, 137, 146, 150, 159–62, 169n, 172–4, 186, 197, 203
division of labor, international 2, 72, 76, 78, *Table* 6.1, 112, 134
drugs (283) *Tables* 2.1, 2.4, 2.5, 4.1, 52, *Fig* 4.2, 163, 166
Duluth 128, *Table* A5.1

educational options 8, 134–5, 144, 147, 151–3, *Table* 9.1, 156, 158, 160–1, 163, *Table* 9.2, 165–8, 169n, 175, 197, *Table* A4.1, 210
electrical equipment (36) 12, *Table* 2.2, 17, 46, 78, 166
electrical industrial apparatus (362) *Tables* 2.1, 2.5, 4.1, *Fig* 4.3, 54
electrical industrial apparatus, NEC (3629) *Tables* 2.6, 3.1, 3.4, 5.1, 5.4, 6.4, 6.5
electrical transmission equipment (361) *Tables* 2.1, 2.5, 4.1, 55, *Fig* 4.4, 58n
electronic capacitors (3675) *Tables* 2.6, 3.1, 3.4, 5.1, 5.3, 5.5, 6.4, 6.5

electronic components and assembly (367) 6, *Tables* 2.1, 2.3–2.5, 31, 49, *Fig* 4.1, *Table* 4.1, 67, 73, 75–6, 103–5, 140–1, 163, 165, 171, 179–80
electronic components, NEC (3679) *Tables* 2.6, 3.1, 29–30, 32, *Tables* 3.2–3.5, 5.1, 5.4, 5.5, 6.2, 6.4, 6.5, 68, 86–7, 103–5
electronic computing equipment (3573) 29–30, 32, *Tables* 2.6, 3.1–3.5, 5.1, 5.3, 6.4, 6.5, 103–5
elevators, moving stairways (3534) *Tables* 2.6, 3.1, 3.4, 5.1, 5.2, 6.4, 6.5
Employment and earnings state and area series (1977) 195
engineering, laboratory instruments and scientific instruments (381) *Tables* 2.1, 2.5, 4.1, *Fig* 4.3, 32, 54, 166
engineering, laboratory, scientific, research instruments (3811) 22n, *Tables* 2.6, 3.1, 3.2, 3.4, 5.1, 5.4, 68, *Tables* 5.5, 6.4, 6.5
engineering and architectural services (891) *Table* 2.1
engineers/engineering technicians 6, 17, *Table* 2.5, 40–1, 46–7, 56–7, 73–4, 76, 117, 122, 134–5, 146–7, 158, 160, 180–1, 183–4
engines and turbines (351) *Tables* 2.1, 2.5, 4.1, 52, *Fig* 4.2, 166
entropy index 80–92, *Tables* 6.4, 6.5, 95, 96n
explosives (2892) *Tables* 2.6, 3.1, 3.4, 5.1, 5.2, 5.5, 6.4, 6.5, 89

fabricated metals (34) *Tables* 2.2, 2.4, 17
Fairchild Semiconductor 73
Fall River-New Bedford 122, *Fig* 7.6, *Table* A5.1
Fantus Corporation 95
Feller, I. 73
fertilizers, mixing only (2875) *Tables* 2.6, 3.1, 3.4, 5.1, 5.4, 6.2, 6.4, 6.5, 81, 86–7, 94
Florida group of SMSAs 7, 125, *Fig* 7.8
Florida state *Tables* 7.1–7.4, *Figs* 7.1, 7.2, 100–1, 112, 130, 172, 179
fluid meters, counting devices (3824) 32, *Tables* 2.6, 3.1, 3.4, 3.5, 5.1, 5.3, 5.5, 6.4–6.6
food and kindred products (20) 12, *Tables* 2.2, 2.4
Fort Collins 125, *Fig* 7.9, *Table* A5.1
Fort Lauderdale 125, *Table* A5.1
Fort Meyers, Florida 173, *Tables* 7.6, A5.1

Fortune 500 headquarters 116, 145, 147, 151, 153, Tables 9.1, 9.2, 156–7, 159–62, 167, 169n, 174–6, 194, 196, Table A4.1, 201–4, 211n
Fortune magazine 194
freeway density 136, 143, 145, 147, 151, Tables 9.1, 9.2, 156, 158–9, 163, 165–7, 175–6, 194, 196, Table A4.1, 204
Fresno Fig 7.3, 119, Table A5.1
Frostbelt 7, 97, 105–6, 116, 130, 134, 172
furniture and fixtures (25) 12, Tables 2.2, 2.4

Galveston 128, Table A5.1
Gary-Hammond, Indiana 166, Table A5.1
general industrial machinery (356) Tables 2.5, 4.1, 52, Fig 4.2, 54, 184
general industrial machinery equipment, NEC (3569) 29, Tables 2.6, 3.1–3.4, 5.1, 5.4, 6.2, 6.4, 6.5, 87
Georgia Table 7.4, 112, 130
Glasgow, Scotland 138–9
Glasmeier, A. 77–8
Gordon, R. and L. Kimball 77
Great Falls, Montana 157, Table A5.1
Greater San Francisco Bay Area 7, 116, 119–22, Fig 7.5, 130, 141, 179
Greeley 125, Fig 7.9, Table A5.1
guided missiles, space vehicles (3761) Tables 2.6, 3.1, 3.2, 3.4, 5.1, 5.2, 5.5, 6.3–6.5, 68, 81, 86, 88, 103–4, 166, 171, 179
guided missiles, space vehicles, parts, NEC (3769) 32, Tables 2.6, 3.1, 3.4, 3.5, 5.1, 5.2, 5.5, 6.3–6.6, 68, 81, 86, 88–9, 96n
guided missiles, space vehicles, propulsion units (3764) 32, Tables 2.6, 3.1, 3.4, 3.5, 5.1, 5.2, 5.5, 6.3–6.6, 68, 86, 88, 94
gum, wood chemicals (2861) Tables 2.6, 3.1, 3.4, 5.1, 5.2, 6.4–6.6, 94

Handbook of labor statistics (1979) 195
Hawaii Fig 7.2, Table 7.4, 111, 113
hoists, industrial cranes, monorail systems (3536) Tables 2.6, 3.1, 3.4, 5.1, 5.4, 6.4–6.6, 92
Honeywell 73
Honolulu, Hawaii 157, Table A5.1
Hoover, Edgar 5, 72, 133
housing prices 76, 134, 144, 147, 151–2, Tables 9.1, 9.2, 156–8, 163–5, 175, 195, Table A4.1, 202–4, 210
Houston, Texas 73, Table 7.6, 116, 128, 141, 173, Table A5.1
Howells, J. R. L. 141

Huntington-Ashland, West Virginia/Kentucky 167, Table A5.1
Huntsville, Alabama 72–3, 140–1, 180, Table A5.1
Hymer, S. H. 75

Idaho Table 7.4, 111–12
Illinois 99, Tables 7.1–7.5, Figs 7.1–7.2, 103–5, 112, 173
Indiana Table 7.1, Figs 7.1, 7.2, 101, Table 7.4, 112
industrial controls (3622) Tables 2.6, 3.1, 3.3, 3.4, 5.1, 5.3, 6.4, 6.5
industrial gases (2813) Tables 2.6, 3.1, 3.4, 5.1, 5.2, 6.2, 6.4, 6.5
industrial inorganic chemicals (281) Tables 2.1, 2.5, 4.1, 54–5, Fig 4.4, 165–6
industrial inorganic chemicals, NEC (2819) Tables 2.6, 3.1, 3.3, 3.4, 5.1, 5.4, 6.2, 6.4, 6.5, 88
industrial instruments for measurement and display (3823) Tables 2.6, 3.1, 3.3–3.5, 5.1, 5.4, 5.5, 6.4, 6.5, 68, 104
industrial organic chemicals (286) Tables 2.5, 4.1, 52, Fig 4.2, 166
industrial organic chemicals, NEC (2869) Tables 2.6, 3.1, 3.2, 3.4, 5.1, 5.3, 6.4, 6.5, 104
industrial patterns (3565) Tables 2.6, 3.1, 3.2, 3.4, 5.1, 5.4, 6.4, 6.5
industrial process furnaces, ovens (3567) Tables 2.6, 3.1, 3.4, 5.1, 5.4, 6.4, 6.5
industrial trucks, tractors, trailers, stackers (3537) Tables 2.6, 3.1, 3.4, 5.1, 5.3, 5.5, 6.4, 6.5
information 42, 72–5, 135; see also business services
innovation 4–5, 8, 10–11, 41–3, 46–9, Fig 4.1, Table 4.1, 54–5, 67, 71–7, 79, 88, 95, 101, 103, 113, 117, 130, 131n, 132, 135–9, 141, 148–9, 156, 162–5, Table 9.2, 172, 178, 180–1, 187, 191, 202
inorganic pigments (2816) Tables 2.6, 3.1, 3.4, 5.1, 5.2. 6.4, 6.5
instruments Table 2.3, 163, 180
instruments, measuring, testing, electrical, electrical signals (3825) 30, 32, Tables 2.6, 3.1, 3.3–3.5, 5.1, 5.4, 6.4, 6.5
internal combustion engines, NEC (3519) Tables 2.6, 3.1, 3.3, 3.4, 5.1, 5.3, 6.4, 6.5, 104–5
Iowa Table 7.4, 112
Israel 76

Index 223

Japan 125, 178, 180
Jersey City 119, *Fig* 7.4, *Table* A5.1
jobs: creation 4, 6, 8, 11–14, *Tables* 2.2, 2.3, 19, 24–39, *Tables* 3.1–3.5, 40–3, 47–58, *Figs* 4.1–4.5, *Table* 4.1, 68, 70, 89, 92, 96, 105–6, *Tables* 7.3–7.5, 110–13, 116–25, 128–32, 137, 171–4, 202–4; location of 1, 3, 79–96, *Tables* 6.4, 6.5, 97–131, *Tables* 7.1–7.6, *Figs* 7.1–7.10, 144–69, *Tables* 9.1, 9.2, 196, *Appendices* 2, 5; loss 4, 6, 11, 24, 32, *Table* 3.4, 38–9, 44, 54–6, 68, 70, 89, 92, 105, 113, 117, 119–25, 128, 131, 137, 171–2
Johnson, President Lyndon 86, 143
Johnson City-Bristol-Kingsport, Tennessee/Virginia 157, *Table* A5.1
Johnstown and Erie, Pennsylvania 166, *Table* A5.1

Kankakee 123, *Fig* 7.7, *Table* A5.1
Kansas 99–100, *Tables* 7.1, 7.2, *Figs* 7.1–7.4, 112–13, 128
Kansas City 128, 140–1, *Table* A5.1
Kenosha, Wisconsin 123, *Fig* 7.7, *Table* A5.1
Kentucky *Table* 7.4, 112
Killeen, Texas 157, *Table* A5.1
Kondratieff, Nikolai 41, 138

labor 4, 6, 46, 72, 74, 78, 87, 89, 92, 95–6, 135; costs 4, 42–3, 46–7, 76, 148, 163; international division of 2, 72, 76, 78, *Table* 6.1, 112, 134; professionals/technicians 6, 17, *Table* 2.5, 40–1, 46–7, 56–7, 73–4, 76, 117, 122, 134–5, 146–7, 158, 160, 180–1, 183–4; supply 5, 7, 11, 72, 74, 87, 98, 132–6, 144, 146, 149–51, 153–4, *Tables* 9.1, 9.2, 156–8, 161–3, 165–7, 202–4, 210
Lafayette, Louisiana 157, *Table* A5.1
Lakeland-Winterhaven *Table* 7.6, 116, *Fig* 7.8, 157, 173, *Table* A5.1
land availability 135, 154, 197, 200n
Laredo, Texas 173, *Tables* 7.6, A5.1
Lawton, Oklahoma 157, 173, *Tables* 7.6, A5.1
leather products (31) *Table* 2.2
Lloyd, P. E. 139
location of high tech industry 2–3, 5, 7–9, 19, 45, 71–96, *Tables* 6.1–6.6, 97–131, 144–69; changes over time 8, 79, 87–96, *Tables* 6.5, 6.6, 144, 148–50, 154, *Table* 9.1, 158–60, 167, 174, 177, 210–11; jobs 1, 3, 79–96, *Tables* 6.4, 6.5, 97–131, *Tables* 7.1–7.6, *Figs* 7.1–7.10, 144–69, *Tables* 9.1, 9.2, 196, *Appendices* 2, 5; patterns 8, 41, 144, 148–50, 154, 157, *Tables* 9.1, 9.2, 160–9, 174, 177–8, *Appendix* 5; plants 3, 71, 79–96, *Tables* 6.4, 6.5, 97–131, *Tables* 7.1–7.6, *Figs* 7.1–7.10, 144–69, *Table* 9.1, 174, 196, *Appendices* 2, 5; l. quotients 98–103, *Tables* 7.1, 7.2, *Figs* 7.1, 7.2, 131n, 141; in SMSAs 7–8, *Table* 6.5, 93–4, 97–8, 113–31, *Table* 7.6, *Figs* 7.3–7.10, 140–69, *Table* 9.1, 172–6, 186, 196–7, *Appendix* 5 in states 7, 97–113, *Tables* 7.1–7.5, *Figs* 7.1, 7.2, 129–30, 145, 172–3, 186–7
location theory 132–43; traditional industrial 5, 7, 87, 133–6
Lockheed 74
London 133, 136
Long Branch-Asbury Park 119, *Fig* 7.4, 201, *Table* A5.1
long-wave theory 41, 138
Los Angeles, California 73, 101, *Table* 7.6, 117–19, *Fig* 7.3, 140–1, 172, 180, *Table* A5.1
Los Angeles Basin 7, 116–19, *Fig* 7.3, 130, 179
Losch, A. 133
Louisiana *Tables* 7.1, 7.2, *Figs* 7.1, 7.2, 104, *Tables* 7.4, 7.5, 111
Lower Great Lakes region 104–5, *Tables* 7.2, 7.5
Lower Lake Michigan 122–5, *Fig* 7.7
Lubbock, Texas 173, *Tables* 7.6, A5.1
lumber and wood products (24) *Table* 2.2

M4 Corridor, England 1, 141
McAllen-Pharr-Edinburgh, Texas 173, *Tables* 7.6, A5.1
machine tool accessories, measuring devices (3545) 29, *Tables* 2.6, 3.1–3.4, 5.1, 5.4, 6.4, 6.5
machine tools, metal cutting types (3541) 29, *Tables* 2.6, 3.1–3.4, 5.1, 5.4, 6.4, 6.5, 104
machine tools, metal forming types (3542) *Tables* 2.6, 3.1, 3.4, 5.1, 5.4, 6.4, 6.5, 104
machinery (35) 12, *Table* 2.2, 163
Maine *Table* 7.4, 111, 113
Malecki, Edward 2, 46, 140–1
Manchester, England 139
market penetration 4, 42–3, 44, 47–9, *Table* 4.1, 52, 54–5, 59, 67, 70–1, 95, 135, 163–6, *Table* 9.2, 169n
market power 6, 41–5, 59–70, *Tables* 5.1–5.5, 76

market saturation 4, 43–4, 48, Table 4.1, 55, Fig 4.4, 54, 59, 66–7, 135, 142, 163–6, Table 9.2, 169n
Markusen, A. 75–6, 139
Marshall, A. 72; Principles of Economics 136
Marx, Karl 136
Maryland 98–9, Tables 7.1, 7.2, Figs 7.1, 7.2, 101, 104, Tables 7.4, 7.5, 131n, 172
Massachusetts Tables 7.1–7.5, Figs 7.1, 7.2, 100, 104–6, 110–11, 172
Massachusetts Division of Employment Security (MDES) 11, Table 2.1
Massachusetts Manpower Development Department 184, 185n
Massey, D. and R. Meegan 89
materials 5, 7, 17, 46, 86–8, 133, 184
mathematicians 17, Table 2.5, 47, 56, 183–4
MCC (Microelectronics and Computer Technical Corporation) research center 129
measuring and controlling instruments (382) Tables 2.1, 2.4, 2.5, 32, 49, Fig 4.1, Table 4.1, 87, 165
measuring, controlling devices, NEC (3829) Tables 2.6, 3.1, 3.4, 3.5, 5.1, 5.4, 5.5, 68, 70, Tables 6.4, 6.5
mechanical power transmission equipment NEC (3568) Tables 2.6, 3.1, 3.4, 5.1, 5.4, 6.4, 6.5
medical and dental supplies (384) 17, Table 2.5, 49, Fig 4.1, Table 4.1, 165, 184
medical, chemical, botanical products (2833) 32, Tables 2.6, 3.1, 3.4, 3.5, 5.1, 5.2, 6.4–6.6
Melbourne-Titusville-Cocoa 73, 125, Fig 7.8, 174, Table A5.1
Mensch, Gerhard 138
metal working machinery (354) Tables 2.5, 4.1, Fig 4.3, 54, 166, 184
metalworking machinery, NEC (3549) 32, Tables 2.6, 3.1, 3.3–3.5, 5.1, 5.4, 6.4, 6.5
Miami 125, Table A5.1
Michigan Table 7.1, Figs 7.1, 7.2, 101, Table 7.4, 112
Middle Atlantic group of SMSAs 7
Midwest industrial heartland 7, 110, 130, 172
military-related sectors see defense
Milwaukee, Wisconsin Table 7.6, 123–5, Fig 7.7, 130, 141, 178, Table A5.1
mining machinery equipment (3532) 32, Tables 2.6, 3.1, 3.4, 3.5, 5.1, 5.3, 6.4, 6.5

Minneapolis-St.Paul, Minnesota/Wisconsin 72–3, Table 7.6, 128, 140, Table A5.1
Minnesota Fig 7.1, 100–1, Tables 7.2–7.4, 106, 110, 112, 130, 131n
Minnesota Mining and Manufacturing (3M) 73
miscellaneous chemicals (289) Table 2.5, 4.1, Fig 4.3, 54
miscellaneous manufacturing (39) Table 2.2
Mississippi Table 7.4, 111
Missouri 101, Table 7.4
model, multiple regression 8, 144–69, 178, Appendix 4
motors, generators (3621) Tables 2.6, 3.1, 3.4, 5.1, 5.3, 6.4, 6.5

Nassau-Suffolk, New York Table 7.6, 116, 119, Fig 7.4, Table A5.1
National Science Foundation (NSF) 12, 46, 183, 195
Nebraska Table 7.4, 113
Nevada Table 7.4, 111, 131n, 172
New Brunswick-Perth Amboy Sayreville 119, Fig 7.4, 167, Table A5.1
New Hampshire Table 7.4, 111, 172
New Jersey 98, Tables 7.1–7.5, Fig 7.1, 103–5, 112, 131n, 172
New Mexico Table 7.4, 111, 113, 130
New York 101, Table 7.6, 116–17, 119, Fig 7.4, 133, 136, 140, 173, 193, 201, Table A5.1
New York state 98–9, Fig 7.1, Tables 7.1, 7.4, 130
New York Times 1
Newark, New Jersey Table 7.6, 116, 119, Fig 7.4, 173, Table A5.1
nitrogenous fertilizers (2873) Tables 2.6, 3.1, 3.4, 5.1, 5.3, 6.4, 6.5
non-electrical equipment (35) 78
non-profit educational, scientific and research organizations (892) Table 2.1
North Carolina Table 7.4, 112, 130, 131n, 166
North Dakota Table 7.4, 112

Oakey, R. P. et al. 141
Occupational Employment Statistics (OES) survey-based matrix (1980) 16–17, 182–3
occupational mix in high tech industries 16–19, Table 2.5, 41, 78, 146, 182–3, Appendix 1
office computing machines (357) Tables 2.1, 2.4, 2.5, 49, Fig 4.1, Table 4.1, 54, 67, 165
office machinery, NEC (3579) Tables 2.6, 3.1, 3.3, 3.4, 5.1, 5.2, 6.4, 6.5

Ohio 99, *Table* 7.1, *Figs* 7.1–7.2, 101, *Table* 7.4, 113, 172
oilfield machinery equipment (3533) 30–2, *Tables* 2.6, 3.1–3.5, 5.1, 5.4, 6.4, 6.5, 87, 89, 103–4
Oklahoma *Tables* 7.1, 7.2, *Figs* 7.1, 7.2, 104, *Tables* 7.4, 7.5, 111, 131n, 166
Oklahoma City *Table* 7.6, 116, *Table* A5.1
Old New England region 103–5, *Tables* 7.2, 7.5, 122, *Fig* 7.6, 172–3
oligopolies 6, 43, 45, 48, 63–5, *Tables* 5.2, 5.3, 67, 75–6, 139
optical instruments and lenses (383) *Tables* 2.1, 2.4, 2.5, 49, *Fig* 4.1, *Table* 4.1, 54, 87, 165
optical instruments, lenses (383) 32, *Tables* 2.6, 3.1, 3.3–3.5, 5.1, 5.4, 5.5, 6.4–6.6, 68
ordnance (348) *Tables* 2.5, 4.1, *Fig* 4.3, 54, 166, 171
ordnance, accessories, NEC (3489) *Tables* 2.6, 3.1, 3.4, 5.1, 5.3, 5.5, 6.4, 6.5, 68, 86
Orlando 125, *Fig* 7.8, *Table* A5.1
orthopedic, prosthetic, surgical applications (3842) 29, *Tables* 2.6, 3.1–3.4, 5.1, 5.4, 6.4, 6.5
Oxnard-Ventura, California *Table* 7.6, *Fig* 7.3, 119, *Table* A5.1

Pacific Southwest region 103–4, *Tables* 7.2, 7.5, 111
paints and varnishes (285) 17, *Tables* 2.5, 4.1, 55, *Fig* 4.4, 165–6, 184
paints, varnishes, lacquers, enamels (2851) 29, *Tables* 2.6, 3.1, 3.2, 3.4, 5.1, 5.4, 6.2, 6.4, 6.5, 87
Palo Alto 74–5
Panama City, Florida 173 *Tables* 7.6, A5.1
paper and products (26) *Tables* 2.2, 2.4
Paterson-Clifton-Passaic 119, *Fig* 7.4, *Table* A5.1
Pennsylvania 99, *Fig* 7.1, 7.2, *Tables* 7.1, 7.4, 112
perfumes, cosmetics, toilet preparations (2844) *Tables* 2.6, 3.1, 3.4, 5.1, 5.3, 6.4, 6.5, 88, 104
pesticides, agricultural chemicals, NEC (2879) *Tables* 2.6, 3.1, 3.4, 5.1, 5.3, 6.4, 6.6, 88
petroleum refining (291) 12, 15, *Tables* 2.2, 2.4, 2.5, 4.1, 55–6, *Fig* 4.5, 166
petroleum refining (2911) *Tables* 2.6, 3.1, 3.2, 3.4, 5.1, 5.3, 6.4, 6.5, 103–4
pharmaceutical preparations (2834) *Tables* 2.6, 3.1–3.4, 5.1, 5.4, 6.4, 6.5, 104–5

Philadelphia *Table* 7.6, 140, 173, *Table* A5.1
Phoenix, Arizona 77, *Table* 7.6, 116, 129, *Table* A5.1
phonograph records, pre-recorded magnetic tape (3652) 32, *Tables* 2.6, 3.1, 3.4, 5.1, 5.3, 6.4, 6.5, 171, 181n
phosphatic fertilizers (2874) *Tables* 2.6, 3.1, 3.4, 5.1, 5.3, 6.4, 6.5, 88
photographic equipment (385/6) *Tables* 2.1, 2.4, 2.5, 4.1, 52, *Fig* 4.2, 166
photographic equipment, supplies (3861) *Tables* 2.6, 3.1–3.4, 5.1, 5.2, 6.4, 6.5, 104–5
photovoltaics *Table* 2.3
Pittsburgh 44, 139, *Table* A5.1
Places rated almanac 194
plants *Table* 3.1, 3.2, 29, 39n; location 3, 71, 79–96, *Tables* 6.4, 6.5, 97–131, *Tables* 7.1–7.6, *Figs* 7.1–7.10, 144–69, *Table* 9.1, 174, 196, *Appendices* 2, 5
plastics and synthetic resins (282) *Tables* 2.1, 2.5, 4.1, 55, *Fig* 4.4, 165–6
plastic materials, synthetic resins (2821) *Tables* 2.6, 3.1, 3.4, 5.1, 5.4, 6.4, 6.5
policy, public 168, 170–81; local 8, 57, 176–8; national 8, 57, 179–81
power, distribution special transformers (3612) *Tables* 2.6, 3.1, 3.4, 5.1, 5.2, 6.4, 6.5
power driven hand tools (3546) *Tables* 2.6, 3.1, 3.4, 5.1, 5.2, 6.4, 6.5, 104
primary metals (33) *Tables* 2.2, 2.4
Principles of Economics (Marshall) 136
printing and publishing (27) 12, *Table* 2.2
printing ink (2893) *Tables* 2.6, 3.1, 3.4, 5.1, 5.3, 6.4, 6.5
product sophistication 11, 13–14, *Tables* 2.1, 2.3, 56
product-profit cycle theory 4–7, 40–58, *Figs* 4.1–4.5, *Table* 4.1, 59, 66–7, 71–2, 75, 77, 88, 95, 131n, 137, 144–5, 163–7, *Table* 9.2, 171–2
profitability 41–3, 45, 47–8
Providence-Warwick-Pawtucket 122, *Fig* 7.6, *Table* A5.1
Provo-Orem 128, *Fig* 7.10, *Table* A5.1
Pueblo 125–6, *Fig* 7.9, 157, *Table* A5.1
pumps, pumping equipment (3561) *Tables* 2.6, 3.1, 3.3, 3.4, 5.1, 5.4, 6.4, 6.5

Racine *Fig* 7.7, 125, *Table* A5.1
radio and TV receiving equipment (365) 6, *Tables* 2.5, 4.1, 55, *Fig* 4.4, 165–6, 184

radio, TV receiving sets, except communication types (3651) 32, Tables 2.6, 3.1, 3.4, 5.1, 5.2, 6.4–6.6, 103, 105, 171
radio, TV transmitting, signal, detection equipment (3662) 29–30, Tables 2.6, 3.1–3.5, 5.1, 5.4, 5.5, 6.2, 6.4, 6.5, 103–4, 165–6
railroad equipment (374) Tables 2.5, 4.1, 55–6, Fig 4.5, 166, 184
railroad equipment (3743) Tables 2.6, 3.1, 3.4, 5.1, 5.2, 6.4, 6.5, 105
rationalization 5, 39n, 44–5, 48, Table 4.1, 54, 66–7, 88, 112, 137, 142, 163–6, Table 9.2, 169n
reclaimed rubber (303) 17, Tables 2.5, 4.1, 55, Fig 4.5, 166, 184
reclaimed rubber (3031) 32, Tables 2.6, 3.1, 3.4, 5.1, 5.2, 6.3–6.5, 171
reconcentration 44–5, 89, 92, 112, 131n
research and development (R & D) 2, 5, 8–10, 13–15, 17, 46, Table 2.4, 56, 73, 76–8, Table 6.1, 132, 139–42, 145, 147–8, 151, 153, Tables 9.1, 9.2, 156–7, 159, 165–6, 169n, 174–5, 177–81, 195–6, Table A4.1; commercial R & D labs Table 2.1, 17, 73, 185
resistors, electric apparatus (3677) Tables 2.6, 3.1, 3.4, 5.1, 5.4, 5.5, 6.4–6.6, 68, 88
resistors for electronic applications (3676) Tables 2.6, 3.1, 3.4, 5.1, 5.3, 5.5, 6.4, 6.5, 88
Richardson, H. 72
Riverside-San Bernardino-Ontario Fig 7.3, 119, Table A5.1
robotics/computer aided manufacturing 2, Table 2.3
Rockford 123, Fig 7.7, Table A5.1
Rogers, E. M. and J. K. Larson 136
rolling mill machinery equipment (3547) 32, Tables 2.6, 3.1, 3.4, 5.1, 5.2, 6.3–6.6
Route 128 (around Boston) 1, 72–3, 77, 100, 147, 174, 176–7
rubber and plastic products (30) 12, Table 2.2

Sacramento Fig 7.5, 122, 201, Table A5.1
St. Cloud, Minnesota Table 7.6, 128, 173, Table A5.1
Salinas-Seaside-Monterey Fig 7.5, 122, Table A5.1
Salt Lake City-Ogden 128, Fig 7.10, Table A5.1
Salt Lake corridor 126–8, Fig 7.10, 130
San Diego, California Table 7.6, 116–19, Fig 7.3, 173, Table A5.1
San Francisco-Oakland, California 74, Table 7.6, 119–22, Fig 7.5, 140–1, 153, 172, Table A5.1
San Jose, California Table 7.6, 116, 119–22, Fig 7.5, 153, 172–3, Table A5.1
Santa Barbara Fig 7.3, 119, 141, Table A5.1
Santa Clara County 73, 78, Table 6.1
Santa Cruz, California 77, Table 7.6, Fig 7.5, 122, Table A5.1
Santa Rosa, California Table 7.6, Fig 7.5, 122, Table A5.1
Savannah, Georgia Table 7.6, 116, Table A5.1
Saxenian, A. 73, 78, Table 6.1, 136
scales, balances, except laboratory (3579 Tables 2.6, 3.1, 3.4, 5.1, 5.2, 6.4, 6.5
Schumpeter, Joseph 4, 41, 136–8
Science Policy Research Unit, University of Sussex 24
scientific instruments (38) Table 2.2, 67
scientific instruments (381) Table 2.4, 7
scientists 40–1, 46–7, 56–7; computer 17, Table 2.5, 183–4; life and physical 6, 17, Table 2.5, 183–4
SDI program 177–9
semiconductors industry 2, 6, 75–6, 87, 95, 171
semiconductors, related devices (3674) 29–30, 32, Tables 2.6, 3.1–3.5, 5.1, 5.3, 5.5, 6.4, 6.5, 70, 88, 103, 191
SIC 6, 22n–3n, 41, 57n, 95, Appendices
Silicon Glen, Scotland 1
Silicon Valley, California 1–3, 72–8, 119–22, 134–6, 141, 146–7, 153, 174, 176–7, 180, 185
small arms (3484) Tables 2.6, 3.1, 3.4, 5.1, 5.2, 5.5, 6.4, 6.5, 86
small arms ammunition (3482) Tables 2.6, 3.1, 3.4, 5.1, 5.2, 6.4, 6.5, 86, 88–9, 104
SMSAs, high tech location in 7–8, Table 6.5, 93–4, 97–8, 113–31, Table 7.6, Figs 7.3–7.10, 140–69, Table 9.1, 172–6, 186, 196–7, Appendix 5
soap, other detergents (2841) Tables 2.6 3.1, 3.4, 5.1, 5.2, 6.4, 6.5
soaps (284) 17, Table 2.5, 4.1, 52, Fig 4.2 184
South Carolina Table 7.4, 112, 130
South Dakota Table 7.4, 112
space vehicles and guided missiles (376) Tables 2.1, 2.4, 2.5, 4.1, Fig 4.3, 54
special cleaning, polishing preparations (2842) Tables 2.6, 3.1, 3.4, 5.1, 5.3, 6.2, 6.4, 6.5

specialty dyes, die sets, jigs fixtures
  industry molds (3544) Tables 2.6,
  3.1–3.4, 5.1, 5.4, 6.2, 6.4, 6.5, 86–7,
  94, 105
speed changers, industrial high drives,
  gears (3566) Tables 2.6, 3.1, 3.4, 5.1,
  5.4, 6.4, 6.5
spin-off of routine functions 42, 72,
  75–6, 79, 112, 116–17, 131–2,
  148–9, 174, 191
Standard Industrial Classification (SIC) of
  the United States Census of
  Manufactures 6, 22n–3n, 41, 57, 95,
  Appendices 1, 2; SIC Manual 11, 57n,
  189, 192n
Standard Metropolitan Statistical Areas
  (SMSAs), high tech location in 7–8,
  Table 6.5, 93–4, 97–8, 113–31, Table
  7.6, Figs 7.3–7.10, 140–69, Table 9.1,
  172–6, 186, 196–7, Appendix 5
Stanford University 2, 73–4, 141, 177
State and Metropolitan Area data book
  (1977) 193–5
states, high tech location in 7, 97–113,
  Tables 7.1–7.5, Figs 7.1, 7.2, 129–30,
  145, 172–3, 186–7
steam, gas, hydraulic turbines (3511)
  Tables 2.6, 3.1, 3.4, 5.1, 5.2, 5.5, 6.4,
  6.5
stone, clay and glass (32) 12, Table 2.2
Strategic Defense Initiative (SDI "Star
  Wars") 177–9
Sunbelt 2, 7, 97, 105–6, 116–17, 130–1,
  134, 139, 160, 172–3, 180
surface active finishing agents (2843)
  Tables 2.6, 3.1, 3.4, 5.1, 5.4, 6.4, 6.5, 88
surgical instruments (384) Table 2.4
surgical, medical instruments apparatus
  (3841) Tables 2.6, 3.1, 3.3–3.5, 5.1,
  5.4, 6.4–6.6
switch gear, switchboard apparatus
  (3613) Tables 2.6, 3.1, 3.4, 5.1, 5.2,
  6.4, 6.5, 104–5
synthetic organic fibers, except cellulose
  (2824) 32, Tables 2.6, 3.1, 3.4, 5.1,
  5.2, 6.4, 6.5, 171
synthetic rubber (2822) Tables 2.6, 3.1,
  3.4, 5.1, 5.2, 6.3–6.5, 87

Tampa-St. Petersburg 125, Fig 7.8, Table
  A5.1
tanks, tank components (3795) 32, Tables
  2.6, 3.1, 3.4, 3.5, 5.1, 5.2, 5.5,
  6.3–6.5, 68, 81, 86, 89, 96n
taxation 139, 145, 154, 197, 200n
telephone, telegraph apparatus (3661)
  29, Tables 2.6, 3.1–3.4, 5.1, 5.2, 6.4,
  6.5, 88, 104–5

Tennessee Table 7.4, 113
Terman, Professor Frederick 2, 73
Texas Tables 7.1–7.5, Figs 7.1–7.3, 101,
  103–6, 111, 128, 130, 153, 166, 179
textiles (22) 12, Tables 2.2, 2.4, 44
Thomas, M. 72
Topeka, Kansas 128, Tables 7.6, A5.1
transportation equipment (37) Table 2.2,
  78
Tucson 129, Table A5.1

unemployment rate 10, 134–5, 144, 146,
  154, Tables 9.1, 9.2, 165, 195, Table
  A4.1
unionization rate 134–5, 144, 146,
  150–4, 158, 161, 163, Tables 9.1,
  9.2, 165–6, 168, 175, 195, Table
  A4.1, 202–4, 210
universities 8–9, 14, 73–4, 116, 122, 135,
  139–41, 147, 151, 153, 156–7, 159,
  Tables 9.1, 9.2, 165–6, 169n, 174,
  177, 180, 195–6, Table A4.1
Utah Tables 7.1–7.4, Figs 7.1, 7.2, 100,
  106, 112–13, 126–8, Fig 7.10, 130,
  179

Vallejo-Fairfield-Napa Fig 7.5, 122, Table
  A5.1
Vermont Table 7.4, 111, 172
Vernon, Raymond 4, 41
Von Thünen, J. H. 135

wage rates 134–5, 142, 144, 146, 150–1,
  154, 158, 163, Tables 9.1, 9.2,
  165–6, 168, 175, 195, Table A4.1,
  201–4, 210
Wall Street Journal 1
Washington, DC Fig 7.2, Table 7.4, 113,
  140–1
Weber, Alfred 5, 7, 133, 135
welding apparatus, electric (3623) Tables
  2.6, 3.1, 3.4, 5.1, 5.3, 6.4, 6.5
West Palm Beach 125, 201, Table A5.1
West Virginia Table 7.4, 113, 130, 166,
  172
Western Gulf region 103–4, Tables 7.2,
  7.5, 111
Wichita 128, 140, Table A5.1
Wisconsin Figs 7.1, 7.2, Tables 7.1, 7.4,
  112
wood (24) Table 2.4
Worcester-Leominster Table 7.6, 116,
  122, Fig 7.6, 172, Table A5.1
Wyoming Table 7.4, 113

Youngstown-Warren, Ohio 166–7, Table
  A5.1